10/26/91

D0992435

PLANT MOLECULAR SYSTEMATICS

PLANT MOLECULAR SYSTEMATICS

MACROMOLECULAR APPROACHES

Daniel J. Crawford

Department of Botany
Ohio State University
Columbus, Ohio

WILEY

A WILEY-INTERSCIENCE PUBICATION

JOHN WILEY & SONS

New York • Chichester • Brisbane • Toronto • Singapore

Copyright © 1990 by John Wiley & Sons, Inc.

All rights reserved. Published simultaneously in Canada.

Reproduction or translation of any part of this work
beyond that permitted by Section 107 or 108 of the
1976 United States Copyright Act without the permission
of the copyright owner is unlawful. Requests for
permission or further information should be addressed to
the Permissions Department, John Wiley & Sons, Inc.

Library of Congress Cataloging in Publication Data:

Crawford, Daniel J.
    Plant molecular systematics: macromolecular approaches / Daniel
J. Crawford.
        p.   cm.
    "A Wiley-Interscience publication."
    Includes bibliographical references.
    ISBN 0–471–80760–5
    1. Plant chemotaxonomy.   2. Plant molecular biology.   I. Title.
QK95.4.C73   1990
581′.012—dc20   89–16702
                                                                          CIP

Printed in the United States of America

10   9   8   7   6   5   4   3   2   1

For Mom and Linda

The application of chemical data to the study of plant systematics represents one of the most—if not the most—active and expanding areas
in the discipline. The hybrid discipline, often called chemosystematics,
traces its origin to the late 1950s and early 1960s, with the pioneering
efforts of dedicated and imaginative workers employing primarily secondary chemistry.

During the past 10 to 20 years the emphasis in plant chemosystematics
has changed from the use of secondary metabolites (flavonoids, terpe-
noids, etc.) to proteins and DNA. The popularity of enzyme electrophoresis (allozymes) has increased rapidly among plant systematists during the past decade. At present, use of DNA represents the most rapidly
expanding (indeed exploding) field in chemosystematics. The popularity
of DNA has increased progressively during the past 5 years. The purpose
of this book is to discuss the use of macromolecular (protein and DNA)
data in plant systematics. Methodology is considered only as needed to
appreciate the nature of the data generated. Likewise, data analysis is
treated to the minimal extent necessary for discussion of how the data
best are applied to systematic problems.

One view that approaches dogma in plant systematics is that the more
data one has about the plants being studied the better with regard to
understanding the plants. At one level this is true because additional
information will allow new insights into *some* aspects of the organisms.
One of the major themes of this book, however, is that particular kinds of
systematic and evolutionary questions are best studied with particular
types of macromolecular data. In each chapter, there is discussion of the
kinds of questions best addressed with the data under consideration, and
examples are given to illustrate the points. Too often, the newest data
may be viewed as a panacea for many problems in plant systematics, yet
this has always proved not to be the case. Thus, this book attempts to
evaluate the strengths and weaknesses of macromolecular data.

The book does not pretend to be comprehensive, either with regard to the topics covered or review of the original literature. Those topics thought to be of most interest and use to practicing plant systematists were included. For any given method or approach, both a discussion of past contributions to plant systematics and probable future contributions are considered.

It is a pleasure to acknowledge those colleagues who read parts of the manuscript, supplied photographs or unpublished manuscripts, explained things to me verbally (i.e, tried to pound something into my thick skull!), or aided in other ways in the preparation of the book. With apologies to those who may have been inadvertently omitted from the list, I thank R.W. Allard, John Doebley, Jeff Doyle, David Fairbrothers, Gerald Gastony, Leslie Gottlieb, Christopher Haufler, Robert Jansen, Uwe Jensen, Richard Lester, Paul Lewis, Gerald Lowenstein, E. Nevo, Jeffrey Palmer, Frank Petersen, James Piechera, Robert Price, Loren Rieseberg, Douglas Soltis, Pamela Soltis, Jonathan Wendel, Charles Werth, Rich Whitkus, and Elizabeth Zimmer. A special thanks to Jeff Palmer for allowing me to spend time in his laboratory. Joe Bruner executed the drawings with great care and with concern for their accuracy. Marilyn Kobayashi word processed the final version of the manuscript; her understanding, kindness, and excellent assistance in this endeavor are very greatly appreciated. Billie Turner's zest for research and his friendship over the years have been a constant source of inspiration. David Giannasi is acknowledged as a good friend and respected colleague. Lastly, Linda Crawford word processed early drafts of the manuscript. More importantly, her support, patience, and devotion over the years made possible the completion of this book.

DANIEL J. CRAWFORD

Columbus, Ohio
December 1989

■■■■■ CONTENTS

Structure of Macromolecules, Rationale for their Use in Plant Systematics, and General Comments on the Phylogenetic Analysis of Macromolecular Data

INTRODUCTION

In his masterful review of the biochemical systematics of plants, Alston (1967, p. 298) indicated that "the future seems to belong to the macromolecules." Viewing this prediction over two decades later, there is no question that Alston was very perceptive. While secondary chemistry has made, and indeed continues to make, important contributions to plant systematics, the most active area of research in plant chemosystematics at present involves macromolecules.

The purpose of this book is to discuss the use of macromolecules in plant systematics and evolution. Emphasis will be on how the molecules can be used for systematic and phylogenetic purposes in plants and not on the evolution of molecules per se. Because the book is concerned with systematics and not population biology, primary emphasis will be on the infraspecific and higher taxonomic categories. Intra- and interpopulational variation will be discussed primarily within the context of its effect on studies of subspecies, species, and so on. Molecular information, like all other data used by plant systematists, cannot be viewed in a vacuum. Rather, it must always be compared and contrasted with all other data, particularly when one is concerned with inferring evolutionary processes or phylogenetic relationships. Emphasis in this book is almost exclusively on vascular plants.

The purposes of this chapter are to provide a general introduction to the kinds of molecules to be discussed in this book, to give the rationale for using macromolecules in plant systematics, and to present some general comments on the analysis of molecular data for phylogenetic purposes. Where appropriate, comments will be provided on how macromolecules differ from micromolecules as data for systematic studies.

Nucleic acids and proteins are the macromolecules of interest. All organisms to be considered have deoxyribonucleic acid (DNA) as their hereditary material; some viruses have ribonucleic acid (RNA), but they will not concern us. The DNA of plants occurs in the nucleus (the nuclear or genomic DNA) and in two organelles, the chloroplasts and mitochondria (organelle DNA). Only the former will concern us at present; chloroplast DNA will be dealt with in a later chapter.

STRUCTURE OF DNA AND RNA

A general description of the composition and structure of DNA and RNA will precede considerations of the functions of each, including the different kinds of RNAs. Both DNA and RNA are polymers with the basic repeating unit of each called the nucleotide. Each nucleotide consists of three distinct parts, namely, a sugar, a nitrogenous base (either a purine or pyrimidine), and a phosphate group (Fig. 1.1). The sugar is a pentose, with deoxyribose present in DNA and ribose in RNA. The sugar is bonded covalently to the base to form a nucleoside (Fig. 1.1). The two common purine bases present in DNA are adenine and guanine, whereas the pyrimidines are cytosine and thymine. RNA differs from DNA in containing the pyrimidine base uracil rather than thymine. The function of the third component of a nucleotide, the phosphate group, is to link together two nucleosides via their sugars. That is, a phosphodiester bond is formed between the 3' carbon of one sugar and the 5' of the other as shown in Fig. 1.2. The sugars and phosphates form the backbone of the molecule, with the bases "hanging" or coming out of this backbone (Fig. 1.2). DNA and RNA molecules are polymers composed of these nucleotide units. The sizes of the molecules cover several orders of magnitude, but naturally occurring, intact DNA molecules may consist of millions of nucleotides.

DNA consists of two polynucleotide chains running antiparallel, that is, in opposite directions. This means that one strand runs in a 3' to 5'

PURINE NUCLEOTIDE

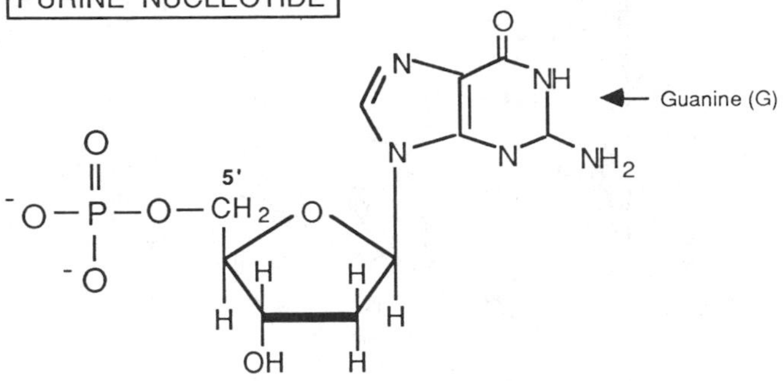

Guanine (G)

Deoxyguanosine 5' - phosphate (dGMP)

PYRIMIDINE NUCLEOTIDE

Cytosine (C)

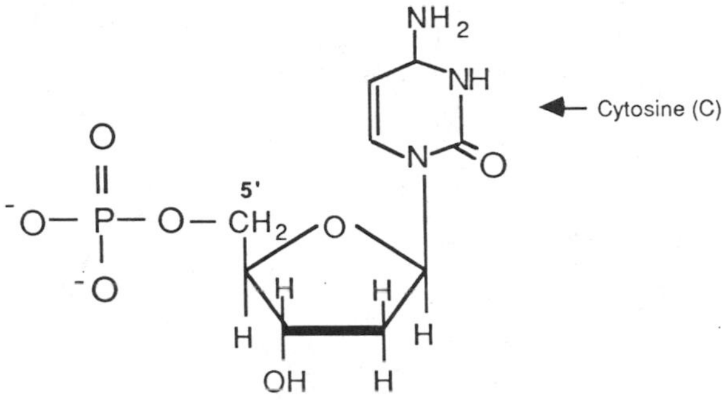

Deoxycytidine 5' - phosphate (dCMP)

Figure 1.1 Structures of a purine and a pyrimidine nucleotide. Each nucleotide is composed of a base, the sugar deoxyribose, and phosphate.

direction while the other is the opposite, the numbering referring to the free sugar group of the terminal nucleotide in each chain (Fig. 1.3). The two strands are held together by hydrogen bonding between nitrogenous bases, specifically between a purine in one strand and a pyrimidine in the other (Fig. 1.3). The pairing is even more precise than this because adenine and thymine are complementary, while guanine and cytosine pair with each other. This precise pairing also occurs between DNA and RNA, with uracil replacing thymine in the latter. The exact base pairing

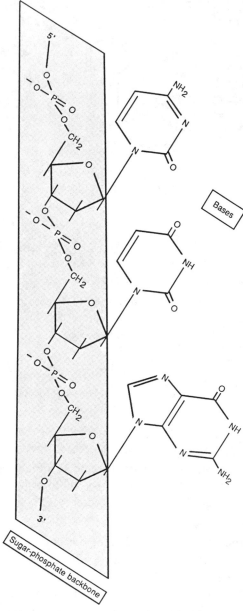

Figure 1.2   A chain of nucleotides linked together by bonds between the sugars and phosphates. The shaded area indicates this sugar–phosphate "backbone" with the bases "hanging out" from it.

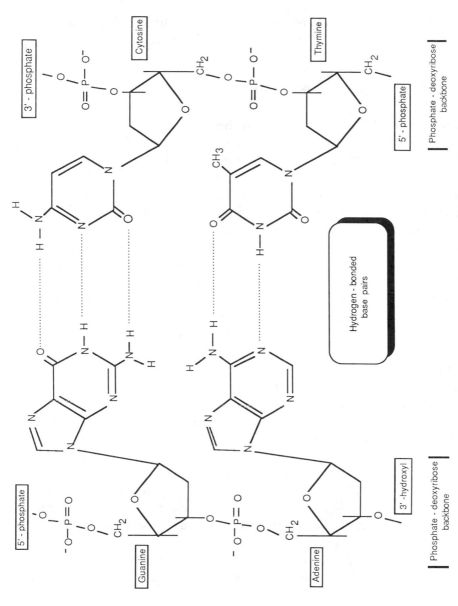

Figure 1.3 Two strands of DNA running antiparallel, with hydrogen bonds holding the strands together. In the DNA molecule, the two strands wind or twist around each other.

is important for DNA replication, and, as will be discussed later, the DNA–RNA specificity is critical for protein synthesis.

DNA, RNA, AND PROTEIN SYNTHESIS

The nuclear DNA contains most of the genetic information for plants (small amounts are present in the organelles) in the form of its base sequences, with three bases (triplets) involved. The processes of transcription and translation transfer the information from DNA into other molecules used by the organism in the multitude of functions involved in growth, development, metabolism, reproduction, and so on. A highly simplified diagram illustrating the results of these two processes is shown in Fig. 1.4, where it may be seen that a sequence of the DNA, making up what may be called a structural gene or cistron, is first transcribed into RNA, and then some of this RNA is translated into protein. No attempt will be made to discuss transcription and tanslation in detail; the reader is referred to several texts for more in-depth yet nontechnical considerations (Lewin, 1987; Fristrom and Clegg, 1988; Goodwin and Mercer, 1983; Ayala and Kiger, 1984; Elseth and Baumgardner, 1984). A simplified diagram of transcription (Fig. 1.5) shows that DNA serves as the template for the synthesis of messenger RNA (mRNA). It is now known that DNA and mRNA are not strictly colinear, but rather the genes may be interrupted (Fig. 1.6); thus the diagram in Fig. 1.5 must be modified. The DNA making up a gene includes certain sequences specifying RNA that is cut out during processing. These sequences are referred to as introns, whereas DNA coding for sequences present in mRNA are called exons. The long RNA molecule produced by transcription of the entire structural gene is designated heterogeneous nuclear RNA (hnRNA), and this molecule then undergoes several alterations (processing) to become mRNA. Certain sequences of bases are added to each end, that is, a "cap" is added to the 5' end of the molecule and a "tail" placed on the 3' end. The former is involved in binding the mRNA to the ribosomes, and

Figure 1.4 Simplified relationship between DNA, RNA, and protein. The information in DNA is transferred to RNA by the process of transcription, and RNA in turn transfers the information to protein via translation.

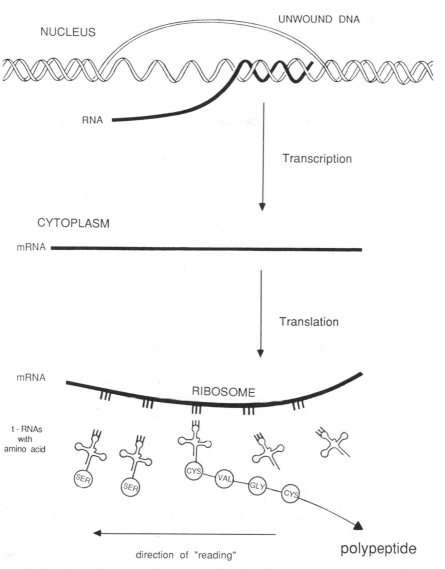

Figure 1.5   Simplified diagram of the processes of transcription and translation to produce a polypeptide or protein. A single strand of DNA forms the template for the messenger RNA (mRNA). The mRNA moves to the cytoplasm, and ribosomes attach to it. Each three-base sequence (codon) in the mRNA is recognized by a complementary three-base anticodon in the transfer RNA (tRNA) molecules. The tRNA molecules with particular anticodons carry a certain amino acid. The base sequences in the DNA, then, ultimately specify the amino acid sequences in the protein.

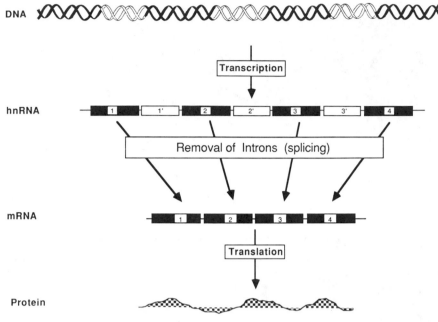

Figure 1.6 Diagramatic representation of transcription and translation showing noncoding regions of DNA (introns). Transcription produces heterogeneous nuclear RNA (hnRNA). The segments specified by the introns are removed, and adjacent segments coded by the exons are spliced together to produce mRNA. Translation then produces the protein. Sequences of nucleotides comprising the introns do not specify amino acid sequences in the protein.

both it and the tail may serve to protect the molecule from attack by nucleases. The hnRNA is then spliced, which involves removal of regions coded by introns followed by the joining together of sequences specified by exons. This processing occurs in the nucleus, and the mRNA then moves through the nuclear envelope into the cytoplasm.

Ribosomes are the sites where mRNA is translated into proteins, and it is here in the cytoplasm that mRNA together with ribosomal RNA (rRNA) and transfer RNA (tRNA) cooperate to bring about protein synthesis. The structures of the latter two types of RNAs will be considered below. Ribosomal RNA and proteins self-assemble to form the ribosome. Ribosomes then move along the mRNA molecules and function in the interaction between mRNA and tRNA to synthesize the protein. Transfer RNA molecules serve as "adapters" between a codon in

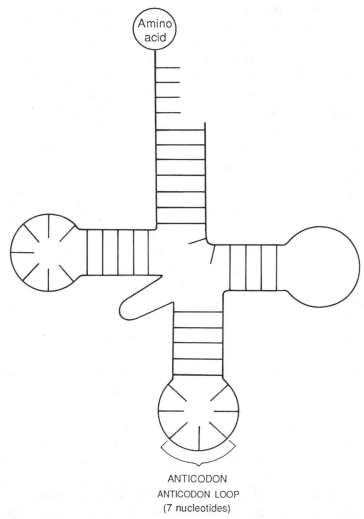

Figure 1.7 Molecule of transfer RNA (rRNA). See text for additional details.

the mRNA and a specific amino acid. The molecules are small, some 75 to 85 nucleotides in length, and have the characteristic cloverleaf form (Fig. 1.7). Complementary base pairing forms the so-called stems, and at the ends of the stems are loops; each stem and loop unit forms an arm (Fig. 1.7). The functional significance of tRNA molecules is that each attaches to a particular amino acid, and in turn the given tRNA can pair with a particular three-base sequence (codon) in the mRNA. That is, a precise sequence of three base pairs in the one part of the tRNA molecule, called the anticodon, is correlated with the ability of the molecule to bind (be charged) with the appropriate amino acid. In this manner, the tRNA functions to translate the base sequences of the mRNA (and indirectly the nuclear DNA) into the amino acid sequence of a polypeptide chain. There are one or more tRNAs for each amino acid. While only a single polypeptide can be synthesized by one ribosome as it moves along the mRNA molecule, more than one ribosome may be moving along a given mRNA, bringing about the simultaneous translation of the mRNA molecule.

A brief consideration of the ribosome from the perspective of the role of rRNA is in order. Eukaryotic ribosomes consist of two subunits that can be dissociated; one of the subunits is about twice the size of the other. The smaller subunit has as its major component the so-called 18S rRNA, whereas the larger subunit is composed primarily of 25S rRNA together with several smaller rRNAs (see below). The genes for rRNA in eukaryotes are present as tandem repeats with each gene present hundreds to thousands of times. The part of the nucleus where rRNA is synthesized is called the nucleolus. Given segments of chromosomes, the so-called nucleolar organizer regions, are associated with the nucleolus. These regions, sometimes referred to as secondary constrictions, contain the rRNA genes (the rDNA). The transcription of rRNA genes is similar to that of mRNA because a large molecule containing the major RNAs (18S and 25S) as well as a small 5.8S is produced initially. The processing of this molecule produces the rRNAs used in ribosome assembly, whereas the so-called spacer regions (transcribed spacer) are removed (Fig. 1.8). Another type of rRNA, 5S, is transcribed from a separate family of repeated sequences not associated with the nucleolar organizer. Ribosomal RNA is the most abundant RNA in the cell, often comprising over 80% of the total RNA.

PROTEIN STRUCTURE

As indicated earlier, the DNA contains the genetic information in the form of its base sequences, that is, these sequences specify or code for particular sequences of RNA, which in turn determine amino acid sequences in proteins. The genetic code refers, then, to the correspondence

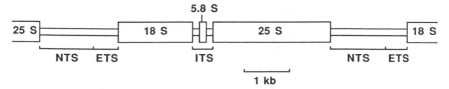

Figure 1.8 Basic repeating unit of genes for 5.8S, 1.8S, and 25S ribosomal RNA (rRNA). An external transcribed spacer (ETS) and internal transcribed spacer (ITS) appear on each side of the 5.85 rRNA gene. The basic units are separated by a nontranscribed spacer region (NTS).

between the sequence of the gene and that of the protein. It consists of different groups of trinucleotides, with each triplet (codon) specifying a particular amino acid. Of the 64 possible three-base codons in RNA, 61 code for specific amino acids (the remaining three function as terminators). Because there are 61 codons for the 21 amino acids, more than one codon must specify at least some of the amino acids. In fact, only two amino acids, methionine and tryptophan, are each specified by a unique codon. For the other amino acids, the number of codons varies from two to six, and the genetic code is said to be degenerate. Degeneracy involves the third base of each triplet in most instances. For 32 of the codons, the third base has no function in specifying the amino acid because all four codons having the same first two bases specify the same amino acid. As examples, CUU, CUC, CUA, and CUG all code for the amino acid leucine. In other cases, the specificity of the third base pertains to whether a purine (A to G) or a pyrimidine (C or U) is produced, and these account for an additional 26 codons. The degeneracy of the genetic code means that base changes (mutations) occurring at the third base of a triplet will not usually cause a change in the amino acid of the resulting protein. The implications of this from a systematic and phylogenetic perspective will be considered later. The genetic code, with rare exceptions, applies to all living organisms, that is, it is nearly universal. The rare exceptions include some genes located in organelles.

  Proteins are similar to nucleic acids in being large polymers, with amino acids comprising the repeating monomeric subunits. There are 20 naturally occurring amino acids, all having the same basic chemical structure. They differ in the nature of the side chain (normally designated by R), with side chains being important in determining various chemical aspects of proteins (Fig. 1.9). The structural similarity of amino acids

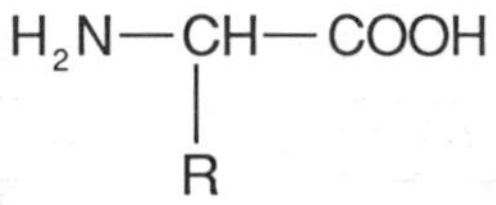

$$H_2N-CH-COOH$$
$$|$$
$$R$$

Figure 1.9 General structure of an amino acid, where R could represent a variety of functional groups.

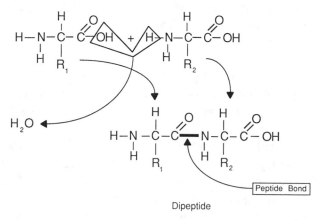

Dipeptide

Figure 1.10 Formation of a peptide bond with release of water.

involves the presence of a carboxyl, –COOH, and an amino, $-NH_2$ group. These give amino acids both acidic (i.e., –COOH loses an H^+ to give COO^-) and basic ($-NH_2$ gains an H^+ to become $-NH_3^+$) properties. The amino group of one amino acid may join to the carboxyl group of another to form a peptide bond, with water given off in the process (Fig. 1.10). The molecule resulting from the formation of a peptide bond between two amino acids is a dipeptide. A chain of amino acid residues linked together by peptide bonds is called a polypeptide. Because the R groups making up the side chains are not involved in forming peptide bonds, they are available to interact chemically with other R groups and are important in determining the structural organization of the polypeptide.

A protein molecule has several different levels of organization. The sequence or arrangement of amino acid residues linked by peptide bonds in a chain is called the primary structure of the protein. The secondary structure of a polypeptide is its three-dimensional shape (or conformation) brought about by hydrogen bonds formed between the oxygen atom of the carboxyl group of one amino acid and the hydrogen atom in the amino group of another amino acid. The helical form, called the alpha helix, represents the most common secondary structure for proteins. Most proteins are not present naturally in helical form, but are further organized into spherical or globular shapes. This common globular conformation is called the tertiary structure of the protein and is the result of various types of bonds, including hydrogen and ionic bonds. Proteins may consist of two or more polypeptide chains aggregated together, these being called multimeric or oligomeric proteins. The quaternary structure refers to how the individual polypeptide chains of a protein molecule are arranged relative to each other. The bonds involved in maintaining the quaterary structure are the same as those producing the tertiary structure except they occur between R groups of amino acids making up different polypeptide chains. An important bond in holding together polypeptides is the disulfide bond.

The polarity of the side chains of amino acid residues is important in determining the conformation (three-dimensional shape) of proteins. In an aqueous environment nonpolar amino acids associate together to the inside or interior of the protein molecule (the amino acids are hydrophobic). By contrast, hydrophyllic amino acids occur on the exterior or surface of the protein. The three-dimensional conformation assumed by a protein is a result of the amino acid sequence or primary structure of the protein.

RATIONALE FOR USING MACROMOLECULES IN PLANT SYSTEMATICS

In previous sections, the structure of the genetic material (DNA) and how information in the DNA is incorporated into RNA and proteins were discussed. This section provides the rationale for using macromolecular data in plant systematics. Harborne and Turner (1984, Chapter 4) presented a lucid, wide-ranging discussion of the rationale for employing chemical data in plant systematics, and it is highly recommended reading. The present discussion will focus more narrowly on using macromolecules as compared with other kinds of data, including secondary chemistry or micromolecules. Certain points made by Harborne and Turner will be commented and/or elaborated upon.

Regardless of what plant systematists prefer to be called (alpha taxonomists, pheneticists, cladists, evolutionary biologists, etc.) the characters they choose for analysis (in whatever manner) are presumed to have a genetic basis. That is, similarities and differences among plants in various characters are inferred to reflect genetic similarities and differences. The more certain one can be that character states are the direct reflection of genetic differences, the stronger the systematic and phylogenetic inferences. From this, it follows that the genetic material itself provides the most basic or fundamental characters that may be employed for purposes of classification and phylogeny. Also, the genetic material is passed from generation to generation, and changes in organisms through time are the result, ultimately, of the transmission of genetic differences to subsequent generations. This forms the basic rationale for the use of DNA in studies of plant phylogeny. From a philosophical (though certainly not practical) point of view, taxonomic classifications at whatever level would best be based on the genetic material. Zuckerkandl and Pauling (1965) were among the first to comment on the differences among molecules regarding information content for phylogenetic purposes. They suggested that molecules occurring in living organisms may be placed into three classes. Semantides are molecules that carry the information of the genes or a transcript thereof. The genes represent the primary semantides in the terminology of Zuckerkandl and Pauling (1965). Previous discussions pointed out that certain sequences of DNA (the introns) in a structural gene are not incorporated into mRNA; thus there is some loss of information in this step. In the terminology of Zuckerkandl and Pauling (1965), mRNA is a secondary semantide (Fig. 1.11).

There is also the potential for considerable loss of information in going from mRNA to polypeptides. The degeneracy of the code means that the complete sequence of mRNA cannot be inferred from knowledge of the amino acid sequence of the protein (see earlier section). Zuckerkandl and Pauling (1965) referred to polypeptides as tertiary semantides (Fig. 1.11). Thus, even if the amino acids of a polypeptide were sequenced, information would still be lost relative to knowing the sequence of its coding structural gene.

Most systematic studies employing macromolecules do not include any type of sequencing. Rather, they probe for differences between molecules employing largely electrophoretic techniques, which are discussed in the next chapter. These electrophoretic studies usually (but not always) involve proteins (either storage proteins or enzymes), and at best they

DNA ⟶ mRNA ⟶ Protein

primary   secondary   tertiary
semantide   semantide   semantide

Figure 1.11 Information content of molecules for systematic and phylogenetic purposes. The genetic material, or DNA, is a primary semantide and has the highest information content, and mRNA is the next highest information molecule, but there is loss of information in the transcription process. Lastly, translation of the message in RNA to protein involves loss of information due to the degeneracy of the code (see text for further discussion).

allow for inferences about allelic differences between individual plants. Sometimes patterns of proteins in gels are used as characters without regard for their genetic bases. These kinds of studies result in considerable information loss relative to sequence information for nucleic acids or proteins. The argument will be made in later chapters that electrophoresis followed by histochemical staining provides very useful systematic and phylogenetic data despite its limitations. Serology will also be considered as another method of ascertaining differences among protein molecules. In this case, particular parts of the protein molecules are probed. Serological data have not been interpreted in genetic terms, but it will be argued that this does not invalidate the method for systematic purposes. The information loss relative to nucleotide sequences that occurs with several commonly used methods does not preclude use of the techniques, it merely reduces the level of precision with which the data may be employed.

A brief consideration of macromolecules versus secondary metabolites (secondary compounds) reveals basic differences in information content, that is, the latter represent a large step away from the genetic material relative to the former. Zuckerkandl and Pauling (1965) referred to molecules whose synthesis is controlled by proteins (tertiary semantides) as episemantic molecules (Fig. 1.12). They suggested that episemantic molecules may be of some use for phylogenetic studies, but contended that such phylogenies need independent confirmation from direct or indirect studies of semantides.

Before using episemantic molecules it should be established whether they are synthesized via the same biosynthetic pathway in different plants. Zuckerkandl and Pauling (1965) asserted that as the number of steps in the biosynthesis of a class of molecules increases, the likelihood

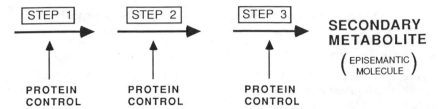

Figure 1.12 Secondary metabolites contain considerably less information compared with macromolecules because their synthesis is controlled by a number of proteins (tertiary semantides). Secondary metabolites are called episemantic molecules.

decreases that their presence results from independent origins in different plants. Thus, at one level, the ability to infer homology of biosynthetic pathways is an important aspect of using episemantides in systematic and phylogenetic studies, but at another level, there is the question of what may be inferred about similarity in primary semantides (DNA) from episemantides, assuming homology of biosynthetic routes has been established. In fact, very little may be inferred about DNA from study of secondary compounds alone, and obtaining the necessary information would require sequencing either the genes encoding the enzymes or the enzymes themselves, which in turn mediate steps in the biosynthetic pathway. Thus, using secondary compounds phylogenetically in a rigorous manner eventually dictates that macromolecules be studied. It was stated earlier that even with relatively "crude" (in the sense of molecular biology) techniques such as gel electrophoresis it is possible to infer allelic differences between plants. This is not the case with secondary compounds in most instances because allelic differences at gene loci specifying enzymes in the pathway may not be reflected in differences in the final product(s). Also, changes at different steps in the pathway could produce the same change(s) in the final product(s). Obtaining maximal phylogenetic value from secondary chemistry involves study of the macromolecules mediating the biosynthesis of the metabolites. This point has been made by several authors (Zuckerkandl and Pauling, 1965; Alston, 1967; Crawford, 1979a; Harborne and Turner, 1984, among others), but it bears repeating because it seems still to be a point of some confusion.

Only a very brief discussion of morphology versus chemistry for inferring genetic differences will be presented. Harborne and Turner (1984, Chapter 4) provided an excellent treatment of the topic and included a table of advantages and disadvantages of chemical versus exomorphic features in systematic research. Many morphological features no doubt result from a complex series of enzyme-mediated steps leading to the final form of a structure. With few exceptions, the genetic bases (in a Mendelian sense) of morphological differences are not known, although it is evident that some features have a relatively simple basis (one or two genes) whereas others are polygenic (Gottlieb, 1984a; Hilu, 1984). Relating morphological differences to DNA differences is, at present, largely only a goal to be highly desired by the plant systematist.

Frelin and Vuilleumier (1979) considered the question of information content in macromolecule versus metabolites and morphology (structures). They called the former genetic systematics and the latter traditional systematics. They argued that ultimately all kinds of data should be incorporated despite the loss of information in certain characters. Frelin and Vuilleumier (1979) indicated that the study of molecules for purely phylogenetic or classificatory purpose may preclude any consideration of the adaptive significance of characters. This is an important point, and while I emphasize the information content of macromolecules for phylogenetic and systematic studies, this does not mean that studies of the other features having less genetic information content yet important to the phenotype of the organism are not critical to the study of plant evolution. Rather, it is a matter of keeping in mind what questions are addressed most effectively with what kinds of data.

ANALYSIS OF MACROMOLECULAR DATA FOR INFERRING PHYLOGENIES

During the past 15 years there has been increased awareness of and emphasis on the analysis of systematic data of all kinds. Of continuing concern and debate has been the issue of phenetic versus cladistic methods for producing hypotheses of relationships among plants. Little purpose would be served by recounting the numerous articles, commentaries, and so on that have appeared in various journals such as *Systematic Zoology* and *Cladistics*. Suffice it to say that, in broadest terms, phenetics is concerned with classifying or grouping organisms on the basis of numerous, equally weighted characters with carefully coded states. In essence, then, phenetics is an attempt to ascertain overall similarity using as many characters as possible. Cladistics, by contrast, is an attempt to determine branching patterns (cladogenesis) in evolution. These patterns are determined by using derived (apomorphic) character states of the characters involved. In cladistics, primitive (plesiomorphic) character states are of no value for producing branching patterns. In fact, only derived character states shared by two or more taxa (synapomorphies) are cladistically informative, while derived character states present in only one taxon (autapomorphies) are not useful for assessing relationships. For those wishing to become more familiar with these schools, general discussions of each may be found in Sneath and Sokal (1973), Clifford and Stephenson (1975), Nelson and Platnick (1981), and Wiley (1981). In addition, reference to recent issues of *Cladistics* will provide one avenue into the literature, and the recent paper by Cronquist (1987) provides a number of references on the use of cladistic methodology in plants.

While there has been some discussion as to whether explicit methods (phenetics and/or cladistics) are superior to more "intuitive" or "narrative" methods (i.e., not explicit with regard to use of characters and methods of analysis) for generating phylogenetic hypotheses when morphological characters are employed, this has never really been an issue with the various kinds of molecular data. That is, it has been accepted that, due in part to the large number of characters involved, one must perform some sort of analysis in order to use the data in a meaningful way to produce phylogenies.

The purpose of the present discussion is to consider in a very general way some of the methods employed for analyzing molecular data because an in-depth critical discussion of assumptions and theories of the various methods is beyond the scope of this book (i.e., beyond my comprehension!). Hopefully, these comments will be useful to the plant systematist and will provide general background information for discussions in later chapters. With regard to the "best way" to construct phylogenetic trees, the recent statement by Hartl and Clark (1989, p. 379) that "There is no universal theory that provides a single optimal way to construct phylogenetic trees . . .' seems most appropriate.

Nei (1987, Chapter 11) and Felsenstein (1988) placed the methods for constructing phylogenetic trees into several categories, including distance matrix methods (including compatibility), parsimony methods, and the maximum likelihood method. Each of these will be considered in turn. There has been considerable debate as to whether distance measures from molecular (or any other) data should be used to produce phylogenetic hypotheses (Farris, 1981, 1985, 1986; Felsenstein, 1982, 1984, 1986,

1988; Nei, 1987, Chapter 11). A variety of distance measures are available; several that have been used for allelic frequency data are presented in Chapter 6. Felsenstein (1982, 1984, 1988) and Nei (1987, Chapter 11) provided a discussion of various distance measures.

Once pair-wise distances have been determined for the operational taxonomic units (OTUs) (these could be individuals, populations, species, or higher taxa), then one attempts to fit a tree to these distances between the OUTs. These methods are phenetic in nature. The tree, by showing the sum of branch lengths between the OTUs, predicts the distance between them (Felsenstein, 1988). The preferred phylogeny from distance data is the one that shows minimal disparity between the distance values in the data matrix and the path lengths in the tree. Felsenstein (1988) emphasized that, while it is widely believed that the use of distance methods assumes a molecular clock (i.e., constant rate of change through time), this is not true.

The questions and debate about the utility of distance measures for generating phylogenies will not be treated in detail; rather, the reader is referred to Farris (1981, 1985, 1986), Swofford (1981), Felsenstein (1982, 1984, 1986, 1988), and Nei (1987, Chapter 11). One source of discussion has been what the branch lengths in the trees really represent. One view is that branch lengths represent the distances between ancestors and descendents in a tree. That is, if one actually had data for hypothetical ancestral taxa at the interior nodes of the tree and not just the extant taxa at the tips of the tree, then it would be possible to ascertain from these imaginary data the actual observed branch lengths in the tree. A problem with this interpretation of branch lengths, as emphasized by Farris (1981), is that often it is necessary to infer distances that could not exist if one attempted to reconstruct evolutionary events. With this interpretation, it is not possible to justify application of distance methods to construct phylogenies. Felsenstein (1984) argued that it is possible to interpret branch lengths in another way. He viewed the branch lengths of the tree as representing expected distances between ancestors and their descendents. That is, for the true tree, there must be an evolutionary process (of a probabilistic nature) that produced the characters one sees for the species in the tree (Felsenstein, 1984). This means that because the branches are expected distances they are not constrained so as to exhibit all the properties of actual distances. Felsenstein (1984) argued that once this view is accepted, then objection from interpreting branch lengths as path lengths do not exist.

Nei (1987, Chapter 11) presented an extended discussion of the various distance measures and the methods for producing trees from the distances. One point emphasized by Nei is that the number of loci used (referring primarily to isozyme loci, but applying to all loci) is a critical factor for producing accurate (as inferred from simulation studies) trees regardless of the methods applied.

While the utility of using distance measures for inferring phylogenetic relationships remains an open question, it seems likely that such measures will continue to be used with data such as alleleic frequencies. As discussed in Chapter 6 and later in this chapter, it is often difficult to use cladistic analyses on allelic frequency data. As long as this situation remains, distance measures will probably be employed even with their potential problems.

It seems fair to say that parsimony represents the most popular method for constructing phylogenies, not only from molecular data, but from all kinds. In simplest terms, parsimony constructs phylogenetic trees by assuming that the observed characters evolved by the minimum number of changes. As Felsenstein (1983) emphasized, parsimony is not "uniquely definable" in the sense that the term does not dictate which events are to be minimized and which allowed.

For molecular data, so-called Wagner parsimony is commonly employed (Felsenstein, 1982; Nei, 1987, Chapter 11). With this method, reversions of character states are allowed, and the tree (or trees) that requires the fewest total changes is the one with maximum parsimony (i.e., is the most parsimonious tree). While Wagner parsimony allows both forward and reverse changes to occur, Dollo parsimony differs in permitting only reverse changes to occur. For any character, consider 0 to be the ancestral state and 1 the derived state. The derived state is viewed as the complex (whether it be morphological or molecular characters) condition and the ancestral condition as less complex. The basic idea of Dollo parsimony is that complex character states arise rarely whereas these character states can be lost much more easily during evolution. Dollo parsimony has been discussed with regard to restriction site gains and losses in DNA (DeBry and Slade, 1985). Most cleavage sites for restriction endonucleases consist of a sequence of six bases. This means that the independent gain of a site at a particular place in a stretch of DNA requires the evolution of the identical six-base sequence, whereas the independent loss of a restriction site in two DNAs could occur by a nucleotide replacement at any of the six bases in the sequence (see Chapter 13 for discussion of restriction sites).

Use of parsimony assumes that evolution occurs in the simplest manner or by the most direct route. As Felsenstein (1983) and others have emphasized, not only is there little evidence that this is invariably the case, but there are indications that evolution often does not take the shortest path. A basic problem, however, is that it is difficult and may require considerable effort to show that evolution in a particular group of plants has not occurred via the simplest path. Thus, in the absence of data to the contrary, the systematist is often forced into an assumption of parsimony.

Felsenstein (1978, 1983) emphasized that there are certain conditions under which parsimony can give incorrect phylogenies. This is particularly true if the rate of change has been very high in certain branches relative to other branches of a tree. The higher rates of character change in branches will increase the probability of parallel changes in these branches. The more characters employed, the higher the probability that changes will be parallel changes. By contrast, branches with much lower rates of change, while containing many fewer parallel character state changes, will nevertheless have many fewer cladistically informative changes. Thus the parallel changes in certain lineages will become much more probable than cladistically informative changes in the branches with lower rates of change.

Compatability differs from parsimony in that it attempts to find the largest number of characters that can be used to produce a tree that has no homoplasy (i.e., no parallelisms or reversals are allowed). With this method, characters that are not compatible with the tree resulting from the largest set of compatible characters are disregarded.

The maximum likelihood method selects a tree that maximizes the probability that the observed data would have occurred (Felsenstein, 1973, 1988). That is, given the data and a model about evolution, one calculates the probability by obtaining the data from a given tree. While Felsenstein has done much to develop maximum likelihood methods for inferring phylogenies, the approach has not yet achieved the popularity of parsimony methods, one of the main reasons being the computational time involved.

Felsenstein (1988) presented extensive discussions of the various approaches mentioned above. The discussions centered on the statistical justifications of different methods. One type of method involves resampling of data, with two of the most widely used methods being the jackknife and bootstrap. The bootstrap method in particular has been employed for resampling in phylogenies. That is, the characters from the original data matrix are sampled (with replacement) until a data set the same size as the original data set is achieved, and the new data set is used to construct tree. The process of sampling and tree construction is repeated many times; then a majority rule consensus tree (see below) may be constructed, and those groups that occur in over half of the trees are considered to be supported by the method (Felsenstein, 1988).

For parsimony and character compatibility methods, one has to determine the ancestral and advanced character states for the characters employed. Although there has been considerable debate about the best methods for polarizing character states, outgroup comparison is the most widely used method for plants, and this is particularly true for certain kinds of molecular data such as restriction site analyses of chloroplast DNA (see Chapter 16). For other types of molecular data such as presence of alleles at isozyme loci, the outgroup method is not as easy to apply for polarizing character states (see Chapter 8). With the outgroup method, the character state present in the taxon used as the outgroup is assumed to represent the ancestral state. One practical problem often encountered when using outgroup comparison in plants is determining the taxon most closely related to the taxon under study. When the outgroup is not at all certain, it is advisable to employ several taxa as outgroups.

One situation that may arise when phylogenetic trees are constructed from different data (or different samples of the same data set) is that different trees may be produced. How to resolve these different phylogenies then becomes a problem. One approach for dealing with different phylogenetic trees is the construction of consensus trees. There are several approaches for producing consensus trees, with the Adams method (Adams, 1972), strict consensus, and majority consensus being the most common (Hillis, 1987). With the Adams consensus tree, clades that are in conflict are collapsed to the node at which the different hypotheses are in agreement (Fig. 1.13c). With an Adams consensus tree, it is possible to identify those taxa that cause the conflicts. However, it may be seen by comparison of Fig. 1.13a–c that the Adams consensus tree may not resemble either of the original trees. A strict consensus tree includes only groups of taxa that occur in all of the trees (Fig. 1.13a,b,d). This method is the most conservative approach for reconciling differences between different phylogenetic trees (Hillis, 1987). A majority consensus tree is constructed by recognizing those groups occurring together in the majority of trees produced. That is, it is similar to strict consensus but requires that only a majority and not all of the trees must have the same groups in order to include them in the consensus tree.

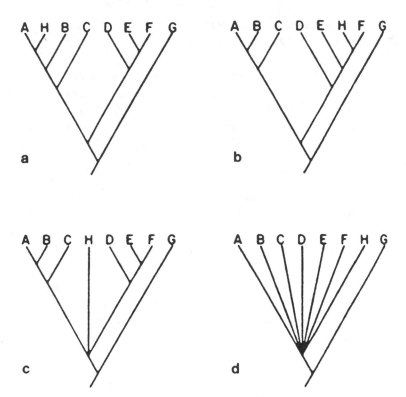

Figure 1.13 Cladograms illustrating Adams consensus and strict consensus trees. (*a,b*) Two conflicting trees. (*c*) Adams consensus tree. (*d*) Strict consensus tree.

The value of consensus trees for resolving conflicting phylogenetic hypotheses is debatable. On the one hand, they do allow one to see the similarities among trees from different data sets, and, conversely, they can indicate where the conflicts reside. In a sense, though, consensus trees do not really solve the problem of *why* there are conflicting trees from different data. Also, a great deal of resolving or descriptive power is often lost with consensus trees (Hillis, 1987). Perhaps one useful purpose of consensus trees is to encourage additional study and analysis of existing data as well as to gather additional data.

It is worth mentioning that phylogenies based on molecular data are usually assumed to represent organismal phylogenies. Nei (1987, Chapter 11) warned that these hypotheses represent the evolutionary history of the genes or other molecules and not necessarily the organisms. This could be of concern to plant systematists, who are interested in organisms and not the molecules or genes per se. Factors such as hybridization and gene duplications could cause problems in distinguishing molecular and organismal phylogenies.

We return to the point made earlier in this discussion, namely, that there is no one best way to construct phylogenetic trees. Felsenstein (1983) emphasized this notion when he indicated that the choice of methods to employ when producing a phylogenetic hypothesis is a biological decision that must be made by the systematist using the method and not by a mathematician or computer programmer. Felsenstein further commented that no one method is preferred in all situations, and he particularly warned that to consider the simplest phylogenetic tree (i.e., the most parsimonious) as the preferred phylogenetic hypothesis leaves one open to several important biological questions. Perhaps the most critical question is why one should always view evolution as occurring by the simplest or shortest route.

Two useful packages of programs for these (and other) phylogenetic analyses are PHYLIP by J. Felsenstein and PAUP by D. Swofford. Because these programs are constantly updated, prospective users are advised to check with Felsenstein and Swofford. Comparisons of the features of these and other programs were given by Fink (1986) and Platnick (1987).

SUMMARY

The DNA, or genetic material, determines the structure of RNA; through RNA, the primary structure of the protein is specified by the DNA. The sequence of bases in the DNA dictates the sequence of bases in messenger RNA. Not all sequences present in DNA are present in messenger RNA. That is, they are "cut out" during the maturation of the RNA. Ribosomal and transfer RNA are essential for translation of the information in the messenger RNA to the amino acid sequence in the protein. The sequence of the amino acids in the protein constitutes its primary structure. Three additional levels of structural organization exist within proteins.

A general discussion of the information content in different molecules serves as a framework for considering macromolecular data for systematic and phylogenetic studies in the following chapters. The practicing plant systematist will recognize that the type of information most desirable for phylogenetic studies, DNA sequences, is rarely available. With the virtually total absence of these data, the advantages and limitations of other methods from a systematic–phylogenetic perspective will be evaluated. Emphasis will be placed on the value of particular methods and molecules for addressing various systematic and phylogenetic questions in plants. Special attention will be given to the taxonomic level at which certain methods and/or molecules are most valid. Specific points will be illustrated and generalizations supported by discussions of particular studies.

CHAPTER 2

Electrophoretic Separation of Proteins: General Comments on Methods and Generation of Data

INTRODUCTION

Electrophoresis of proteins is a commonly employed technique for generating systematic data from macromolecules, and the method has become increasingly popular among plant systematists. Because the next several chapters will be devoted to electrophoretic studies, this one will provide a general consideration of the method, hopefully furnishing adequate background information for discussion of results in the following chapters, and preventing repetitive comments on methods in each chapter. Certain more specialized aspects, such as histochemical staining following separation of enzymes, will be considered in the appropriate chapters.

THE ELECTROPHORETIC METHOD

Many biological molecules (amino acids, proteins, nucleic acids, etc.) have groups that can be ionized so the molecules occur in solution with an electrical charge, that is, as an anion ($-$) or a cation ($+$). These charged molecules will migrate or move in an electrical field, in essence forming the basis of electrophoresis. During electrophoresis, cations ($+$ net charge) will move to the cathode ($-$) while anions ($-$ net charge) migrate to the anode ($+$). Both the samples to be electrophoresed and the supporting medium (such as a gel) must be dissolved in a buffer in order to conduct the electric current. The buffer is also critical for maintaining the same ionization conditions of the molecules in the samples because a change in pH may alter the charges on the molecules. The current is actually maintained because electrolysis occurs at the electrodes, which are emersed in tanks containing buffer. Electrolysis produces hydroxyl ions and hydrogen at the cathode, and oxygen and hydrogen ions at the anode as follows:

$$\text{Cathode} \quad 2e + 2H_2O \dashrightarrow 2OH^- + H_2$$

$$\text{Anode} \quad H_2O \dashrightarrow (2H)^{2+} \tfrac{1}{2}O_2 + 2e$$

Hydroxyl ions generated at the cathode cause dissociation of the weak acid (HA) portion of the buffer, thus causing the formation of more anion (A^-) to conduct current to the anode. At the anode, the anions combine with H^+ ions to form HA again, and the electrons go into the electric current. The majority of the current between the electrodes is conducted by ions from the buffer, with the samples providing a very small proportion of the ions. If the electric field is removed before the ions in the samples reach the electrodes, the ions in the mixture are separated according to electrophoretic mobility. The net charge on molecules in a sample is an important factor affecting mobilities. Electrophoresis, in essence, represents a form of electrolysis.

Electrophoretic separation of proteins is normally carried out in some type of inert supporting medium such as gel, paper, or cellulose acetate. The two most widely employed supporting media for electrophoresis of plant proteins in systematic studies are starch and polyacrylamide gels. Use of a supporting medium causes ions in a sample to migrate as zones, which is why the method is called zone electrophoresis. With zone electrophoresis, the net charge affects relative mobility, with an increase in charge causing the relative migration distance to increase. Gels also contain pores that act as molecular sieves; thus both sizes of molecules and their shapes (or conformations) will affect their relative mobilities because of the sieving effects of the gels.

The net charge of a protein molecule at a particular pH will depend on the sum of the negative and positive charges on the side chains of the constituent amino acids. Aspartic acid and glutamic acid have carboxyl groups (which could carry a negative charge), whereas arginine, lysine, and histidine have side chains carrying a positive charge.

Although "routine" electrophoresis involves a combination of net charge as well as size and/or conformation for separating proteins, other methods use only one of these features. For example, the technique of isoelectric focusing (or electrofocusing) involves placing proteins in a pH gradient created in a gel by the incorporation of substances called ampholytes. The proteins then migrate in the gel until they reach the pH at which they no longer have a net charge, that is, the proteins reach their isoelectric points. This is a very useful and sensitive method for separating mixtures of proteins. In electrofocusing, the pores in the gels exert very minimal sieving effects.

Another electrophoretic method involves using a buffer system that dissociates proteins into their constituent polypeptide subunits. The ionic detergent sodium dodecyl sulfate (commonly designated SDS) is usually employed to dissociate the proteins. A thiol reagent, such as dithiothreitol or 2-mercaptoethanol, is needed to break disulfide bonds. Proteins are denatured by heating at 100°C with the SDS and thiol reagent. The constituent polypeptides bind the SDS at a constant weight ratio, and thus the polypeptides have the same negative charge densities. This means that the polypeptides will migrate in gels (polyacrylamide is used rather than starch) according to their sizes. This method is also very useful for ascertaining the sizes of polypeptides.

The methods of electrofocusing and SDS–polyacrylamide gel electrophoresis can be combined to separate mixtures of proteins. In two-dimensional electrophoresis, proteins are often separated in the first dimension in a pH gradient (i.e., by electrofocusing), and then SDS–polyacrylamide electrophoresis is employed for separation in the second dimension.

A detailed consideration of electrophoretic methods is beyond the scope of the present discussion. Gordon (1975), a series of papers in Hames and Rickwood (1981), and Chapter 1 in Dunbar (1987) provide good general discussions of starch and polyacrylamide electrophoresis.

GENERATION OF SYSTEMATIC DATA

This topic will be discussed in detail in the following chapters; some very cursory comments will suffice for the present. Proteins that consistently migrate differentially in a gel under a given set of conditions are different in some way, whether it be in net charge, size, shape, or a combination of these factors. In turn, it may be inferred (with some exceptions) that the plants producing these different proteins are different genetically. In some instances, these genetic inferences may be made at the level of allelic variation of structural genes. In other cases, such inferences are not possible, and it can only be said that genetic differences of some sort must exist.

While it is easy to demonstrate differences between proteins electrophoretically, demonstrating similarity (i.e., identity) is a more difficult matter. As will be considered in some detail later, proteins may differ in amino acid sequences yet these differences will not result in alteration of electrophoretic mobility under a given set of conditions, or even under two or more conditions. This may be especially true if amino acid differences do not result in changes in net charge of the protein. The method of choice is to subject the proteins to various electrophoretic conditions, such as gel size and pH, and see if different mobilities are detected. If none is found, then one simply reports that the proteins are indistinguishable under these conditions, and assumes they have the same structure. It is recognized, of course, that this may not be the case in all instances, but the greater the number of conditions that fail to separate the proteins, the higher the probability they are identical.

SUMMARY

Gel electrophoresis is an effective method for separating proteins. Proteins migrate differentially in a gel because of differences in net charge, size, shape, or a combination of these factors. Electrofocusing may be employed to separate proteins on the basis of differences in net charge. Alternatively, SDS–polyacrylamide gel electrophoresis resolves proteins according to size rather than charge or conformation. Electophoresis cannot demonstrate conculsively the identity of two proteins; it can only show differences. The greater the number of electrophoretic conditions under which proteins comigrate, the higher the probability that they are identical.

CHAPTER 3

Electrophoresis of Seed Storage Proteins

INTRODUCTION

For the past 25 years electrophoretic profiles of seed storage (and to a lesser extent pollen) proteins have been employed in systematic studies. The method involves extracting the proteins (the proteins obtained will depend on the extracting medium), subjecting the mixture to electrophoresis (usually in acrylamide gels), and then visualizing them in the gels with a general protein stain such as Coomassie Blue or Imido Black. This produces a pattern of stained bands in the gels, and these bands represent the data for the systematist (Fig. 3.1).

Seeds have been used most commonly as the sources of proteins because they represent a well-defined stage in the life history of a plant. Variables associated with development of a leaf, for example, limit the value of protein profiles for taxonomic purposes. In addition, seeds represent a rich source of proteins and provide sufficient quantities for electrophoresis.

Historically, solubility has been used as an important criterion for classification of proteins, and while there are problems associated with strict use of this criterion, it is an important consideration for the systematist. This is particularly true with seed storage proteins because use of proper extracting medium is critical for obtaining proteins for electrophoresis. Albumins, by definition, are soluble in water, globulins are soluble in dilute saline solutions, and prolamins are soluble in alcohol. A class of proteins called glutelins are not easy to solubilize, and because they usually are of high molecular weight, they are often viscous in solution. Dilute alkali or acids are often required to bring them into solution. The solubility of proteins is dependent largely upon the nature

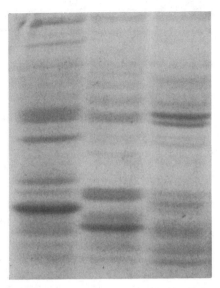

Figure 3.1 Seed storage proteins from *Chenopodium* separated by polyacrylamide gel electrophoresis and stained with Coomassie Blue.

of their predominant amino acids. For example, albumins are rich in ionizable amino acids such as arginine, lysine, and glutamic acid, making them soluble in water.

Knowledge of the predominant type of protein present in the seeds of particular groups of plants is important because solubility dictates the extracting medium to be used. The primary type of storage protein may be constant for a particular taxon, for example, globulins are the storage proteins for the Leguminosae, whereas in the grasses (Gramineae) the major seed proteins are (with few exceptions) prolamins and glutelins (Larkins, 1981; Miege, 1982).

ELECTROPHORETIC PATTERNS OF SEED PROTEINS: RATIONALE FOR USE IN SYSTEMATIC STUDIES

The basic rationale for the use of electrophoretic profiles of seed storage proteins in systematics is that proteins are relatively direct gene products (see earlier discussion of protein biosynthesis). The profiles, then, presumably represent some measure of genetic similarities and differences among the plants being compared. A related concern is how the proteins are inherited, that is, what is the genetic basis of differences encountered in profiles visualized by general protein staining following electrophoresis. Most evidence has come from studies of legumes and grasses, a not too surprising situation given the economic significance of the seeds (or actually fruits in the grasses) of these two families. The focus of certain of these studies has not been systematic or phylogenetic in nature but rather primarily toward elucidating protein components of the economic quality of the seed (Payne et al., 1979, 1980, 1981; Payne, 1987; duCros et al., 1983; Kitamura et al., 1984; Lafiandra et al., 1984; Schroeder and Brown, 1984, as examples).

Simple genetic control with codominance has been shown for seed proteins from controlled crosses, with maternal gene dosage effects noted for endosperm proteins (Shewry et al., 1978, 1980; Meacham et al., 1978; Lawrence and Shepherd, 1981; Mahmoud and Gatehouse, 1984; Salcedo et al., 1984; Leonard et al., 1988, as representative examples). Results of genetic studies have also included nonadditive profiles in F_1 hybrids (Dhaliwal, 1977; Kim et al., 1979; Brown et al., 1981, Kraljevic-Balalic et al., 1982, as examples). Other phenomena such as the presence of "novel" bands in the hybrids (Kim et al., 1979; Burnouf et al., 1983) and the possible role of regulator genes (Salcedo et al., 1978; Branlard, 1983) have also been reported. The occurrence of multiple genes for seed storage proteins has been documented for a variety of different plant species (Payne, 1986; Murphy and Thompson, 1988, pp. 81–88, for discussions). These factors suggest that caution is needed when attempting to infer genetic differences from complex protein profiles.

The use of seed protein profiles in systematic studies is based on the assumption that proteins from different individuals, populations, species, and so on are homologous if they have the same mobility in a gel, and produce bands of similar relative intensity and width following staining. The individual bands are treated as discrete characters, with these characters assumed to be relatively direct gene products (keeping in mind the loss of information in going from coding gene to protein). A common means of assessing protein similarities among populations, taxa, and so on is to employ some sort of similarity measure (Johnson, 1972a; Altosaar et al., 1974; Crawford and Julian, 1976; Sahai and Rana, 1977; Booth and Richards, 1978, as examples). Simply computing a/b where a represents the number of shared bands and b the total number of bands found in two populations, taxa, and so on is a frequent method for expressing protein similarities. It is important to stress, however, that this technique does not yield an estimate of genetic distance (see later discussion under enzyme electrophoresis) but rather represents a "fingerprinting" technique.

Consider first several potential technical and interpretive problems when employing protein profiles in systematic studies. One possibly serious source of error is the inference of protein homology from migration equivalence using only one set of electrophoretic conditions, which is how most systematic studies have been carried out. Sometimes mixtures of protein extracts from two taxa are run to ascertain whether bands separate or comigrate. This procedure may allow for a slightly more refined assessment of comigration compared with running the separate extracts side by side in a gel slab. The problem of the possibility of two different proteins comigrating remains, however, if the same electrophoretic conditions are employed in both instances (i.e., whether run side-by-side or as a mixture). A more effective method would be to fractionate the proteins under different gel concentrations, pH values, and so on to see if they may be distinguished; this topic will be considered in greater detail in the discussion of enzyme electrophoresis. While inferences of homology are on sounder grounds if proteins comigrate under a variety of electrophoretic conditions, assumptions of identity are still questionable without genetic tests.

Seed protein profiles may be composed of 20 or more individual bands (Fig. 3.1). The complexity of banding patterns can cause problems of interpretation, and it may become difficult to compare a number of different variable profiles. Also, the problem of establishing comigration of bands becomes more onerous the greater the numbers of bands and populations or taxa being compared.

Despite the limitations mentioned above, protein profiles (primarily from seeds) have been used in various types of systematic studies. Boulter et al. (1966) provided an early overview of the use of electrophoretic profiles, and Ladizinsky and Hymowitz (1979) presented a more recent brief discussion of seed protein electrophoresis in taxonomic and evolutionary studies. The latter authors discussed certain advantages of seed protein profiles including the stability of mature seed storage proteins in the light of environmental and seasonal fluctuations, and the age of the seed. Ladizinsky and Hymowitz (1979) also suggested that protein profiles are often species-specific. This is the case in many instances, but exceptions do exist (see below), and the constancy of the characters must be established for each species (as with any feature).

INTRAPOPULATIONAL STUDIES

In most studies involving seed protein electrophoresis, seeds from the same or different plants are pooled to obtain sufficient protein for the profiles. Exceptions include investigations of plants having seeds of sufficient size to produce profiles from single seeds (Payne, 1976; Shewry et al., 1978; Comas et al., 1979; Nevo et al., 1983, as examples). This means that unless plants have relatively large seeds it is difficult to determine whether there is any spatial substructuring of seed protein variation within a population. This, however, does not limit the use of protein profiles at the interpopulational level or higher.

Nevo et al. (1983) examined seed storage proteins (hordeins) electrophoretically in *Hordeum spontaneum* (wild barley). They collected seeds from individual plants along two transects in a population, and carried out electrophoresis on single seeds from plants. The hordeins visualized in the gels are specified by two gene loci (actually two groups of linked loci) designated *hor-1* and *hor-2* (Fig. 3.2). A total of 15 different *hor-1* and six *hor-2* phenotypes was found, with seven *hor-1* and five *hor-2* common patterns (Nevo et al., 1983). These workers detected significant differentiation of both *hor-1* and *hor-2* phenotypes associated with soil type and topography. They interpreted their results as indicating that some of the seed protein variation found in wild barley may be of adaptive value and that selection may occur over extremely short distances. Regardless of whether the interpretation of the causes of variation is correct, this study showed that seed protein profiles may be used for demonstrating patterns of genetic variation within populations.

Payne (1976) presented preliminary results showing extensive variation of protein profiles among individual seeds of *Ambrosia trifida* (ragweed). While the methods were not spelled out in detail, it appears that appropriate steps were taken to prevent variation caused by immaturity or other developmental differences among the seeds, and the protein variation ostensibly reflects genetic variation within the population of ragweeds.

Levin and Schaal (1972) used individual seeds to study protein variation within and among populations of *Phlox pilosa*. Populations of this species sometimes contain an additional band, with the frequency of the "novel" protein approaching half of the seeds examined in certain populations, whereas it was not detected at all in others. This novel protein comigrates with one occurring in another species, *Phlox glaberri-*

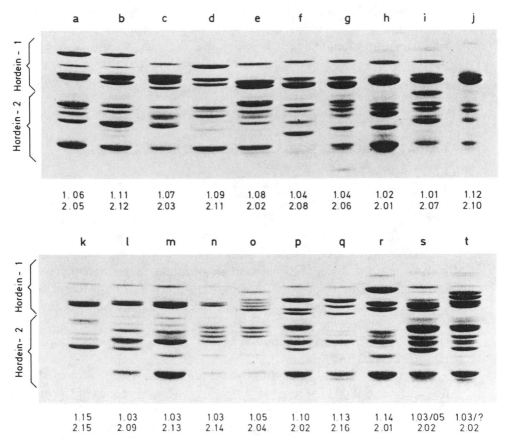

Figure 3.2 Twenty phenotypes (labeled a–q) of hordein storage proteins from *Hordeum spontanlum*. The banding patterns are controlled by two "compound" loci. Numbers below patterns designate phenotypes at each "locus." (With permission from E. Nevo et al., Theor. Appl. Genet. **64**: 123–132. 1983, Springer-Verlag, Inc. Photograph courtesy of E. Nevo.)

ma, suggesting it is the same in each species. These workers hypothesized that the protein occurs in *P. pilosa* because of introgressive hybridization with *P. glaberrima*. Evidence in support of this suggestion comes from the more frequent occurrence of the protein in those populations of *P. pilosa* with the highest possibility for gene exchange with *P. glaberrima*, that is, where incompatibility and seasonal and spatial barriers are the weakest. This study represents one of the very few (if not the only) examples of employing seed protein profiles for detecting introgression. The complexity of the patterns and the attendent difficulties in interpreting homologies of bands in variable profiles are probably major reasons for the paucity of such studies. Additionally, it is imperative that single seeds be used for producing the profiles if one is concerned with the frequencies of particular bands in populations, and often adequate protein is not present for this sort of analysis.

VARIATION AMONG POPULATIONS OF THE SAME SPECIES

A number of studies have focused on intraspecific (interpopulational) variation of protein profiles. Edmonds and Glidewell (1977) found specific banding patterns with little or no variation in those species of *Solanum* that are relatively uniform morphologically. By contrast, species exhibiting higher levels of morphological variability in general displayed greater interpopulational variation in seed proteins. It appears that genetic variation in these species of *Solanum* is reflected in both morphological and protein features. However, in three of the variable species, *S. nigrum, S. villosum*, and *S. americanum*, variation in profiles is not concordant with morphological variability as recognized by intraspecific taxa (Edmonds and Glidewell, 1977).

Crawford (1974, 1976) and Crawford and Julian (1976) examined intraspecific variation of seed protein profiles in several species of *Chenopodium*. Two species, *C. fremontii* and *C. incanum*, exhibited extensive interpopulational variation with similarities between pairs of populations often being as low as 60% (i.e., $a/b \times 100$ where a = number of bands shared by two populations and b = collective number of bands in the two populations). In *C. incanum*, there is little correlation between geographic origin of populations and their profile similarities (Crawford, 1974). By contrast, in *C. fremontii* there is a much higher concordance between geographic distribution and seed proteins as compared with *C. incanum* (Crawford, 1976). Study of several additional species of *Chenopodium*, such as *C. atrovirens, C. hians, C. leptophyllum*, and *C. pratericola*, revealed little or no interpopulational variation (Crawford and Julian, 1976). These species, like *C. fremontii* and *C. incanum*, are diploid, weedy, annual plants widely distributed in the western United States. The reasons for the large differences in variation of seed protein profiles within different species of *Chenopodium* are unknown. Taxa with uniform profiles are not consistently less variable in ecological preferences or morphology than are the two species with variable seed proteins. These results do indicate the necessity for assessing variation within species before making comparisons between species.

Ladizinsky and Waines (1982) examined seed protein variation in *Vicia sativa*, a species variable in chromosome number (and with different karyotypes within each number), morphology, and ecological preferences. Despite these differences, the chromosomal variants can be hybridized to produce fertile F_2 segregants. The same authors found extensive interpopulational variation in *V. sativa*, but the variation was not correlated with intraspecific taxonomic assignment, chromosome number, or the karyotypes of each number. The authors suggested that the polymorphism detected in the profiles could result from germination fitness of particular proteins in given habitats, that is, there has been selection for particular protein components of the seed. They also implicated hybridization among different karyotypes as another cause of variability.

The results from studies of intraspecific variation in seed protein profiles indicate that wide differences may exist even among closely related species in the same genus. This means that if the data are to be used for interspecific comparisons then variation (or lack thereof) must be established for each species. The majority of studies assessing intraspecific variation has found little or no variation among populations, but this cannot be assumed a *priori* for any species.

INTERSPECIFIC VARIATION AND TAXONOMIC UTILITY

Protein electrophoresis has been used most frequently in studies at the interspecific level within a genus and for investigating the progenitors of polyploid taxa. Comas et al. (1979) examined seed proteins (probably largely albumins because of extraction with water) in seven of the eight species in the genus *Bulnesia* (Zygophyllaceae). The genus is restricted to South America, with two extensive geographic disjunctions between certain species (Comas et al., 1979). Morphological, geographical, and ecological differentiation is very pronounced in *Bulnesia*, with habit ranging from large trees to shrubs.

Results of the electrophoretic studes are shown in Fig. 3.3, where the phenogram based on distance among the taxa is shown. Data from seed proteins, in general, support strongly the inferences of relationships obtained from other information. For example, the two tropical arboreal

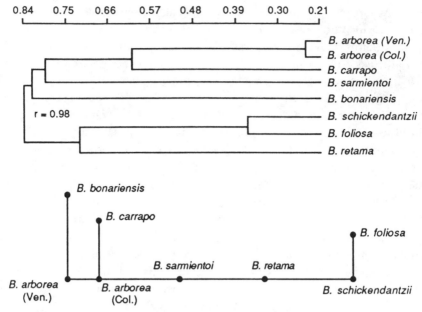

Figure 3.3 (Top) Phenogram of eight OTUs based on the UPGMA cluster analysis of OTU × OTU distance matrix constructed from electrophoretic banding patterns of seed proteins from species of *Bulnesia*, cophenetic correlation coefficient. (Bottom) Prim network for the 8 OTUs based on the eight OTU × OTU distance matrix. See text for additional details. (Redrawn with permission from C.I. Comas et al., Biochem. Syst. Ecol. 7:303–308. 1979, Pergamon Press, Ltd.)

species *B. arborea* and *B. carrapo* have similar yet different protein profiles (Fig. 3.3), a result concordant with other data suggesting they are closely related yet distinct species. The protein similarity of the species pair *B. foliosa*–*B. schickendantzii* agrees with results of a numerical taxonomic study based on 43 morphological features (Fig. 3.3). The morphological study placed *B. bonariensis* near *B. arborea* and *B. carrapo*; in the phenogram based on seed protein data, the former species comes out fairly close to the latter two taxa (Fig. 3.3). Previous studies by Crisci et al. (1979) suggested that these three species represent primitive elements in the genus, and from this perspective, *B. bonariensis* represents somewhat of an intermediate element between the northern, tropical, arboreal species (i.e., *B. arborea* and *B. carrapo*) and the southern taxa (*B. schickerndantzii*, *B. foliosa*, and *B. retama*).Numerical analyses of morpholoical features revealed the same two basic groupings, with *B. bonariensis* intermediate (Crisci et al., 1979). In the "southern" group of three species, *B. foliosa* and *B. schickendantzii* are clearly much more similar in seed proteins than either is to *B. retama* (Fig. 3.3), and this is concordant with the morphological data (Crisci et al., 1979). Both seed proteins and morphology point to an isolated position for *B. sarmientoi* within *Bulnesia* (Fig. 3.3) (Crisci et al., 1979). The study of seed protein profiles in *Bulnesia* provided strong confirming data for interspecific relationships inferred from morphological features, with the relative levels of interspecific divergence in protein profiles and morphological characters paralleling each other very closely.

Hill (1977) carried out seed protein electrophoresis on 11 species of *Mentzelia* (Loasaceae) occurring in the northern Rocky Mountain region and found each species highly differentiated in protein profiles, with a percentage similarity of 78 being the highest for any two species. In certain instances concordance was found between protein similarity and relationships inferred from other data, whereas in other cases this was not so. For example, *M. speciosa* and *M. sinuata* are viewed as closely related, perhaps recently diverged species, and they show higher protein similarity than any other pair of species (Hill, 1977). By contrast, *M. rusbyi* and *M. multiflora* appear closely related on the basis of other data yet exhibit low protein similarities. Hill (1977) interpreed the low protein similarities among the species of *Mentzelia* as the result of divergence following reproductive isolation. In certain instances relative rates of morphological and ecological differentiation between species appear to parallel divergence at gene loci specifying seed proteins, whereas this does not seem to be so in other cases.

Hammel and Reeder (1979) studied seed proteins in 32 populations of the three species of the Old World grass genus *Crypsis* that are naturalized in the United States. Two of the three species, *C. schoenoides* and *C. vaginiflora*, had often been treated as conspecific, whereas the third species (*C. alopecuroides*) is highly distinct morphologically (Hammel and Reeder, 1979). Seed protein profiles clearly distinguished the three species, with *C. schoenoides* and *C. vaginiflora* readily separable (Fig. 3.4— the latter species designated as "compacta"); no intraspecific variation was encountered among the 32 populations examined. Seed proteins were useful evidence in augmenting morphological and chromosomal data, lending support to the recognition of three species of *Crypsis* in the United States.

C. alopecuroides   C. schoenoides   "compacta"

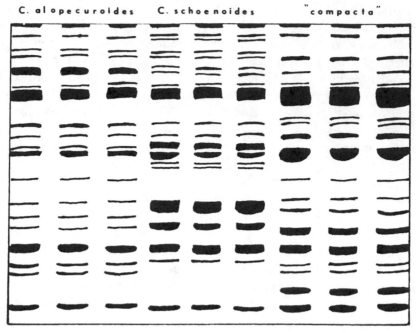

Figure 3.4   Drawing of seed protein profiles from species of *Crypsis* occurring in the United States. "Compacta" is treated as specifically distinct; the correct name is *C. vaginiflora*. (With permission from B. Hammel and J.R. Reeder, Syst. Bot. 4:267–280. 1979, American Society of Plant Taxonomists.)

Crawford and Julian (1976) examined seed protein profiles in 69 populations of a diploid complex of *Chenopodium* occurring in the western United States. The complex had been accorded various taxonomic treatments, the most common being recognition of the seven species *C. atrovirens, C. desiccatum, C. hians, C. icognitum, C. leptophyllum, C. pratericola,* and *C. subglabrum.* Only *C. subglabrum* is a rare and little-collected taxon. Taxonomic difficulties result primarily from phenotpyic plasticity, although hybridization had been suggested as a factor blurring species boundaries. Percentage protein similarity was calculated for all pair-wise comparisons of the 69 populations and a phenogram constructed from these data (Fig. 3.5). It may be seen from Figure 3.6 that each species (with the exception of *C. incognitum*; see below) has little or no interpopulational variation. The three species *C. atrovirens, C. desiccatum,* and *C. pratericola* have identical or nearly identical seed protein

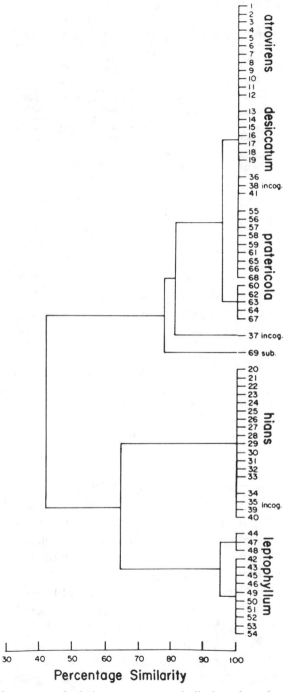

Figure 3.5 Phenogram depicting percentage similarity of seed protein profiles among populations of *Chenopodium*. Numbers refer to populations. (With permission from D.J. Crawford and E.A. Julian, Am. Jo. Bot. **63**:302–308. 1976, Botanical Society of America.)

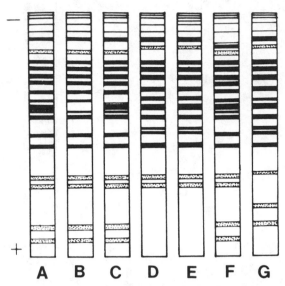

Figure 3.6 Drawings of seed protein profiles in species of *Chenopodium*. (*A*) All populations of *C. atrovirens* and *C. desiccatum*, populations 36, 38 and 41 of *C. incognitum*, and populations 55–59, 61, 65, 66, and 68 of *C. pratericola*. (*B*) Populations 60, 62–64, and 67 of *C. pratericola*. (*C*) Population 37 of *C. incognitum*. (*D*) Populations 42, 43, 45, 46, and 49–54 of *C. leptophyllum*. (*E*) Populations 44, 47, and 48 of *C. leptophyllum*. (*F*) *C. subglabrum*. (*G*) All populations of *C. hians* and populations 34, 35, 39, and 40 of *C. incognitum*. Population numbers same as in Fig. 3.5. (With permission from D.J. Crawford and E.A. Julian, Am. Jo. Bot. **63**:302–308. 1976, Botanical Society of America.)

profiles. Other information suggested that these taxa are distinct yet very closely related (see discussion in Crawford and Reynolds, 1974; Crawford and Julian, 1976; Reynolds and Crawford, 1980). Plants previously included under *C. incognitum* exhibited two electrophoretic profiles, one identical to *C. hians* and the other very similar to or the same as the *C. atrovirens–desiccatum–pratericola* profile (Fig. 3.6). Other information likewise indicated that *C. incognitum* represented two biological entities (Crawford, 1975). *Chenopodium hians* and *C. leptophyllum* had sometimes been viewed as doubtfully distinct, yet they have protein profiles with less than 65% similarity (Fig. 3.5), and the differences in seed proteins correlate with morphological and chromosomal distinctions (Crawford and Reynolds, 1974; Crawford, 1975) in arguing for the recognition of two species. Lastly, the rare *C. subglabrum* had been treated by certain workers as a variety of *C. leptophyllum*, but the protein profile of the one population examined was more similar to proteins of the *C. atrovirens–desiccatum–pratericola* complex. The morphology of *C. subglabrum* likewise suggested a much closer relationship to the latter three taxa than to *C. leptophyllum*. In summary, seed protein profiles were useful for assessing relationships and for helping to circumscribe species in *Chenopodium*.

Bulinska-Radomska and Lester (1985) compared results from electrophoresis of seed proteins with morphology and intercompatibility for examining relationships in five species of the grass genus *Lolium*. Three species are outcrossers, and two are selfers; the former form natural hybrids, have similar karyotypes, and intergrade morphologically. The selfing species do not form fertile hybrids when crossed artificially and are distinct karyotypically. The outcrossing and selfing species are in turn corss-incompatible and show a consistent difference in karyotype. The results of seed protein electrophoresis parallel cross compatibility, karyotype, and morphology in demonstrating two distinct groups; the protein similarities were low for the selfing species when compared as a group with the outcrossing taxa. Also, the two selfing species are more distinct each other in seed proteins as compared with the three outcrossers.

SEED PROTEIN PROFILES AND THE STUDY OF HYBRIDS AND POLYPLOIDS

Seed protein profiles have been used for inferring the 2N progenitors of polyploid species. The method is of considerable value if each putative diploid progenitor produces unique protein bands, because the polyploid derivatives often display simple additive patterns (Vaughan and Waite, 1967; Johnson and Thien, 1970; Cherry et al., 1970; Levin and Schaal, 1970; Edmonds and Glidewell, 1977, as examples).

Levin and Schaal (1970) used seed protein profiles to examine phylogenetic relationships among several species of *Phlox*. Other data had suggested a pattern of reticulate evolution revolving around *P. pilosa* (Fig. 3.7). The three "key" diploid species *P. carolina*, *P. pilosa*, and *P. drummondii* have diagnostic protein profiles. From these profiles, proteins of the tetraploid taxa may be evaluated in terms of their presumed diploid progenitors. The seed of the two tetraploid species *P. aspera* and *P. villosissima* each exhibit several of the diagnostic protein bands from

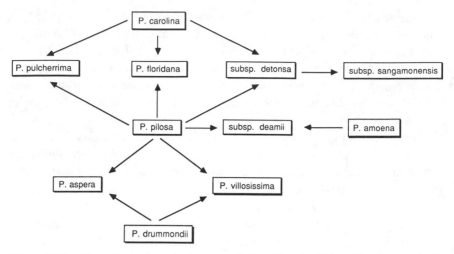

Figure 3.7   Hypothesized evolutionary relationships in *Phlox*. Those taxa indicated only by subspecies belong to *P. pilosa*. (Redrawn with permission from D.A. Levin and B.A. Schaal, Am. Jo. Bot. **57**:977–987. 1970, Botanical Society of America.)

their putative progenitors *P. pilosa* and *P. drummondii*. Even more interesting is that each of these tetraploid species exhibits a distinctive array of protein bands while at the same time being additive relative to the two diploid species. In other words, each inherited several different diagnostic proteins from the same diploids. The presumed allotetraploid derivatives of *P. carolina* and *P. pilosa*, namely, *P. floridana* and *P. pulcherima*, have almost completely additive protein profiles (Fig. 3.7). Thus, in all instances, seed protein profiles support the hypothesized genomic constitutions of the tetraploid species.

At the diploid level, seed proteins provided insight into the origin of *P. pilosa* subsp. *sangamonensis*, a taxon endemic to Illinois whose relationships had been the subject of considerable discussion (see Levin and Schaal, 1970). Briefly, the electrophoretic profiles provided evidence that subsp. *sangamonensis* is closely related to (derived from?) *P. pilosa* subsp. *detonsa*, and that both of these taxa are derivatives of hybridization between *P. pilosa* and *P. carolina* (Fig. 3.7). *Phlox pilosa* subsp. *deamii* had been considered a hybrid derivative of *P. pilosa* and *P. amoena* based on morphological and ecological intermediacy. The seed protein profiles of subsp. *deamii* represent the summation of the profiles

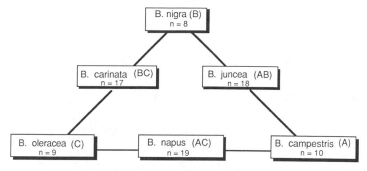

Figure 3.8 Proposed origins of three species of allotetraploids from three diploid species of *Brassica*. Letters in parentheses indicate genomes present; haploid chromosome numbers are given.

of these two species, providing evidence for its hybrid nature (Fig. 3.7).

In addition to using protein profiles for testing hypotheses of reticulate evolution in *Phlox*, Levin and Schaal (1970) examined proteins in synthetic hybrids as well as autotetraploids and found the seed of synthetic diploid bybrids to have additive profiles. Also, autotetraploids of *P. drummondii* and *P. subulata* have the same profiles as the diploids. These results lend increased validity to phylogenetic inferences made from the electrophoretic data.

Seed proteins have been examined in the three diploid species *Brassica nigra*, *B. oleracea*, and *B. campestris*, and their three presumed allotetra-ploid derivatives *B. carinata*, *B. juncea*, and *B. napus* (Fig. 3.8) (Vaughan et al., 1966, 1970; Vaughan and Denford, 1968; Yadava et al., 1979). In the first paper, Vaughan et al. (1966) stated that "The acrylamide gel separation of the seed proteins indicates a greater similarity between *B. campestris* and *B. nigra* than between *B. oleracea* and *B. nigra*. . . ." This evaluation was based on data from albumins and globulins. In a later paper, Vaughan and Denford (1968) examined albumins and globulins of the same three diploid species together with their presumed allopoly-ploid derivatives and gave tables of percentage similarities of albumins and globulins for the three diploid species. In these tables, *B. nigra* has a higher similarity to *B. oleracea* than it does to *B. campestris* (60% versus 25%) in seed protein globulins. With regard to seed alubmins, the simi-larity of *B. nigra* and *B. campestris* is 28%, while the former exhibits a 23% similarity to *B. oleracea*. By contrast, the similarity between *B. campestris* and *B. oleracea* is given as 50%. Based on these data, the conclusion from the earlier study that *B. nigra* is more similar to *B. campestris* than the former is to *B. oleracea* would not appear to hold, and the diagram given in Vaughan and Denford (1968) suggests a closer relationship between *B. campestris* and *C. oleracea.* Vaughan and Denford (1968) emphasized that the number of species-specific bands is limited and that taxa are distinguished primarily by the permutation of bands.

In most instances, the allopolyploid species displayed diagnostic bands from their putative diploid progenitors, that is, *B, napus* has bands from its presumed parents (*B. campestris* and *B. oleracea*). The same applies to the allopolyploid *B. carinata* except that two bands presumed to be the same as in *B. campestris* are present. This would not be expected because *B. campestris* is not considered a parent of *B. carinata* (Fig. 3.8). Vaughan and Denford (1968) also interpreted a protein band in *B. juncea* as identical to one in *B. oleracea*, with the latter not generally viewed as a genome donor to the former (Fig. 3.8). The study of Yadava et al. (1979) of the same diploid and polyploid *Brassica* species included staining for several enzymes in the seeds as well as stains for general proteins. These authors concluded, apparently on the basis of enzyme as well as general protein banding patterns, that *B. campestris* and *B. nigra* are more similar to each other than either is to *B. oleracea* (the same conclusion reached by Vaughan et al., 1966). They also suggested that the parental genomes of *B. carinata* are *B. nigra* and *B. campestris* (recall that Vaughan and Denford, 1968, reported *B. campestris* bands in *B. carinata*) rather than *B. nigra* and *B. oleracea,* as is commonly thought. The implication of this interpretation is that both *B. carinata* and *B. juncea* have the same diploid parents. Yadava et al. (1979) interpreted their data to support *B. campestris* and *B. oleracea* as the parents of *B. napus*, which is the generally held hypothesis.

The question arises as to why two studies (Vaughan et al., 1966; Vaughan and Denford, 1968) gave different interpretations of protein similarities among the three diploid species of *Brassica*. In addition, it seems puzzling that protein bands presumably specific to a certain diploid species apparently occur in those allopolyploid taxa that, from the preponderance of data, are not viewed as containing the genome of the given diploid. It may be that problems in the interpretation of banding patterns account for differing inferences of protein similarity among the diploid species, and the same explanation could apply with regard to banding patterns in the allotetraploids. The *Brassica* situation will be discussed in

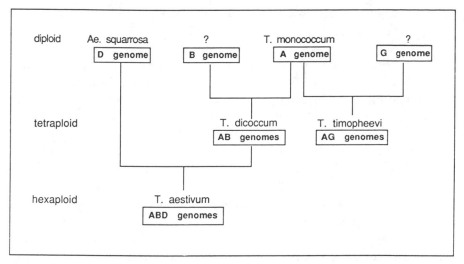

Figure 3.9   Hypothesized origins of polyploid wheats. See text for additional details.

future chapters, and will be summarized as one of the case studies in Chapter 17.

Seed protein electrophoresis has been employed to address the question of the genomic constituents of polyploid wheats (Johnson, 1967, 1972a,b, 1975; Caldwell and Kasarda, 1978). The outline of a commonly proposed phylogeny is shown in Fig. 3.9 along with the genome designations. Certain species are variously treated taxonomically as members of either *Aegilops* or *Triticum*, and either the wild or cultivated species may be included in particular studies. While this grouping may cause some confusion for those not familiar with various taxonomic concepts in wheat, the basic phylogenetic questions remain the same. Johnson (1972a) examined protein profiles in the hexaploid wheat *Triticum aestivum* and interpreted the results as concordant with the hypothesis that it contains the genomes of the tetraploid cultivar *T. dicoccum* and the wild diploid species *Aegilops squarrosa*. The hexaploid displayed banding patterns representing essentially a summation of bands found in the latter two species. As shown in Fig. 3.9, this agrees with the widely accepted view of the donors of two of the three genomes to the hexaploid.

Determining the diploid species that donated the B genome to *T. dicoccum* and thus also to *T. aestivium* has been a long-standing and difficult problem (see discussions in Johnson, 1972b, 1975; Caldwell and Kasarda, 1978). Suffice it to say that various species of *Triticum* and *Aegilops* have been suggested as the donor, with the evidence coming from a variety of sources including morphology and cytogenetics (Johnson, 1975; Caldwell and Kasarda, 1978). Johnson (1972a,b) used seed protein profiles as evidence for a species of *Triticum*, probably *T. urartu*, as the B genome donor. Johnson (1972b, 1975) interpreted the protein data to mean that a species of *Aegilops* had not donated this genome; *Ae. longissima* was among the taxa included in these studies. Caldwell and Kasarda (1978) interpreted banding patterns as indicating *T. urartu* as one genome donor to *T. dicoccum*, but said that they could not dismiss the possibility that the genome from a species of *Aegilops* occurs in the two polyploids. Like Johnson, however, they did indicate that *Ae. longissima* is probably not a genome donor to *T. dicoccum*. The following chapter on rubisco (fraction I protein) presents evidence on the origin of the B genome donor. In addition, the discussions of nucleic acid hybridization (Chapter 14) and chloroplast DNA (Chapter 16) will consider additional data for these same species, and all the information will be brought together as one of the case studies in Chapter 17. It is worth mentioning that Vittozi and Silano (1976) employed data from amylase inhibitors to suggest *Ae. longissima* as the possible B-genome donor.

Johnson (1975) and Caldwell and Kasarda (1978) detected differences in seed protein banding patterns between *T. dicoccum* and another tetraploid, *T. timopheevii*. The hypothesis had been that the two tetraploids *T. timopheevii* and *T. dicoccum* are monophyletic in the sense that they have the same basic genomes (Fig. 3.9). Johnson (1975) suggested that different biotypes of *T. urartu* donated the B genome to both *T. dicoccum* and *T. timopheevii*. Caldwell and Kasarda (1978) did not view their data as conclusive on this point, indicating that neither *Ae. apeltoides* nor *T. urartu* could be ruled out. Data from amylase inhibitors likewise suggest that the *Ae. speltoides* genome is present in *T. timopheevii* (Vittozzi and Silano, 1976). In summary, electrophoretic studies of seed proteins by Johnson (1972a,b, 1975) generated results that were interpreted to implicate *T. urartu* as the probable B-genome donor to polyploid wheats. Caldwell and Kasarda (1978), while suggesting that their results could be interpreted to agree with Johnson on the genomes in polyploid wheats, viewed their data as less than definitive. These results will be discussed further in Chapter 17.

Johnson and Thein (1970) and Cherry et al., (1970) examined electrophoretic profiles of seed proteins in *Gossypium*, and inferred evolutionary relationships from them. Similar, but not identical, techniques were employed for extracting proteins from the seeds, although Johnson and Thein (1970) defatted the seeds first and also lyphilized the material before extracting proteins with distilled water. Cherry et al. (1970) extracted seeds with tris glycine buffer (pH 8.3). Similar types of proteins (albumins) would be extracted with these two procedures. Cherry et al. (1970) did not describe their methods in detail, but by consulting the paper to which they referred, it was ascertained that they used the basic method of Davis (1964) with a 7.5% acrylamide running gel. Johnson and Thein (1970) employed very similar electrophoretic methods, but used an 11.5% acrylamide running gel. It appears, then, that the methods were roughly comparable though not identical. Presumably at least some of the same seed stocks were used because both groups acknowledged the same person for supplying seeds. Johnson and Thein (1970) analyzed their data in a quantitative fashion by computing correlation coefficients for pairwise comparisons of the species, and a phenogram was constructed based on these coefficients. Cherry et al. (1970) provided a narrative of bands shared (or not shared) by different species.

The interpretations of protein similarities by the two groups of workers were in general agreement for certain species, but for other taxa they were different. For example, in the A-genome taxa, both agreed that *Gosypium arboreum* and *G. herbaceum* exhibited very similar electrophoretic profiles. By contrast, in the B-genome group, *G. anomalum* and *G. triphyllum* were interpreted by Cherry et al. (1970) as having very similar seed protein profiles, whereas Johnson and Thein (1970) found them to be quite dissimilar. The latter workers interpreted the low seed protein similarity as concordant with other data suggesting that *G. triphyllum* is remotely related to other B-genome species. Other examples of concordance and nonconcordance between protein profile interpretations by Cherry et al. (1970) and Johnson and Thein (1970) could be cited. The important point is that variations in methodology and/or interpretation of banding patterns can sometimes cause two different laboratories to draw rather disparate conclusions. The reader perhaps could gain an appreciation of the difficulties involved in banding pattern interpretation by consulting the photographs of gels in the two original papers, this being particularly true of gels shown by Cherry et al. (1970).

SUMMARY AND FUTURE PROSPECTS

The routine use of seed (or pollen) protein profiles produced by one set of electrophoretic conditions (one buffer system, one gel concentration, etc.) will probably have limited use in future systematic studies because of the questionable validity of assuming that bands with similar mobilities under these conditions represent homologous proteins. The method is best applied to studies of variation within species or among closely related species. But even in such instances, it is desirable to carry out electrophoresis under several different conditions. The use of electrophoresis, then, involves continual attempts to demonstrate differences among bands, and failure to do this under a variety of conditions (see discussion of sequential electrophoresis in Chapter 5) strengthens the inference of homology.

Drawing conclusions about the genomic constituents of polyploids, particularly if the polyploids are old (providing the opportunity for divergence at gene loci specifying seed proteins in both the diploid progenitors and the polyploids) requires considerable caution because bands from the diploids and polyploids migrating to similar positions in the gels may not represent homologous proteins. In other words, through time one or more proteins in the diploid or polyploid that comigrate could have evolved, that is, there has been convergence for electrophoretic mobility. When such evolutionary significance is assigned to bands in gels, it is imperative that considerable effort be spent attempting to demonstrate that two bands are not the same, thus providing stronger evidence for inferring their similarity.

Seed protein electrophoresis is useful for addressing a variety of specific questions in plant systematics, but for the reasons discussed earlier, it is doubtful the method will enjoy increased application. Indeed, it is significant that the majority of studies discussed in this section were done during the 1970s or earlier. The increased popularity of enzyme electrophoresis with its attendant advantages probably represents a major reason for the decline in seed protein studies.

CHAPTER 4

Electrophoretic Studies of Rubisco: Contributions to Plant Systematics and Phylogeny

INTRODUCTION

The previous chapter dealt with the use of electrophoresis of seed storage proteins in systematic and phylogenetic studies, and the next several chapters are concerned with enzyme electrophoresis. Material included in this chapter could "fit" into these other chapters because the methods bear some similarity to seed protein electrophoresis in that a general protein stain is employed and electrophoretic mobilities are used as the data in most instances. On the other hand, the molecule is an enzyme with a known (and important) function. These considerations, plus the numerous systematic and phylogenetic applications of the molecule, argue for inclusion of this material in a separate chapter. Later chapters will deal with other contributions of rubisco to the study of plant phylogeny.

FEATURES OF THE ENZYME AND METHODS OF STUDY

The most abundant protein present in green plants has as many interesting aspects for study by the plant systematist as it has names. This enzyme, which often comprises half of the total soluble protein in green plants, has been known (among other names) as fraction I protein, carboxydismutase, ribulose 1,5 bisphosphate carboxylase-oxygenase, RuDPCase, RuBP carboxylase, and rubisco. In this chapter it will be referred to as rubisco because this is a simple and convenient designation (Wildman, 1983).

The enzyme is important because in photosynthesis it catalyzes the step that fixes carbon dioxide in the Calvin-Benson (C_3) pathway. Rubisco functions in the reaction of carbon dioxide with ribulose 1,5 bisphosphate to form two molecules of 3-phosphoglyceric acid. It also catalyzes the initial reaction in photorespiration. The enzyme occurs only in the chloroplasts, and because of its presence in such large quantities, it has been purified to homogeneity from a wide variety of different plant species.

Rubisco is a large molecule with a molecular weight of about 550,000. It is an oligomeric protein consisting of eight large subunits (LS) having a molecular weight of 55,000 each and eight small subunits (SS) each of molecular weight 13,000. The large subunits of the enzyme provide the catalytic sites; the function of the small subunits is not known.

From a systematic perspective, the most interesting aspect of rubisco is the mode of inheritance of its different subunits. It was first demonstrated in the genus *Nicotiana* that the small subunits are inherited biparentally whereas the large ones are maternally inherited (Chan and Wildman, 1972; Kawashima and Wildman, 1972; Sakano et al., 1974; see also reviews by Gray, 1980; Wildman, 1983). This pattern of inheritance has been documented for several additional genera of plants (Gray, 1980). It has been established that the large subunit is encoded by a chloroplast gene and the small subunit is specified by a nuclear gene; in a number of species small gene families (some tightly linked) rather than a single gene encode the small subunits (Dean et al., 1985a, 1987; Krebbers et al., 1988; Murphy and Thompson, 1988, pp. 142–144 and references therein).

The different patterns of inheritance for the rubisco subunits are of potential value to the plant systematist, particularly for studies of the origin of species via hybridization. Since evolution of new species via allopolyploidy represents a very common mode of plant speciation involving hybridization, rubisco data are of potential value for documenting the origin of polyploids in many plant groups. In this chapter, the systematic and evolutionary value of electrophoretic studies of rubisco will be discussed and assessed.

The electrophoretic method most commonly employed in systematic studies of rubisco is isoelectric focusing or electrofocusing (see Chapter 2). The oligomeric protein is dissociated into its component subunits, and the individual subunits are separated on the basis of their isoelectric

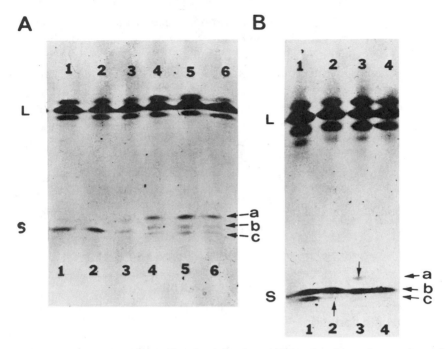

Figure 4.1 Large and small subunit polypeptides of rubisco in species of *Lycopersicon* resolved by electrofocusing. L designates large and S small subunit. (*A*) 1, *Solanum pennellii* (used as standard); 2, *L. hirsutum*; 3, *L. chilense*; 4, *L. cheesmanii*; 5, *L. pimpinellifolium*; 6, *L. esculentum cerasiforme*; (*B*) 1, *L. peruvianum*; 2, *L. parviflorum*; 3, *L. chimielewskii*; 4, *Solanum pennellii*. Arrows point to polypeptide bands that are faint yet consistent and reporducible. (With permission from H. Uchimiya et al., Biochem. Genet. **17**:333–341. 1979, Plenum Publishing Corp.)

points. The polypeptides are then visualized with a general protein stain, and the data consist of the patterns of the bands in the gels (Fig. 4.1) Polypeptides of the small and large subunits are usually separable by this method, with the former often migrating to the region of pH 5.5, whereas the latter migrate to the 6.3 region (Gray, 1980). Considerable variation may be found in the isoelectric point of small subunit polypeptides, and on occasion they may overlap with a larger subunit of the same plant. Gray (1980) presented a critical discussion of the generation of bands of the large subunits and indicated that they are most likely artifacts of the methods employed, but he also emphasized that these are reproducible artifacts and once understood cause no problems of interpretation. Lin et al. (1986) suggested that posttranslational modification of protein encoded by one gene generates the bands. Three bands are invariably found for the large subunit in all species examined.

The small subunits sometimes show more lightly staining "minor" bands in addition to the more heavily stained ones (Fig. 4.1). The "minor" bands were once thought to represent artifacts, but this is apparently not the case because variation in the methods thought to produce the "extra" bands caused no change in their presence in gels (Melchers et al., 1978; also see discussion by Gray, 1980). Even more important, genetic analyses have demonstrated that the lighter staining bands are not artifacts. The "minor" bands may reflect the differential expression of several different genes for the small subunit (Fluhr et al., 1986; Dean et al., 1985a; Sugita et al. 1987).

SYSTEMATIC AND PHYLOGENETIC STUDIES

Intraspecific Variation

Rubisco shows little variation within species, although in most studies sampling has not been very extensive. Over 100 plants of *Nicotiana tabacum* (tobacco) were examined, including a rather wide range of morphological forms comprising several cultivars, but no variation was detected in electrophoretic patterns (Chen et al., 1975; Uchimiya et al., 1977). Likewise, no variation was found in rubisco among many different seed accessions of another species of *Nicotiana, N. glauca* (Chen et al., 1975). Uchimiya et al. (1979) examined 10 accessions of *Lycopersicon esculentum* and five of *L. pimpinellifolium* without finding variation in the banding patterns in either species.

Uchimiya et al. (1979) detected variation among the three accessions of *Lycopersicon peruvianum*, with one accession having two and the others three polypeptides for the small subunit. Uchimiya et al. (1977) reported variation in small subunit composition within populations of *Nicotiana suaveolens*. Three different banding patterns were observed, with one of them containing each of the bands unique to the other two patterns. The more complex additive pattern no doubt resulted from hybridization between plants with the two similar patterns. Bosbach (1983) reported no intraspecific variation in several species of *Erysimum*, but sample sizes were not given. Available evidence suggests that rubisco

TABLE 4.1. Small Subunit Composition of Rubisco in
Lycopersicon **and** *Solanum Penellii*

Taxon	Subunit
S. penellii	b
L. hirsutum	b
L. chimielewskii	a,b
L. chilense	a,b(c)[a]
L. parviflorum	b,c
L. peruvianum	b,c
L. cheesmanii	a,b,c
L. esculentum	a,b,c
L. pimpinellifolium	a,b,c

From Gatenby and Cocking (1978a) and Uchimiya et al. (1979).

[a] Gatenby and Cocking (1978a) reported a, b, and c for this species, whereas Uchimiya et al. (1979) reported a, b subunits.

variation within species may be minimal. However, the lack of adequate sampling in all but a handful of species dictates caution in any statements on the matter, but if further work reveals little or no variation in rubisco then the enzyme may well prove to be a useful species marker.

Use at Higher Taxonomic Levels

Uchimiya et al. (1979) and Gatenby and Cocking (1978a) examined rubisco in the genus *Lycopersicon* (and *Solanum penellii*) and found no differences in large subunit polypeptides among the species. Both laboratories reported the same single small subunit (designated b) polypeptide in *Solanum penellii* and *Lycopersicon hirsutum* (Table 4.1). It should be pointed out that *S. penellii* is closely related to *Lycopersicon* and is perhaps best treated as a member of the genus (Rick, 1979); in fact, D'Arcy (1982) has formally transferred the species to *Lycopersicon*. *Solanum* is often considered ancestral to *Lycopersicon*, that is, it represents the outgroup in cladistic terminology. Both groups of workers reported the same subunit composition for all species except *L. chilense*, in which Gatenby and Cocking (1978a) reported polypeptides a,b,c and Uchimiya et al. (1979) found a,b, in the species.

The rubisco data are interesting when viewed against systematic and phylogenetic concepts in *Lycopersicon* generated from other information. The identical large subunit composition (as determined from electrophoresis) of *S. penellii* and *Lycopersicon*, and the presence of small subunit b in all these species, provides further evidence that *S. penellii* should be treated as a species of *Lycopersicon*, as has now been done. Drawings in Gray (1980) indicate that another species of *Solanum* (*S. meglongena*) has electrophoretic patterns distinct from *Lycopersicon*, and thus unlike *S. penellii*. Considerably more survey work in this huge genus (*Solanum*) would be desirable. The grouping of *S. penellii* (*L. penellii*) with *L. hirsutum* on the basis of the single polypeptide b in common agrees with chloroplast DNA data suggesting a close relationship between the two (Palmer and Zamir, 1982) (also see discussion in Chapter 16). The presence of all three small subunit polypeptides in *L. cheesmanii*, *L. esculentum*, and *L. pimpinellifolium* suggests a close evolutionary relationship among the three, which is concordant with other data (Rick, 1979; Palmer and Zamir, 1982). On the basis of crossability, *Lycopersicon chilense* and *L. peruvianum* are considered to be closely related (Rick, 1979), and data from chloroplast DNA (Plamer and Zamir, 1982) (see Chapter 16) also support this view. It may be seen from Table 4.1 that the rubisco data are not in agreement with this hypothesis regardless of whether *L. chilense* is considered to have a, b or a, b, and c subunits. rick et al. (1976) presented considerable data indicating that while *L. chmielewskii* and *L. parviflorum* should be recognized as distinct species, they are very closely related evolutionarily, with the latter having evolved from the former. The small subunit composition of rubisco supports their taxonomic separation because the two species have different subunits (Table 4.1); however, given the conservative nature of the small subunit, one might not expect plants this closely related to have different subunits. Gatenby and Cocking (1978a) indicated that the presence of polypeptide a in *L. chmielewskii* may be a reflection of its origin as a hybrid between *L. pimpinellifolium* and *L. hirsutum*. The extensive data of Rick et al. (1976) do not suggest this origin for the species. Also, if one were to invoke *L. pimpinellifolium* as the source of a subunit in *L chmielewskii*, then one must also assume that the c polypeptide from the former was not incorporated into the latter.

Both Gatenby and Cocking (1978a) and Uchimiya et al. (1979) viewed species with fewer subunits to be primitive and those with more to be advanced. No reasons are given why this should invariably be true. In the case of *L. hirsutum* (with only one polypeptide) (Table 4.1), this trend in subunit number does not coincide with its relationships based on other data, that is, the species has been viewed as derived relative to species with more subunits. Another problem with always reading evolution in one direction is the lack of knowledge of the mechanism by which additional polypeptides are generated. Two likely processes are gene duplication followed by divergence of the polypeptides and hybridization between species with different subunits. In certain instances, postulating hybridization for the presence of subunits in a species necessitates assuming that the event was so ancient that other evidence of hybridization has been lost. While this hypothesis cannot be dismissed, it isn't very appealing, for example, for explaining the subunit composition of *L. chmielewskii*. Gene duplication followed by divergence seems a feasible explanation for differences in the number of small subunits seen for rubisco. As discussed earlier, in a number of species several genes encode the small subunit (see Wimpee et al., 1983; Dean et al., 1985a,b, 1987; Krebbers et al., 1988, as examples, and discussion in Murphy and Thompson, 1988, pp. 142–144). The loss of subunits could occur by deletions or silencing of genes for a subunit, for example, so there is no reason to postulate that the evolutionary trend is always toward increase. Also, differential expression of genes could affect the number of bands detected in gels.

Gray (1980) cautioned that inferring relationships from electrofocusing patterns alone may not be valid because species with apparently idential patterns can be distinguished by additional studies of the rubisco subunits such as serology and tryptic digests. As with many other characters, chemical or otherwise, the problem with electrofocusing of rubisco is not demonstrating differences but rather inferring identity with confidence.

Extensive studies have been carried out on rubisco in *Nicotiana*, with 63 species surveyed (Chen and Wildman, 1981). Four different clusters of large subunit polypeptides (designated A, I, II, and III types) occur among these species. A total of 13 different small subunit polypeptides was detected in the 63 species. Large subunit type A is restricted to Australian species except for two species in South America. The presence of the type A subunit in the South American plants was viewed as a relic of the rubisco present in the members of the genus that gave rise to Australian species of *Nicotiana* (Chen and Wildman, 1981). In addition, the two South American species contain a small subunit polypeptide otherwise restricted to California. For a complete discussion of the distribution of subunit types in *Nicotiana* together with comments on biogeography, the reader is referred to Chen and Wildman (1981).

Perhaps the best-known example of using rubisco in *Nicotiana* involves the origin of cultivated tobacco, *N. tabacum*, which has a chromosome number of $2n = 48$. Morphological and cytogenetic data suggested that the species originated via chromosome doubling following hybridization between *N. sylvestris* ($2n = 24$) and either *N. otophora* or *N. tomentosiformis* (Gray et al., 1974; Kung, 1976; Uchimiya et al., 1979). Gray et al. (1974) used electrofocusing to show that *N. tabacum* originated from a cross with *N. sylvestris* as the egg parent and *N. tomentosiformis* as the pollen parent. Tobacco contains the same large subunit polypeptides as *N. sylvestris* (which would be expected with maternal inheritance), whereas it combines small subunits characteristic of *N. sylvestris* and *N. tomentosiformis*. Data from rubisco effectively dismissed *N. otophora* as one of the parents and also indicated which species served as the egg parent. The tobacco study demonstrates an important advantage of employing rubisco for determining the origin of allopolyploids. The small subunits, which are encoded by nuclear genes, can be used to indicate the two parental taxa, whereas the maternally inherited large subunits indicate the direction of the cross.

The origin of polyploid wheats was discussed earlier in Chapter 3 with regard to electrophoresis of seed proteins, and will be considered in subsequent chapters. The basic question on the origin of hexaploid wheat concerns the donor of the B genome; it is generally accepted that *Triticum monococcum* provided the A genome, while *Aegilops squarrosa* donated the D genome (see Fig. 3.9) (see note in Chapter 3 concerning names of taxa). *Triticum urartu* and *Ae. speltoides* have been suggested as possible donors of the B genome to the tetraploid *T. dicoccum* and thus to the hexaploid *T. aestivum* as well. As discussed in Chapter 3, Johnson (1972a,b, 1975) concluded from seed protein electrophoresis that *T. urartu* is the probable source of the B genome. Chen et al. (1975) and Hirai and Tsunewaki (1981) carried out electrofocusing studies of rubisco in *T. monococcum, T. boeoticum, T. dicoccum, T. dicoccoides, T. timopheevii, T. aestivum, T. urartu, Ae. speltoides,* and *Ae. squarrosa*. They found the same single band for the small subunit in all taxa, but two patterns were detected for the large subunits (Fig. 4.2). One pattern occurs in *T. aestivum, T. dicoccum, Ae. speltoides, T. timopheevii,* and *T. dicoccoides,* while the other pattern was found in *T. boeoticum, T. urartu, T. monococcum,* and *Ae. squarrosa*. Viewing these data in light of the proposed relationships of diploid and polyploid taxa, one must infer that the maternal diploid parent of *T. dicoccum* donated the B genome to it and,

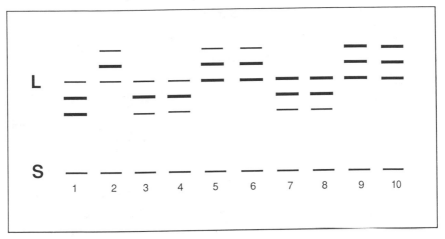

Figure 4.2 Drawing of patterns of large and small subunits of rubisco in species of *Triticum* and *Aegilops*, and reciprocal interspecific hybrids as resolved by electrofocusing in polyacrylamide gels. L designates the large and S the small subunit. 1, *Ae. squarrosa*; 2, *Ae. speltoides*; 3, *T. urartu*; 4, *T. monococcum*; 5, *T. dicoccum*; 6, *T. aestivum*; 7, *T. boeoticum*; 8, Hybrid with *T. boeoticum* the egg parent and *T. dicoccoides* the pollen parent; 9, Hybrid with *T. dicoccoides* the egg parent and *T. boeoticum* the pollen parent; 10, *T. dicoccoides*.

further, that *T. dicoccum* is the maternal parent of *T. aestivum*. These inferences follow because the large subunits of rubisco are maternally inherited (Fig. 4.2). Since *T. timopheevii* has the same electrophoretic pattern for large subunits as *T. dicoccum* and *T. aestivum*, Chen et al. (1975) and Hirai and Tsunewaki (1981) reasoned that the former must have the same B genome donor as the latter two species. The rubisco data implicate *Ae. speltoides* rather than *T. urartu* as the source of the B genome. Chen et al. (1975) pointed out that only one accession of each species was examined, and B.L. Johnson (1976) emphasized this point further by questioning the validity of either accepting *Ae. speltoides* or dismissing *T. urartu* (or *T. boeoticum* for that matter) as B genome donors because the sample sizes did not preclude the possibility that rubisco variation could exist within one or more of the species. Hirai and Tsunewaki (1981) examined a number of lines of several species without detecting infraspecific variation. Gray (1980), citing a personal communication from K. Chen, indicated that further accessions of the species had been examined without any evidence of infraspecific variation for the large subunit of rubisco.

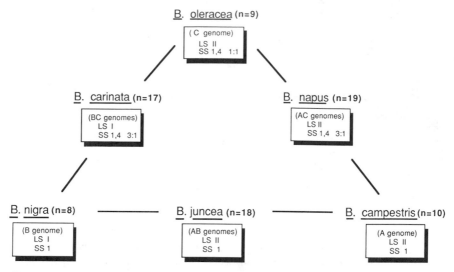

Figure 4.3 Proposed phylogenetic relationships among species of *Brassica*. Chromosome numbers are given in parentheses after each species name. Each box contains the genome designation and the large and small subunit polypeptide composition for each species. The ratios following the small subunit designations for three of the species indicate the dosages of the respective polypeptides present in each species. See text for further discussion and details.

Electrophoretic banding patterns for the large subunit of rubisco clearly implicate *Ae. speltoides* as the B genome parent of the teraploid *T. dicoccum* and hexaploid *T. aestivum*) while arguing against *T. urartu* as the source of the B genome. The results of Chen et al. (1975) suggest that *Ae. speltoides*, in addition to donating the B genome to *T. dicoccum*, also is the source of the G genome to *T. timopheevii* because the banding patterns are identical for all three taxa. These data will be contrasted with other molecular information available for wheat in Chapter 17.

Electrofocusing of rubisco in diploid species of *Brassica* and their presumed allopolyploid derivatives has been carried out by a number of different workers (Gatenby and Cocking, 1978b; Uchimiya dn Wildman, 1978; Robbins and Vaughan, 1983). A wealth of data, primarily from morphology and cytogenetics, suggested the relationships shown in Fig. 4.3. Uchimiya and Wildman (1978) and Robbins and Vaughan (1983) documented the distribution of large subunits among the species of *Brassica* (Fig. 4.3). From these patterns it follows that *B. nigra* must have been the egg parent and *B. oleracea* the pollen parent of the polyploid *B. carinata*. Likewise, *B. campestris* was the maternal and *B. nigra* the paternal parent of *B. juncea* (Fig. 4.3). the direction of the cross involving *B. oleracea* and *B. campestris* to produce *B. napus* is not discernible from the rubisco data because both of the presumed parental species have the same large subunit (Fig. 4.3).

Gatenby and Cocking (1978b) determined relative amounts (via densitometry) of small subunit polypeptides in *B. oleracea, B. carinata,* and *B. napus*, with the ratios being equal in the former and three SSI to one SS4 in the latter two species (Fig. 4.3) Since the basic chromosome number of *Brassica* is $x = 5$, Gatenby and Cocking (1978b) argued that the species in Fig. 4.3 with the lower ploidy level are in fact polyploids (perhaps old ones). The presence of both types of small subunit polypeptides in *B. oleracea* is, therefore, a reflection of its hybrid origin involving species with each of the types. If one accepts this, then the three to one ratio of the two polypeptides in the polyploid species *B. carinata* and *B. napus* may be explained as follows. If all low ploidy level species are ancient polyploids, then their genomes would contain a minimum of two coding genes (with two alleles each) for small subunits. Furthermore, if *B. oleracea* has one gene with both alleles encoding one type of polypeptide and both alleles the other gene encoding the alternative subunit, then its gametes would contribute one allele for each of the polypeptides, and when combined with the gametes of either *B. nigra* or *B. campestris* (containing two alleles for SS1), the resulting progeny would contain three alleles encoding SS1 and one specifying SS4. With this model, chromosome doubling presumably has not altered the relative amounts of the two polypeptides. This scenario to account for the relative amounts of rubisco subunits assumes that at least two structural genes are encoding proteins in the lower ploidy level species and that the products of each of the genes are transcribed and translated equally. Also, given that several tightly linked genes for the small subunit are probably present even in the lower chromosome number plants, it must be assumed that these clusters of genes are acting as single genes with regard to encoding the proteins.

In summary, both the large and small subunits of rubisco have been of some value for interpreting relationships in *Brassica*. The large subunit patterns indicate the maternal parents of two of the three polyploids, but could not resolve the situation for the third polyploid. With regard to the diploid species, the rubisco data suggest a close relationship between *B. campestris* and *B. oleracea* if one considers only the large subunits (Fig.

4.3). Evidence with regard to the small subunit is not definitive. These data will be considered with other information for *Brassica* in Chapter 17.

Chen and Wildman (1981) employed electrofocusing of rubisco to examine variation among 19 species in the genus *Gossypium*. They found four different large subunit patterns and eight different small subunit polypeptides among the species. The subunits found in the two cultivated amphidiploid species *G. barbadense* and *G. hirsutum* support the hypothesis that they originated by hybridization between one diploid species with the A genome and one with the D genome. All D genome species contain a unique large subunit pattern designated LSII and the two small subunits designated 5 and 7, whereas A genome species contain LSIV (otherwise found only in *G. longicalyx* with the F genome) and small subunits 4 and 6 with traces of 5. The two polyploid species have large subunit IV, meaning that the A genome species was the egg parent in the original cross. Small subunits 4–7 occur in *G. barbadense* and *G. hirsutum*, which is the additive profile expected if the A and D genomes are present.

The origins of polyploids in other genera have been studied with electrofocusing of rubisco. Steer and Thomas (1976) and Steer and Kernoghan (1977) examined species in *Avena*, and found a particular large subunit pattern in diploid species with the A genome and a different pattern in C genome diploids. *Avena murphyi*, a tetraploid with the A genome, showed the same pattern as diploids with the A genome. The hexaploid *A. sativa* exhibits the large subunit pattern typical of A genome species, which means that if C genome diploids were involved in the origin of *A. sativa*, they must have been the pollen parents. The small subunits were of no value in studying *Avena* because they had the same isoelectric point in all species.

Gatenby and Cocking (1978c) examined the origin of the European potato by comparing its rubisco profiles with several species thought by various workers to be involved in the parentage of the tetraploid species. One hypothesis is that the European cultivated potato, *Solanum tuberosum* subsp. *tuberosum*, originated from doubling of the genome of *S. stenotomum*, its presumed closest relative. This would mean that it is an autopolyploid. The rubisco data are not concordant with this view because the two species differ in both large and small subunit polypeptides, and presumably doubling of chromosomes would not change the kinds of polypeptides produced. Another hypothesis is that subsp. *tuberosum* was formed by hybridization between *S. stenotonum* and *S. sparispilum*. The rubisco data, however, do not support this view because the two species have identical large subunit polypeptides, which in turn differ from the subunit composition of subsp. *tuberosum.* Also, *S. sparsipilum* contains a small subunit polypeptide not found in subsp. *tuberosum. Solanum tuberosum* subsp. *andigena,* the Andean cultivar, has the same rubisco pattern as *S. stenotomum,* which would be concordant with the origin of the former via chromosome doubling of the latter. As Gray (1980) pointed out, studies of additional species of *Solanum* and synthetic hybrids may help clarify the origin of cultivated potatoes. In addition, the activities of plant breeders have no doubt complicated the situation.

SUMMARY AND GENERAL CRITIQUE OF SYSTEMATIC VALUE OF RUBISCO

Electrofocusing of rubisco subunit polypeptides has made significant contributions to systematic and phylogenetic studies of a range of plant groups. The most widely investigated problem has been the origin of polyploid species, particularly allopolyploids. As mentioned earlier (and pointed out by Gray, 1980 and discussed by Wildman, 1981), assuming homogeneity of two polypeptides only on the basis of isoelectric point may not always be valid. Additional analyses based on other properties of the polypeptides have revealed heterogeneity in certain cases. Presumably, amino acid substitutions could occur, but if they do not affect the isoelectric point of the polypeptide, one might conclude that the two are identical. Lin et al. (1986) demonstrated that different large subunit patterns in tobacco result from a single changed residue difference.

Chen and Wildman (1981) and Wildman (1983) discussed the use of rubisco electrofocusing patterns for inferring the ages of groups of angiosperms; in particular, they attached significance to the number of large subunit patterns and small subunit polypeptides. They used the number of polypeptides in the Lemnaceae, in which fossil evidence can be used to estimate the age of the family, as a "clock" or reference point for estimating the ages of other taxa such as *Nicotiana, Gossypium,* and *Lycopersicon.* They also attempted to correlate morphological diversity of a taxon with both observed diversity of rubisco and assumed age, and they suggested that on the whole the data correlate well.

Several assumptions made by Chen and Wildman (1981) appear of questionable validity, and they may have applied rubisco electrofocusing patterns to problems for which they are not appropriate. Using rubisco diversity in Lemnaceae as a standard for inferring the ages of other taxa is questionable because it assumes a uniform mutation rate for amino acid substitutions affecting isoelectric points. Gray (1980) correctly indicated the potential fallacy of the argument. It is also of questionable validity to make general assessments of the relative degree of differentiation among particular taxa, and it is perhaps even less advisable to infer the relative ages of taxa from these assesments. Chen and Wildman (1981) did this in several instances, stating, for example, that *Nicotiana* and *Gossypium* are as strongly differentiated as Lemnaceae, and then concluding that the three groups probably have been evolving for about the same period of time.

The problem with inferring divergence in rubisco employing isoelectric point as the only criterion was discussed earlier, but it is worth further consideration in the context of the age of plant groups. Use of this single property for measuring rubisco divergence in essence disregards amino acid differences not affecting isoelectric point. The most accurate measure of divergence between two rubiscos would be obtained by sequencing the genes encoding them, or failing that, determining the amino acid sequence of the proteins (see discussion in Chapter 12). Since it is not feasible to use either of these methods in surveys of large numbers of species, additional tests of the homogeneity of polypeptides with the same isoelectric points should be carried out. Wildman (1981) and Gray (1980) discussed a number of these additional procedures. Suffice it to say that broad phylogenetic and biogeographical inferences for plants are not justified from electrofocusing patterns of rubisco subunits. This limitation, however, does not detract in any way from the many significant contributions that this method has made to plant systematics and phylogeny.

CHAPTER 5

Enzyme Electrophoresis: Basic Methods and Interpretation of Banding Patterns

INTRODUCTION

The last two chapters dealt with basic electrophoretic techniques and the usc of general protein staining following electrophoresis as methods for generating systematic data. This chapter discusses staining for specific enzymes following electrophoretic separation. While the method is similar in some respects to electrophoresis of seed proteins, the kinds of data produced and their analyses are quite different from the comparisons of banding patterns seen with a general protein stain. The purpose of this chapter is to discuss methods for generating and interpreting data from enzyme electrophoresis.

The use of starch gel for electrophoresis (as first described by Smithies, 1955) combined with histochemical staining for particular enzymes allowed Hunter and Markert (1957) to demostrate the existence of multiple molecular forms of enzymes within a single organism. Prior to this demonstration, there had been suggestions of molecular heterogeneity, but multiple forms were often dismissed as artifacts or seen as a reflection of possible shortcomings of the biochemist doing the isolations (Markert, 1977). The zymogram technique (i.e., zone electrophoresis combined with histochemical staining) provided a gentle yet sensitive method for the resolution of enzymes in cell extracts. Markert and Møller (1959) demonstrated different forms of lactate dehydrogenase associated with particular tissues and with ontogeny within individuals. They also documented the existence of species-specific patterns. Their results lcd to the development of the isozyme concept: that living organisms often produce multiple forms of enzymes, that is, different molecular forms catalyzing the same reaction may be found within an individual. The term isozyme (or isoenzyme) was originally coined with reference to these multiple forms of enzymes at a time when their genetic bases were unknown. As more information became available on the factors creating multiple forms of enzymes, attempts were made to provide more precise terminology. In the following discussion, different molecular forms of an enzyme encoded by different alleles of the same gene locus will be referred to as *allozymes*. If different molecular forms are encoded by different loci, they will be referred to as *isozymes*, but with some indication that their genetic bases are known. Should nothing be known about the genetic bases of the multiple molecular forms, they will simply be designated as isozymes.

ELECTROPHORESIS AND HISTOCHEMICAL STAINING

The zymogram technique involves electrophoretic separation of enzymes followed by histochemical staining. Various plant parts, such as leaves, roots, or floral buds are ground in cold extracting buffer, and this crude extract containing many different proteins is subjected to an electric current. With the many proteins present in the gel following electrophoresis, it is necessary to locate specific enzymes. This is accomplished by incubating gel slices in a buffer containing the necessary substrate and cofactors for a specific enzyme together with some type of dye. In essence, the enzymes in the gel cause changes in the substrate, which in turn cause a color to develop in the gel where the enzymes are located. In other words, the activities of the enzymes cause them to reveal their locations in the gel. In contrast to a process using Coomassie Blue or some other general protein stain, the zymogram technique involves staining for enzymes with specific activities. Methods will not be considered in detail, but an example showing the general procedure for staining for dehydrogenases (oxidoreductases) will suffice (Fig. 5.1). Examples of substrates for which dehydrogenases are often assayed in plants include malic acid, glutamic acid, ethyl alcohol, glucose-6-phosphate, 6-phosphogluconic acid, skikimic acid, isocitric acid, and glyceraldehyde-3-phosphate. Thus a considerable number of enzymes can be studied with the same basic system by substituting different substrates. Additional enzymes often assayed in plants will be enumerated later.

While methodology will not be discussed in detail, we stress that the

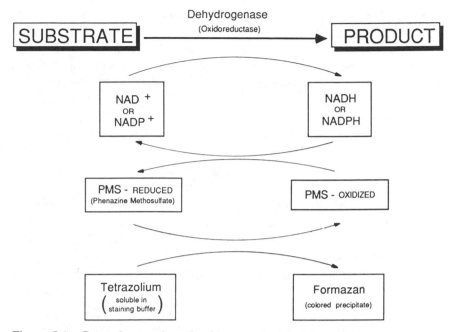

Figure 5.1 General procedure for histochemical staining for dehydrogenases. The soluble tetrazolium dye, upon being reduced, forms the colored formazan that precipitates in the gel to identify the location of enzyme activity.

extracting, gel, and electrode buffer systems suitable for animal tissues are not adequate for plants. The large number of secondary compounds and cell walls are no doubt troublesome factors. The conditions producing good band resolution for one species or group of species may be unsatisfactory for other species; therefore extracting, gel, and electrode buffers must be determined empirically for different groups of species. Several especially useful compilations of recipes are by Soltis et al. (1983b), Shields et al. (1983), Vallejos (1983), Wendel and Stuber (1984), Werth (1985), and Morden et al. (1987).

INTERPRETATION OF BANDING PATTERNS

Several factors, which must be understood for proper interpretation of patterns, determine the number of bands detected in an individual plant when stained for a particular enzyme. These factors include whether the organism is homozygous or heterozygous at each gene locus, the active subunit composition of the enzyme (whether one, two, or more polypeptides combine to form the active enzyme, i.e., whether they are monomers, dimers, etc.), the number of gene loci specifying different forms of the enzyme, whether polypeptides encoded by different loci combine to form active enzymes, and whether posttranslational modifications have occurred.

Consider first allozymes, that is, different molecular forms encoded by alternative alleles of the same gene locus. Most allozymes probably result from point mutations where one or more nucleotide replacements in the structural gene cause amino acid substitutions, which in turn alter the electrophoretic mobility of the enzyme. In many instances the amino acid substitutions probably change the net charge on the enzyme molecules, but there is evidence that electrophoretic mobility can be altered by other factors (see below).

Allozymes are inherited in a codominant manner, making it possible to distinguish between plants that are homozygous and heterozygous at a gene locus (Fig. 5.2). If the active enzyme consists of a single polypeptide (i.e., is monomeric), then plants homozygous at the locus encoding this enzyme will display a single band, whereas heterozygous individuals will produce two bands (Figs. 5.2, 5.3). From Figs. 5.2 and 5.3 it may be seen that when an enzyme consists of two associated polypeptides (a dimeric enzyme), plants homozygous at the locus will produce a single band, the same as a monomeric enzyme. Heterozygous individuals, however, will exhibit three bands because of random association of polypeptides to produce the active form of the enzyme (Figs. 5.2, 5.3). A tetrameric enzyme consists of four polypeptides (or subunits), and therefore a heterozygous plant will have five bands (Fig. 5.2). In the three-banded pattern typical of an individual heterozygous at a locus encoding a dimeric enzyme, the middle band represents the active enzyme composed of two different subunits (the heterodimer). It will typically stain darker (and wider) than the other two bands composed of two identical subunits (the homodimers) (Fig. 5.3). This is so because with random association of subunits, there will be twice as many enzyme molecules formed between unlike polypeptides as there are with the same polypeptides. A comparable pattern is usually found with tetrameric enzymes, although the ratio of staining intensities of the heterodimers to homodimers becomes more complex.

From study of Fig. 5.2, it is apparent that a plant heterozygous at a

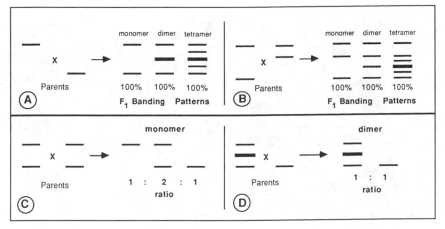

Figure 5.2   Enzyme banding patterns of parents and progeny for enzymes with different active subunit compositions. (*A*) Parents homozygous for different alleles at the same locus. All progeny are heterozygous, but if the enzyme is monomeric there are two bands, if dimeric there are three, and if tetrameric five bands are present. (*B*) Parents are homozygous for the same allele at one gene locus and for different alleles at the other locus. The situation is the same as in A except for the additional band resulting from the homozygous locus. (*C*) Parents are each heterozygous for the same two alleles coding for a monomeric enzyme, and the progeny exhibit the characteristic 1:2:1 ratio. (*D*) One parent is heterozygous and the other homozygous for one of the same alleles at a gene locus specifying a dimeric enzyme. The progeny will exhibit a 1:1 ratio for each of the parental types.

gene locus encoding a tetrameric enzyme will differ from a homozygous individual by four bands, with the corresponding difference being two bands for a dimeric enzyme and only one band for a monomer. Gottlieb (1977a) emphasized that simple comparisons of banding patterns with use of some sort of similarity measure (as has been used for seed protein profiles; see Chapter 3) could lead to erroneous conclusions about genetic similarities between plants. For example, the percentage similarity (bands in common/total bands) for two individuals homozygous and heterozygous at a gene locus specifying a tetrameric enzyme is 20%, while the similarity is 33% for a dimeric enzyme and 50% for a monomeric enzyme. Yet the genetic similarity is the same in each case!

The number of structural gene loci present for a given enzyme will affect the number of colored bands appearing in the gel (Fig. 5.4). Thus both heterozygosity at each locus and an increasing number of gene loci

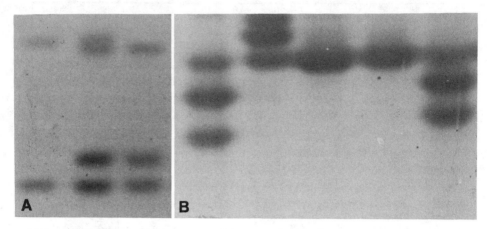

Figure 5.3 Photographs of gels showing segregation for different banding patterns. (*A*) Phosphoglucomutase (PGM), a monomeric enzyme, from *Coreopsis*. The faster migrating forms at the top occur in the chloroplasts, with plants homozygous for different alleles on each side in the photograph and the heterozygous plant with a two-banded pattern in the middle. The slower migrating forms occur in the cytosol; the plant on the left is homozygous and the other two are heterozygous. (*B*) Phosphoglucoisomerase (PGI), a dimeric enzyme, from *Coreopsis*. Only the cytosolic form is shown. Three different three-banded patterns are represented together with two plants displaying the same single band. All plants have one band in common and thus share the same allele. Plants exhibiting the three-banded pattern are heterozygous, whereas the two plants with the same single band are homozygous for the same allele.

will increase complexity of banding patterns for a given enzyme. For multimeric enzymes, subunits (polypeptides) specified by different loci could combine to form what may be called interlocus or isozymic heteromers, but these heteromers do not form between isozymes present in different subcellular compartments (Fig. 5.4).

Given the importance of interpreting enzyme banding patterns in a genetic context, it is desirable to carry out some sort of genetic analysis. Several procedures exist for doing this, with the method of choice dictated to some extent by the plants. With self-incompatible plants easily cultivated in the greenhouse or growth chamber, crosses between individuals with known banding patterns followed by examination of the progeny (including both the F_1 and backcross or F_2) is a feasible approach (Fig. 5.2). Failing this, highly self-pollinating plants are best analyzed by

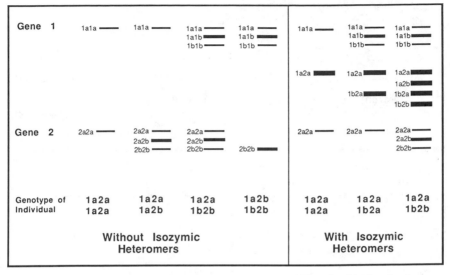

Figure 5.4 Drawing of enzyme banding patterns with and without the formation of heteromers between isozymes (polypeptides encoded by different loci). Genotypes are indicated by using a number to designate the gene and a letter to represent the allele. The same designations are employed for the polypeptide subunit composition of the enzymes. The enzyme is assumed to be dimeric. It may be seen that the formation of isozymic heteromers complicates the patterns considerably.

determining the phenotypes of various enzymes in individual plants, allowing the plants to self-pollinate and then examining the progeny (Fig. 5.5). A last method involves examining the progeny of seeds collected from individual plants in nature (Brown et al., 1975). The basic rationale of the approach is that all progeny from an open-pollinated individual will have alleles from that plant (the maternal parent) in common. Studies of the segregation patterns in the progeny thus allow one to ascertain which bands are specified by alleles of the same genetic loci and which are controlled by different loci (Fig. 5.6).

Weeden and Gottlieb (1979) showed that using pollen as the source of enzymes simplifies interpretation of the banding patterns of multimeric enzymes. Each pollen grain, being haploid, contains only one allele of each locus encoding an enzyme, and, as a result, allozymic heteromers do not form in the pollen of heterozygous plants. For example, the intermediate band (heterodimer) normally present in the leaf or stem of a

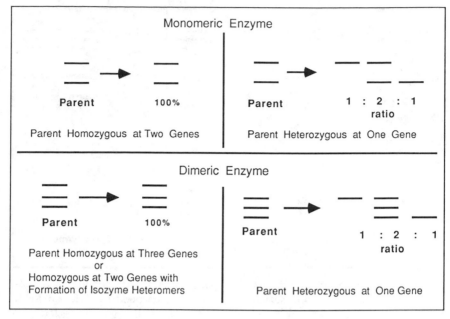

Figure 5.5 Drawings of banding patterns of progenies resulting from self-pollination of parents with known patterns. These progeny tests may be used to infer the genetic bases of patterns seen in the parental plants. In all cases, homozygous plants will yield identical banding patterns, whereas segregation will occur upon selfing of heterozygous plants.

heterozygous plant does not form when pollen is used as the enzyme source (Fig. 5.7). By contrast, subunits (polypeptides) encoded by different gene loci (but only if products of the two loci are localized in the same places in the cell; see later discussion) may form heteromers because each pollen grain contains one allele for each locus (Fig. 5.7). Using pollen, it is possible to interpret some of the chemical complexities of protein banding patterns genetically. Because multimeric enzymes often produce the most complex banding patterns, it is fortunate that pollen is useful for interpreting these enzymes.

Conifers offer a special advantage for interpreting the genetic bases of enzyme banding patterns. Individual megagametophytes (which are haploid) may be removed from the seeds gathered from individual trees and used as the sources of enzymes for electrophoresis (Mitton, 1983). These tissues contain many enzymes in high activity, and running several gametophytes from a tree allows for inferences about the genotype of the tree.

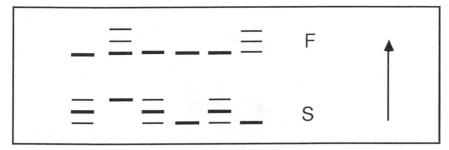

Figure 5.6 Banding patterns of a hypothetical dimeric enzyme from progeny of an open-pollinated plant. Segregation patterns suggest that two gene loci are expressed because variation in the two zones is independent. The maternal parent must be homozygous for the allele encoding the slower form of the enzyme in the F zone because all progeny have it. The maternal parent must be heterozygous for the two alleles whose products are expressed in the S zone. By examining a sufficient number of progeny for a variety of enzymes, it is possible to infer the genotype of the maternal parent and likewise to infer the genetic bases of the banding patterns for the enzymes.

In addition, when the embryo of a seed is run together with the mega-gametophyte from the same seed, it is possible to tell directly the allelic contributions of the maternal and paternal parents. These observations allow for ready determination of the genetics of enzyme banding patterns in conifers.

Homosporous ferns are even more amenable to genetic analysis than are conifers. Fern gametophytes are the derivatives of individual meiotic products (spores), yet because they are multicellular it is possible to carry out electrophoresis on single gametophytes. This allows one to interpret the complex banding patterns of the sporophytes by looking at segregation patterns in their resulting haploid products (Gastony and Gottlieb, 1982).

"HIDDEN" VARIATION AND GEL SIEVING

One consideration with the use of enzyme electrophoresis is whether all variation is being detected with the methods employed. That is, could a single band in a gel represent more than one form of an enzyme? Or, could two bands from different individuals migrating to the same place in a gel represent different forms of an enzyme? Only about 30% of the

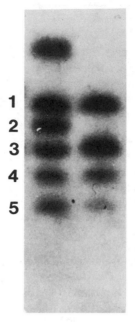

Figure 5.7 Patterns of the dimeric enzyme phosphoglucoisomerase (PGI) in the stems (left) and pollen (right) of *Clarkia dudleyana* (anode toward top). The most anodal band occurs in the chloroplast and is not present in the leachate from the pollen. The slower migrating bands occur in the cytosol and are specified by two gene loci. The bands in this region are designated 1 through 5, with 1 being the fastest (most anodal) and 5 the slowest band. This plant is homozygous at one locus that encodes band 5. It is heterozygous at the other locus, and bands 1 and 3 in the stem extract represent the homodimers, while band 2 is the heterodimer (actually, band 3 also contains another form of the enzyme; see below). Note that band 2 is absent from the pollen leachate. Band 4 of the stem extract and the corresponding band from the pollen leachate represent the isozymic heteromer. That is, the same subunits that comprise bands 3 and 5 associate to produce this active form of the enzyme. The isozymic heteromer formed between subunits comprising bands 1 and 5 is present in band 3 and overlaps one of the homodimers, as discussed above. Comparison of pollen leachate and stem tissue allows the cytosolic and chloroplast forms of PGI to be distinguished. Also, it is possible to ascertain which bands represent allozymic versus isozymic heteromers because the former will be absent from pollen since each pollen grain (being haploid) has only one allele. Isozymic heteromers occur in pollen because both gene loci are expressed in the haploid tissue, and the resulting heteromer will be present. (With permission from N.F. Weeden and L.D. Gottlieb, Biochem. Genet. **17**:287–296. 1979, Plenum Publishing Corp. Photograph courtesy of L.D. Gottlieb.)

nucleotide replacements in a gene will result in the substitution of an amino acid with a different net charge, and thus change the net charge on the polypeptide (see discussion in Chapter 1 on protein structure). If electrophoretic mobility of a molecule were based solely upon net charge, then obviously much of the variation would go undetected by electrophoresis, leading to the question of whether differences in electrophoretic mobility result entirely (or primarily) from charge differences.

G.B. Johnson (1976, 1977a,b) suggested from electrophoretic data that some of the electromorphs (bands in gels) detected in butterflies result from differences in conformation rather than net charge on molecules. Unless the amino acid sequences are known for the molecules, however, it is not possible to determine with certainty the reasons for differences in mobility. Ramshaw et al. (1979), in a novel approach to the question, used electrophoresis to examine different variants of hemoglobin with known amino acid sequences and with known three-dimensional structures. Sequential electrophoresis was carried out employing different conditions of pH and gel concentration, that is, if two hemoglobins were not separable under one set of conditions (pH and gel concentration) they were run under a second set of conditions. If the first two systems did not resolve the molecules, they were fractionated under a third system. With a random sample of 20 hemoglobin variants fractionated under three different conditions, 17 electromorphs were detected in the gels, that is, 85% of the different molecules were separated. With one system, eight variants were separated in the gel. A second series of experiments compared hemoglobin variants with the same amino acid substitutions, but in different parts of the molecule, and sequential electrophoresis separated 90% of these variants. The final experiment examined hemoglobin variants with chemically different but charge-equivalent amino acid substitutions at the same sites in the molecules; 80% of these variants were distinguished by sequential electrophoresis.

Ramshaw et al. (1979), while cautioning that their results from hemoglobin should not be applied a *priori* to all molecules, concluded that factors other than net charge affect the mobility of molecules during electrophoresis. Their results also indicated that amino acid substitutions occurring in the interior of the molecule (when considering three-dimensional structure) still result in electrophoretically detectable differences, though they have less of an effect than substitutions on the surface. One class of substitutions not studied by Ramshaw et al. (1979) was that of one neutral amino acid replacing another in the interior of the molecule. This type of change might be expected to produce the least change in electrophoretic mobility.

Several studies, particularly with *Drosophila*, have detected additional variation when employing sequential electrophoresis. These studies include demonstrations of increased variation within species (Singh et al., 1976; Singh, 1979; McDowell and Prakash, 1976; Coyne et al., 1978; Choudhary and Singh, 1987, as examples) or between species (Coyne, 1976; Coyne et al., 1979).

One question facing the plant systematist employing enzyme electrophoresis concerns how much of the variation is being detected and how the undetected variation is affecting systematic and phyletic inferences. Most studies employing sequential electrophoresis within species (such as with *Drosophila*; see above) have detected additional variation at gene loci previously shown to be the most variable (i.e., to have a large number of alleles) when subjected to one condition of electrophoresis. This means that the amount of additional variation detected at a gene locus with sequential electrophoresis is proportional to the amount found under one set of conditions (see discussion by Gottlieb, 1981a; Ayala, 1982). Thus, within a species, sequential electrophoresis will most likely reveal additional alleles at variable loci, but the alleles present in highest frequency will remain the same as under the original condition, that is, the "new" alleles detected will occur in low frequency. Genes found to be invariant originally will probably remain so after sequential electrophoresis. Thus it is unlikely that sequential electrophoresis will alter the results for populations of the same species if one were using a measure of similarity such as Nei's (1972) genetic identity. This notion is true because low-frequency alleles have little affect on these values. If a cladistic analysis were employed, then additional low-frequency alleles could effect results.

Sequential electrophoresis may be more critical in comparisons among species, as illustrated by the work of Coyne et al. (1979) in which variation in the enzyme alpha-glycerophosphate dehydrogenase from 71 species of *Drosophila* was examined under different electrophoretic conditions. Their analyses detected no additional alleles for this enzyme within species, but considerable additional variation was documented among species of *Drosophila*. In fact, the variation found among the species is concordant with accepted taxonomic relationships in the genus (Coyne et al., 1979). Gottlieb (1979) used several electrophoretic conditions (pH and polyacrylamide gel concentration) to test for additional variation in *Stephanomeria malheurensis* but detected none beyond that originally found with starch gels. Additional plant studies in which electrophoresis under several conditions was employed (Gottlieb, 1984b; Crawford and Smith, 1984; Lowrey and Crawford, 1985, as examples) failed to detect additional variation either within or among taxa. By contrast, Stuber and Goodman (1983) showed that additional allozymes of phosphoglucomutase (PGM) could be resolved by a change in pH of the gel and electrode buffers.

Despite results of several studies showing no additional variation among species of plants with varying electrophoretic conditions, it is advisable to use sequential electrophoresis before inferring identity of electromorphs from congeneric species (Gottlieb, 1981a). As mentioned previously, electrophoresis may be used to detect differences among proteins; identity of proteins (in terms of amino acid sequence) can only be inferred from the inability to demonstrate electrophoretic differences. The more conditions under which differences fail to be detected, the stronger the inference of identity.

LOCALIZATION OF ISOZYMES AND CONSERVATION OF ISOZYME NUMBER

Gottlieb (1982) summarized data showing that in plants isozymes may occur in the cytosol or be localized in subcellular compartments. The number of isozymes of a given enzyme often is dependent upon the number of subcellular compartments in which the same reaction must be catalyzed. The two pathways in which this phenomenon has been documented best are glycolysis and the oxidative pentose phosphate pathway (Gottlieb, 1981a, 1982). Isozymes of enzymes in these two pathways occur in the cytosol and plastids because the reactions occur in both places in the cell (Gottlieb, 1981a, 1982; Weeden, 1983 give discussions and listings of enzymes). Other enzymes such as aspartate aminotransferase and glutamate dehydrogenase are highly conserved in plants. The number of isozymes reported in diploid plants for various enzymes commonly examined electrophoretically together with their active subunit compositions is given in Table 5.1.

For many plants, it is relatively easy to isolate chloroplasts and determine which isozymes occur in them (Weeden and Gottlieb, 1980a; Mills and Joy, 1980; Gastony and Darrow, 1983; Pichersky et al., 1984). Plastid

TABLE 5.1. Conserved Number of Isozymes in Diploid Plants and Their Active Subunit Composition for Enzymes Commonly Examined Electrophoretically in Plants[a]

Enzyme	Number of Isozymes	Subunit Composition
Alcohol dehydrogenase	2 or 3	Dimer
Aspartate aminotransferase	3 or 4 (1 or 2)	Dimer
Fructose-bisphosphate aldolase	2	Tetramer
Glucose-6-phosphate dehydrogenase	2	Dimer
Glutamate dehydrogenase	Single enzyme	Hexamer
[NAD]glyceraldehyde-3-phosphate dehydrogenase[b]	2	Tetramer
[NADP]isocitrate dehydrogenase[b]	2 (1 or 2)	Dimer
Leucine aminopeptidase[b]	2 (1)	Monomer
Malate dehydrogenase	3 or 4	Dimer
6-Phosphogluconate dehydrogenase	2	Dimer
Phosphoglucomutase	2	Monomer
Phosphoglucoisomerase	2	Dimer
Shikimate dehydrogenase[a]	2 (1)	Monomer
Triosephosphate isomerase	2	Dimer

[a] Common isozyme numbers for ferns are given in parentheses when they differ from other plants.

[b] One isozyme often reported.

isozymes are encoded by nuclear genes (Weeden and Gottlieb, 1980a,b), and ascertaining which isozymes occur in the same organelle allows comparison of genes specifying the plastid forms in different plants. Weeden and Gottlieb (1980a,b) demonstrated that soaking pollen in a buffer for 4 hours resulted in release of isozymes from the cytosol into the buffer, whereas the plastid forms were not released (Fig. 5.7). A possible difficulty with this procedure is obtaining sufficient quantities of pollen. Also, leaching the pollen for a prolonged period of time will cause the plastid isozyme to appear in the leachate. Both isolation of chloroplasts and leaching of pollen (if done carefully) may provide valuable complementary information on the localization of isozymes in plants.

Although isozymes may be localized in different organelles, as discussed above, they are encoded by nuclear genes, showing there has been conservation of gene number in the nuclear genome of diploid plants; as indicated by Gottlieb (1982), this conservation encompasses a vast taxonomic array of plants including conifers, monocots, and dicots. One important systematic aspect of this conservation is that it provides some assurances that gene loci encoding the same isozymes (i.e., same subcellular localization) in different plants are homologous. Also, the occurrence of an "expected" number of isozymes in diploid plants helps to simplify the genetic interpretation of banding patterns. Lastly, any deviation from the "normal" number of isozymes is of potential systematic and/or phylogenetic interest; this topic will be discussed in more detail in Chapter 10.

SUMMARY

Zone electrophoresis combined with histochemical staining permits the visualization of different forms of enzymes as colored bands in gels. Different forms encoded by alternative alleles of the same gene locus are termed allozymes, whereas the term isozyme refers either to forms of an enzyme encoded by different loci or when the genetic basis of the enzyme differences is not known. Genetic interpretation of banding patterns is necessary for extracting maximal value from the data, and several procedures are available that facilitate these interpretations. Electrophoresis undoubtedly does not detect all the variation in plants, but available evidence suggests that undetected variation is not a serious problem for the plant systematist.

CHAPTER 6

Generation and Analysis of Data from Enzyme Electrophoresis

INTRODUCTION

The last chapter discussed the basic methods of enzyme electrophoresis and how to interpret the resulting banding patterns. This chapter will consider procedures for generating and analyzing electrophoretic data for systematic and phylogenetic studies. The discussion will include the sampling of enzymes and plants as well as methods of data analysis and application to problems in plant systematics.

SAMPLING WITHIN AND AMONG POPULATIONS OF PLANTS

In many electrophoretic studies, the sources of enzymes are progeny of plants from natural populations; seeds are collected from plants and germinated, and the resulting seedlings are used for electrophoresis. This method is used when it is not feasible to collect in the field and send plants to the laboratory; it is more desirable to sample plants from natural populations, however. Hamrick and Loveless (1986) enumerated the advantages of natural sampling for the study of electrophoretic variation in tropical species, and their comments apply to plant species in general. Random sampling from a natural population alleviates the problem of seed samples not being representative of the population. Differential germination of seeds could bias results. For perennial species, sampling can be done for different age classes, which is not possible if seeds are used. Hamrick and Loveless (1986) also pointed out that using material directly from populations makes it feasible to collect over a greater period of the year, and not just when the plants have mature seed. They also noted that if mature seed is dispersed rapidly or is subject to predation it may be difficult to obtain good samples. Germination of seed may be a problem with some plant groups, and seed viability may be relatively short. Both of these problems are circumvented with the collection of leaves from populations.

Hamrick and Loveless (1986) freeze-dried leaf samples in the field, shipped them to the laboratory, and stored them in an ultracold freezer until they were used for electrophoresis. This procedure allowed 15 to 20 enzymes to be assayed, which represents a rather large number for plants, but it requires availability of a freeze drier in the field. Sytsma and Schaal (1985a) collected leaf material in the field in Panama, placed it on ice, and shipped it back to the laboratory for enzyme extraction. This method worked well for resolving banding patterns despite a delay of several days.

If seeds are used it is important to collect them from individuals throughout a population; one method would be to collect along a transect. The number of individuals from which seeds should be gathered will vary with the size of the population. It is preferable not to combine or mix seeds from different individuals because of the possibility that seeds from only one or a few plants will germinate and thus give an inaccurate estimate of allozyme variation in the population. If plants are growing in different microhabitats within a population, seeds should be collected from these areas and the differences noted.

The number of plants that need to be examined from each population, once the aforementioned factors are considered, will depend on how concerned one is with detecting low-frequency alleles. This is not a serious concern if only genetic similarities are computed among populations or species, because low-frequency alleles have little effect on these values. However, small samples could lead to erroneous conclusions on whether particular low-frequency alleles are present in one population, species, and so on, but lacking in another. Such conclusions, in turn, could affect phylogenetic interpretations. Marshall and Brown (1975) indicated that for a "worst case" situation in which 20 alleles, each in a frequency of 0.05, are present at a locus in a population, 30 diploid individuals (60 independent gametes) should provide 95% probability of detecting all alleles. Brown and Weir (1983) provided an excellent discussion of the use of isozymes for measuring genetic variation in plant populations.

Populations of a species should be examined throughout the known geographic distribution of the taxon, although no hard and fast rules exist as to how many populations should be included. It is prudent to sample populations growing in different habitats such as at different altitudes, and on different soil types. In effect, the same general rules apply for sampling electrophoretically as for stuides of morphology, chromosome numbers, and so on. However, it seems fair to say that sampling within populations in electrophoretic studies has been more extensive than in investigations employing other characters.

Breeding system is also a factor for consideration when sampling populations of a species. In outcrossing plants, a given population is often highly similar to all other populations in having the same alleles in the highest frequencies at most loci (see Gottlieb, 1981a; Crawford, 1983 for reviews). Such similarity may not occur in outcrossing species that are animal-pollinated, but it is common in most wind-pollinated plants with high levels of interpopulational gene flow (see Hamrick, 1987 for discussion). By contrast, populations of selfing plants often display one of two patterns of interpopulational variation at isozyme loci. They may be monomorphic at almost all loci over their geographic ranges, that is, there is minimal intra-and interpopulational variation. Another pattern occurs when different populations exhibit large differences in allelic frequencies at several loci, including populations that are monomorphic for different alleles. Examples of these patterns may found in Nevo et al. (1979), Crawford and Wilson (1977, 1979), Schoen (1982), Roberts (1983), Gottlieb (1984b), Crawford et al. (1985), and D.E. Soltis et al. (1987). While extensive interpopulational sampling needs to be done for all plants, it is probably even more critical for selfers than outcrossers.

SAMPLING ENZYMES

An important question in any electrophoretic study concerns the numbers and kinds of enzymes to be included. There is no correct answer as the question being addressed in a given study as well as other factors of a more technical nature (which enzymes can be resolved electrophoretically from the plants) will influence the decision.

Consider first studies directed at assessing genetic divergence and similarity among populations of the same species or among cogeneric species. If a measure such as the genetic identity statistic (see next section) is to be employed for assessing the allelic frequency data, then the general rule is that the more gene loci included the better, as is the case with phenetic studes utilizing morphological characters. Certainly 20 or more loci are to be desired.

Another concern is the selection of enzymes, which, as indicated above, may be dictated by technical considerations. These factors aside, as many enzymes with known natural substrates (and hence conserved number of isozymes) as possible should be included. This makes it easier to interpret the genetic bases of the banding patterns and to assess homologies between isozymes from different plants than is possible with enzymes having nonspecific substrates because the latter often have more isozymes expressed and thus more complex banding patterns. However, peptidases, which are relatively nonspecific, usually exhibit rather simple patterns and thus represent an exception to this generalization. Given the complex patterns of several "nonspecific" enzymes, genetic tests for inferring the bases of the banding patterns are particularly critical. In effect, it is not suggested that these enzymes be excluded from systematic studies, but rather that obtaining optimal value from their use may require more time and effort than is necessary for other enzymes. The elegant studies carried out by Rick and collaborators (Rick and Fobes, 1975; Rick et al., 1976, 1977, 1979; Rick and Tanksley, 1981; Tanksley and Rick, 1980, as examples) demonstrated the systematic and phyletic utility of enzymes with nonspecific substrates when detailed genetic tests are employed. The investigations of Wilson (Wilson, 1976, 1981; Wilson and Heiser, 1979; Wilson et al., 1983) documented the value of leucine aminopeptidase, a relatively nonspecific enzyme, in systematic studies when thorough genetic tests are performed.

The numbers and kinds of enzymes included in electrophoretic studies are important for comparable results from different plant taxa. For example, one laboratory may report a mean genetic identity for populations of species X and Y in a particular genus as 0.74, whereas another laboratory may calculate the mean identity to be 0.56 for species A and B in another genus. Unless the enzymes included in both studies are the same (or very similar) in kind and number, the results are not comparable. As an illustration, if fewer enzymes were examined in species A and B as compared with X and Y, and the enzymes examined in the former pair were primarily the more variable ones with nonspecific substrates, then the reported differences in genetic identities may have little or no biological significance.

There is a potential side benefit from examining as many enzymes with specific substrates and conserved isozyme numbers as possible when conducting electrophoretic surveys. The possibility exists for detecting instances of increased isozyme number for certain enzymes; these cases are worthy of further study because of their potential phylogenetic significance (see discussion in Chapter 10).

Increasing the number of enzymes often necessitates the use of additional buffer systems in order to achieve adequate resolution, and this involves additional time, labor, and expense. By contrast, once buffer systems for certain enzymes have been worked out for a group of plants, it is relatively simple to examine large numbers of plants. As a consequence, most systematic studies employing electrophoresis are probably deficient in number of enzymes rather than number of plants examined (Crawford, 1983). As anyone who has practiced enzyme electrophoresis on a variety of taxa is aware, the plants themselves are quite variable with regard to the numbers and kinds of enzymes that resolve well with given buffer systems. Ascertaining the best methods for a new plant taxon is to a large extent an empirical exercise. Certain taxa yield superior activity and resolution for a large number of enzymes with as few as two buffer systems, whereas other plants may require considerable effort and several buffer systems to resolve the same enzymes. In other words, the plants may be as important as the expertise and ingenuity of the investigator.

GENETIC IDENTITY, GENETIC DISTANCE, AND GENOTYPE IDENTITY

From previous discussion of enzyme electrophoresis it is clear that the data generated differ from most other information used by the plant systematist. The basic difference is that the banding patterns in the gels allow for inferences to be made on the number of gene loci included for each enzyme, the number of alleles present at each locus (and thus whether an organism is homozygous or heterozygous at a gene), as well as which alleles occur. From these data for individual plants one may then estimate allelic frequencies at each locus for plants in a population, in a species, and so on. In addition, one may ascertain the distribution of individual alleles among populations or congeneric species provided that sampling is adequate. Another question that may be addressed with allelic data is how genetic variation is partitioned within and among populations or species.

Several measures of genetic variation in a population, species, and so on have been employed for allelic data. The more common ones will be commented upon briefly. One measure is the percentage of polymorphic loci in a population or species. A gene locus is considered to be polymorphic if the most common allele is present in a frequency of less than 0.99 or 0.95. That is, two criteria of polymorphism have been employed, and thus the percentage of polymorphic loci will vary depending on the criterion applied. Another measure of variation is the mean number of alleles per locus or per polymorphic locus. This measures variation by indicating the number of alleles present rather than frequency of alleles. Another commonly employed measure is Nei's (1973) gene diversity statistic, which is expected heterozygosity in a panmictic population.

Brown and Weir (1983) provided a cirtical discussion of the various single-locus measures that have been used to quantify variation within populations. While the reader is referred to this paper for more detailed considerations, several salient points are worthy of comment. Brown and Wier (1983) emphasized that two basic concepts can be recognized with regard to genetic variation that is detected electrophoretically. These include the richness of alleles (number of alleles detected) and their evenness (distribution of alleleic frequencies).

The percentage of polymorphic loci represents at best a rather crude indicator of the level of genetic variation in a plant population (Brown and Wier, 1983). Problems with this statistic include the arbitrary definition of polymorphism, dependence on the size of the sample, and kinds of enzymes sampled. This measure combines parts from both the evenness and richness concepts. The mean number of alleles per locus is one measure of allelic richness, but its value depends on sample size which causes problems in comparing samples. The gene diversity statistic of Nei (1973) is a measure of the evenness of allelic frequencies; it represents expected heterozygosity under random mating. This is a good measure of genetic variation, but one problem is that its value is determined primarily by the two most frequent alleles.

Several different measures of genetic similarity and distance have been proposed and will be mentioned briefly. An advantage of these measures is that they incorporate a considerable amount of data into a form feasible for visualizing relationships among populations, species, or other taxa, that is, the methods take allelic frequencies or other data from a number of different loci and reduce them to a single value of similarity or difference (distance) between populations. One potential problem with the approach is that useful information from individual gene loci may be lost. It is always desirable, therefore, to consider the distribution of particular alleles within or among populations or taxa.

One commonly employed measure of similarity for allelic data is the genetic identity statistic (or the genetic identity) of Nei (1972). Consider two plant populations A and B; J is a given locus, and i different alleles were found at the locus in the two populations. The frequencies of the alleles in populations A and B may be represented as a_1, a_2, a_3, ..., a_i, and b_1, b_2, b_3, ..., b_i, respectively. The genetic similarity between populations A and B at locus J is measured by the genetic identity, I_j, *in the following manner:*

$$\Sigma \, a_i b_i \, / \, \sqrt{a_i^2 b_i^2}$$

where $a_i b_i$ the products of the frequencies $a_1 b_1$, $a_2 b_2$, and so on from populations A and B, a_i^2 refers to the squares of the frequencies a_1^2, a_2^2, and so on in population A, and b_i^2 denotes the squares of the frequencies for alleles from population B. The genetic identity statistic is used to calculate the normalized probability that two alleles taken at random from each population will be identical. The identity value at a single locus for two populations may vary from 0 to 1. If the two populations have no alleles in common the value would be 0, as one may see from substituting values in the formula. If the two populations exhibit the same alleles in identical frequencies, their genetic identity would be 1. Various values between these lower and upper limits could exist depending on the occurrence and frequency of alleles.

In practice, of course, the plant systematist includes more than one locus when comparing plant populations. This is done by using the arithmetic means of $\Sigma a_i b_i$, Σa_i^2, and Σb_i^2 over all gene loci. Thus let *Iab*, *Ia*, and *Ib* represent these averages; then the genetic identity, I, between the two populations will be calculated as follows:

$$Iab \, / \, \sqrt{Ia \; Ib}$$

The genetic distance D between two populations is given by

$$D = -ln \; I$$

Genetic distance is employed to estimate the number of allelic substitutions per locus that have occurred during the separate evolution of two populations or species. The values for genetic distances of two populations may range from 0, where there has been no allelic change, to infinity (this latter value is possible only because this is not a true distance measure in the mathematical sense). Nei's genetic identity and genetic distance are commonly employed measures for expressing genetic similarity (or divergence) among populations or taxonomic entities. Various discussions and methods have been presented for assessing the impact of small sample sizes and for making statistical inferences (e.g., calculation of 95% confidence intervals) on these measures (see Nei, 1978; Mueller, 1979, as examples). Programs written by Whitkus (1985, 1988) are very useful for computing several of the genetic statistics from allelic frequency data.

Another method for calculating genetic distance from allelic frequency data was proposed by Rogers (1972). The formula for estimating distance (Rogers' distance, R) between populations A and B at a locus is given by

$$R = \sqrt{0.5 \; \Sigma_1^n \; (P_{Ai} - P_{Bi})^2}$$

where n = number of alleles

P_{Ai} = frequency of ith allele in population A

P_{Bi} = frequency of ith allele in population B

The value of R is 0 when two populations are genetically identical, and it is 1 when the populations are fixed for different alleles. Calculation of R over all loci for two populations is usually done by computing the arithmetic mean of the R values for each locus (Rogers, 1972). In practice, estimates of genetic divergence using the methods of Nei and Rogers are highly correlated. Useful discussions of these and other measures are provided in Chpater 5 of Richardson et al. (1986).

Hedrick (1971) suggested that using the probability of genotype identity rather than allelic identity has certain advantages. The method is similar to genetic identity because it represents the probability of getting the same genotypes when drawing from two populations or species as compared with the average probability of getting identical genotypes from the same populations on two successive independent draws (see Hedrick, 1983, pp. 73–74 for discussion). The values obtained for single gene loci may then be averaged over all loci. This measure is applicable to electrophoretic data from polyploid plants when it is not possible to determine allelic frequencies from gels.

PHENETIC ANALYSES

Once allelic frequencies, genetic distances, or identities have been calculated for plant populations, it is useful to analyze the data by phenetic methods. Techniques such as cluster analyses and principal component analyses are valuable for determining relationships among populations of the same and different species.

While many examples of such analyses of isozyme data exist, the studies of *Zea* by Doebley et al. (1984, 1985, 1986) will serve to illustrate the utility of these methods. Figure 6.1 shows a cluster analysis (average linkage) of isozyme data for species and races of *Zea*. Modified Rogers' distances were calculated from allelic frequency data and were employed to construct the phenogram. Cluster analysis is useful because it allows one to visualize similarities among taxa by the levels at which they group together.

Doebley et al. (1984) employed principal components analysis to assess relationships among collections of maize and its wild relatives the teosintes. Each allele was treated as a character, the value of which could vary from 0 to 1, that is, its frequency in a collection. It may be seen in Fig. 6.2 that the first two principal components allow visualization of

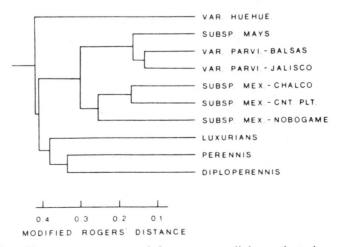

Figure 6.1 Phenogram constructed from average linkage clustering of species and races of *Zea* using modified Rogers' distances calculated from allelic frequency data of allozymes. (With permission from J.F. Doebley et al., Syst. Bot. **9**:203–218. 1984, American Society of Plant Taxonomists.)

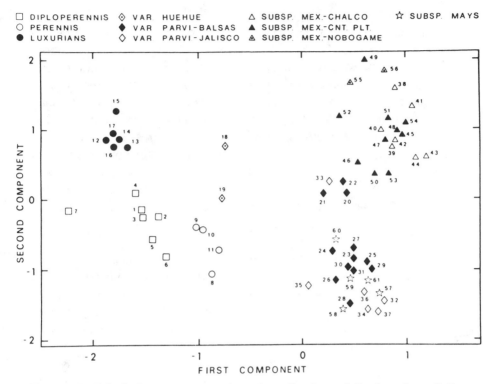

Figure 6.2 Principal component analyses for collections of *Zea* based on allelic frequency data. (With permission from J.F. Doebley et al., Syst. Bot. **9**:203–218. 1984, American Society of Plant Taxonomists.)

the distinctiveness of collections of the various taxa of *Zea*. Analyses of isozymic data beyond calculation of mean genetic identities and/or distances for pair-wise comparisons of populations are desirable because they allow for additional insights into relationships. These phenetic analyses should be used more in future studies than they have been in the past.

CLADISTIC ANALYSES

Cladistic analyses of electrophoretic data from plants using the Hennigian approach have rarely been employed. An excellent discussion of the problems associated with such analyses of allelic data is given in Chapter

12 of Richardson et al. (1986). The following treatment is based largely on that discussion. One problem with cladistic analyses is ascertaining which allozyme encoded by a locus represents the ancestral condition. The preferred method is to employ an outgroup of one or several species that are related to, but clearly not within, the taxon under study. Presumably, the allozyme present in the outgroup represents the ancestral state in the taxa under study. One potential problem with this method is that the outgroup may contain none of the allozymes present in the species being studied. For example, the species under investigation may contain alleles a, b, and c at a given locus. If the outgroup contains only alleles d, e, and f, then it is not possible to ascertain the ancestral allele (or allozyme). Because cladistic analysis is based on shared, derived character states (synapomorphies), it is critical that the correct ancestral condition be determined.

Another critical factor in using allelic data for cladistic purposes is the proper interpretation of transition series. Assume that allozymes a, b, and c occur in the taxa being investigated, *and* that it has been possible to determine via outgroup analysis that a represents the ancestral condition (Fig. 6.3). One is still faced with the questions of whether b and c were derived independently from a; or a gave rise to b, which then gave rise to c; or a gave rise to c, and the latter in turn gave rise to b. Richardson et al. (1986) explained that determining the proper transition series for the three alleles is critical to the construction of the correct phylogeny. If b and c were derived independently from a, then taxa containing each of these would form a monophyletic assemblage (Fig. 6.3). In the second situation in which a -> b -> c, all taxa with c would represent one monophyletic group, and those with b would include a larger monophyletic assemblage. Where species with b should be placed relative to those with c is not known from the data (Fig. 6.3). If a gave rise to c, which in turn gave rise to b, then those taxa having b in common are monophyletic, but it is not possible to ascertain where they should be placed relative to the larger group of taxa with a or c (Fig. 6.3).

Richardson et al. (1986) emphasized that in many cladistic studies of electrophoretic data once the ancestral condition has been determined all other alleles at a locus are treated as shared derived character states (synapomorphies) uniting groups of species as monophyletic assemblages. The problems associated with this approach are evident in Fig. 6.3.

Richardson et al. (1986) also discussed the problems of using the presence–absence of an allele as the character instead of employing the

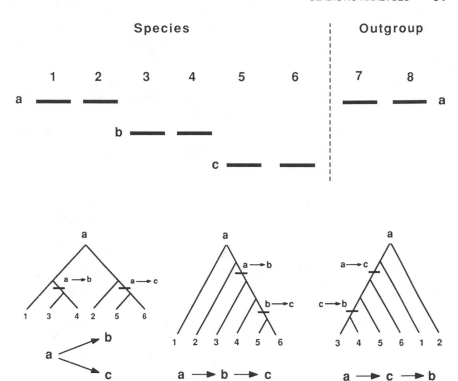

Figure 6.3 (Top) Allozymes present at a locus in six hypothetical species and two species serving as the outgroup. (Bottom) Cladograms constructed from the assumption of different transition series among the allozymes. See text for additional details and discussion.

gene locus as the character with alleles as the character states. From Fig. 6.3 it can be seen that the presence of allele a would be treated as the ancestral condition because it is present in the outgroup and its absence would be viewed as derived. The opposite condition applies to alleles b and c, with their presence treated as the derived and their absence the ancestral condition. If these assumptions were accepted, then species 3 and 4 would be considered a monophyletic group because they share allele b, and 5 and 6 would likewise be treated as a monophyletic assemblage because they possess allele c. Also, species 3, 4, 5, and 6 would be viewed as monophyletic, with species 1 and 2 excluded. It is apparent from Fig. 6.3 that these conclusions on phylogenetic relationships are not justified from the data.

Richardson et al. (1986) argued that the more desirable approach is to combine observations from more than one locus. For example, assume that the taxa in Fig. 6.3 are examined at a second locus and it is found that species 1, 2, and 3 have allozyme a, while species 4, 5, and 6 have allozyme b. Furthermore, assume that the outgroup has only a. This means that a is the ancestral condition and that b unites species 4, 5, and 6 as a monophyletic assemblage, with species 1, 2, and 3 being ancestral. It is then possible to go back to the original locus and use species 1, 2, and 3 as the outgroup for 4, 5, and 6, and to infer that allele b is the ancestral (plesiomorphic) condition for species 4, 5, and 6. This in turn indicates that the presence of allele c in species 5 and 6 is the derived condition and represents a synapomorphy uniting them as a monophyletic group. Therefore, comparative data from another locus suggest that a -> b -> c represents the proper transition series. Although it may be possible to elucidate a transition series at a locus by examining other loci, the task will be quite complex when a large number of alleles occur at loci.

Swofford and Berlocher (1987) discussed the use of allelic frequency data for constructing phylogenies. The trees are constructed by finding the most parsimonious tree (or trees) with regard to Manhattan distance. Doing this requires that two different problems be addressed. The first of these tasks involves assigning allele frequencies to loci in hypothetical ancestral taxa such that the length of the tree is minimal for that tree topology. Swofford and Berlocher (1987) emphasized that the frequencies for each locus in the hypothetical ancestors must add up to one. In essence, then, the process involves finding ancestors whose allelic frequencies are such that a minimal amount of change is required to produce a tree for taxa with the observed allelic frequencies. Swofford and Berlocher (1987) suggested that this problem is solved by using linear programming. This procedure optimizes the value of a linear function given certain linear constraints on the variables.

The second part of the problem of obtaining phylogenies from allelic frequency data involves finding from all possible trees the tree topologies that allow global minimization of the tree length. Swofford and Berlocher (1987) indicated that this is more difficult than the first problem of assigning allelic frequencies to hypothetical ancestors. They suggested heuristic methods for obtaining approximations. The reader is referred to the original paper for suggested methods.

Use of allelic frequency data for constructing phylogenies offers some advantages over possible problems with using only presence–absence of alleles or attempting to determine transition series for alleles at a locus. The method advocated by Swofford and Berlocher (1987) assumes that changes in allelic frequencies occur in the most parsimonious way during evolution. Whether indeed this is always the case is open to question.

It seems fair to conclude that the generation of phylogenies from allelic frequency data is subject to problems of both a biological and methodological nature. Perhaps this accounts for the very few studies of plants in which cladograms were constructed solely from allelic data.

SUMMARY

Enzyme electrophoresis allows for the estimation of the number of alleles and their frequencies at gene loci in populations and taxa of plants. These allelic data may then be employed to estimate the genetic variation within populations or taxa; several different measures of variation are employed including percentage of polymorphic loci, mean number of alleles per locus, and expected heterozygosity with random mating (gene diversity of Nei, 1973). Several measures of genetic similarity (or divergence) may be calculated from allelic frequencies, including Nei's genetic identity and genetic distance, and Rogers' distance. A similar measure that may be useful is one based on genotypic similarity rather than allelic similarity. Cluster and principal component analyses are valuable techniques for showing relationships among populations. While cladistic analyses of allelic data may provide insights into phylogenetic relationships among species, care must be taken to assure that assumptions about the ancestral and derived condition as well as the transition series for allozymes are valid.

CHAPTER 7

Allelic Data: Studies of Populations and Infraspecific Taxa

INTRODUCTION

This chapter is the first of three concerned with employing allelic data generated from enzyme electrophoresis in plant systematics. Genetic identity will be employed as the primary measure of similarity among populations, infraspecific taxa, and species. Electrophoretic variation within and divergence among populations of the same taxa and infraspecific taxa (whether recognized taxonomically as subspecies or varieties) will be considered. Emphasis is on the latter because the focus of this book is systematics rather than population biology.

In this and the following chapters, the tacit assumption is made that in most instances different allozymes encoded by a locus are neutral or nearly neutral. If selection were acting strongly on allozymic diversity, then allozymes would not be nearly as useful systematically as they have proved to be. It is not argued, however, that all allozymes are always selectively equivalent.

POPULATIONS OF THE SAME TAXON

Gottlieb (1977a), reviewed electrophoretic data in plant systematics and summarized available information indicating high genetic identity (ca. 0.95) among plant populations belonging to the same taxon. Later summaries of allozyme data from many additional genera (Gottlieb, 1981a; Crawford, 1983; Giannasi and Crawford, 1986) provided general support for the initial conclusion drawn by Gottlieb (1977a). For readers interested in correlations between the genetic structure of plant populations (as revealed by enzyme electrophoresis) and ecological aspects of the plants, the reviews by Loveless and Hamrick (1984) and Hamrick (1987) are recommended. Data gathered since the initial publication by Gottlieb in 1977 indicate that mean genetic identities for conspecific populations of plants from a wide variety of taxonomic groups with different life histories (herbaceous and woody plants, selfers and outcrossers, annuals and perennials, ferns, gymnosperms and angiosperms, etc). are above 0.90 (see Gottlieb, 1981a; Crawford, 1983; Giannasi and Crawford, 1986 for reviews). More recent studies of additional and taxonomically highly diverse genera include: *Abies* (Jacobs et al., 1984); *Adiantum* (Paris and Windham, 1988); *Aeschynomene* (Carulli and Fairbrothers, 1988); *Agastache* (Vogelmann and Gastony, 1987); *Antennaria* (Bayer, 1988; Bayer and Crawford, 1986); *Asplenium* (Werth et al., 1985a); *Blechnum* (P.S. Soltis and D.E. Soltis, 1988b); *Bommeria* (Haufler, 1985); *Botrychium* (D.E. Soltis and P.S. Soltis, 1986b); *Cabomba* (Wain et al., 1983); *Camassa* (Ranker and Schnabel, 1986); *Camellia* (Wendel and Parks, 1985); *Chenopodium* (Walters, 1988; Wilson, 1988a,b,c); *Cirsium* (Loveless and Hamrick, 1988); *Coreopsis* (Crawford et al., 1984; Crawford and Whitkus, 1988); *Cucurbita* (Decker and Wilson, 1987; Wilson, 1989); *Gaillardia* (Heywood and Levin, 1984); *Galvesia* (Elisens, 1989); *Gymnopteris* (Ranker, 1987); *Helianthus* (Wain, 1982, 1983); *Hemionitis* (Ranker, 1987); *Hydrocharis* (Scribailo et al., 1984); *Lasthenia* (Crawford et al., 1985; Crawford and Ornduff, 1989); *Lens* (Pinkas et al., 1985; Hoffman et al., 1986); *Limnanthes* (McNeill and Jain, 1983; Kesseli and Jain, 1984); *Mabrya* (Elisens and Crawford, 1988); *Pellaea* (Gastony and Gottlieb, 1985); *Picea* (Yeh et al., 1986); *Pinus* (Wheeler and Guries, 1982; Wheeler et al., 1983; Ledig and Conkle, 1983; Loukas et al., 1983; Furnier and Adams, 1986; Conkle et al., 1988; Millar et al., 1988); *Plectritis* (Carey and Ganders, 1987); *Populus* (Weber and Stettler, 1981; Hyun et al., 1987); *Quercus* (Manos and Fairbrothers, 1987); *Solanum* (Whalen and Caruso, 1983); *Tellima* and *Tolmiea* (Rieseberg and Soltis, 1987a); *Tillandsia* (Soltis et al., 1987); *Triticum* (Nevo et al., 1982); *Zea* (Doebley et al., 1984, 1985, 1986); and *Zizania* (Warwick and Aiken, 1986). This list is not intended to be exhaustive.

Earlier evidence indicated that a greater range of variation in genetic identities is found among populations of predominately self-pollinating plants as opposed to outcrossers (Gottlieb, 1981a; Crawford, 1983; Giannasi and Crawford, 1986). Such a range may occur because different populations are fixed for alternative alleles at several loci in selfers whereas outcrossers exhibit the same alleles in similar frequencies. To put it differently, in selfers, a greater proportion of gene diversity (see Nei,

1973) within a species exists among its component populations as compared with outcrossers, in which it occurs within populations. Schoen
(1982) demonstrated that within one species, *Gilia achilleifolia*, there is a
lower mean genetic identity among populations of selfers as compared
with outcrossers, and likewise, a greater proportion of the genetic diversity exists among populations of selfers as opposed to the outcrossers. A
rather dramatic example of the same situation was provided by Gottlieb
(1984b), who detected no variation at 40 gene loci within each of two
populations composed of selfing plants of *Clarkia xantiana*, but the two
populations were monomorphic for different alleles at 8 of the 40 loci. By
contrast, in populations of outcrossing plants of *C. xantiana*, nearly half
the loci were polymorphic. Rick et al. (1977) found strong positive
correlations between outcrossing and allozyme variation in populations of
Lycopersicon pimpinellifolium. Similar patterns have been found among
congeneric species differing in breeding systems (*Phlox*, Levin, 1978;
Limnanthes, Brown and Jain, 1979; *Lasthenia*, Crawford et al., 1985;
Agastache, Vogelmann and Gastony, 1987; *Plectritis*, Carey and Ganders,
1987; *Tillandsia*, D.E. Soltis et al., 1987, as examples). An extreme
example of differences among populations of a selfing species is provided
by *Bidens discoidea*, in which pairs of populations may exhibit identities
as low as 0.688 (Roberts, 1983). By contrast, other populations may have
an identity exceeding 0.99. A group of plants in which lowered mean
identities have been found among population is the homosporous pteridophytes. Ranker (1987) reported mean identities as low as 0.78 for
populations in species of *Hemeonitis*, with pairs of populations having an
identity of 0.52. Ranker (1987) reported values nearly as low for populations of a species of *Gymnopteris*.

In addition to patterns of variation seen in genetic identities averaged
over all loci, it is advisable to examine the distribution of individual
alleles among populations. Wilson (1981), for example, demonstrated
interesting patterns of geographic variation at two gene loci for leucine
aminopeptidase in tetraploid members of *Chenopodium* in South
America.

INFRASPECIFIC TAXA: SUBSPECIES OR VARIETIES

Infraspecific variation, when recognized taxonomically, usually involves
naming entities as subspecies or varieties. There is not always agreement
on which category should be used, and in reality the two are often used to

designate the same situation regardless of the name applied. In certain instances both categories are employed, with different varieties of a subspecies recognized. Subspecies or varieties usually designate geographically discrete groups of populations that are consistently separable by several features. The groups may occasionally come in contact and produce fertile hybrids. It is thought that varieties or subspecies sometimes represent a stage in the process of speciation (Grant, 1981). The studies discussed in this chapter are somewhat arbitrary in the sense that different taxonomic treatments could recognize entities as species rather than infraspecific taxa (and thus would be considered in the next chapter).

In an earlier review (Crawford, 1983), it was shown that populations treated as distinct at the infraspecific level often exhibit mean genetic identities as high as conspecific populations in which no taxonomic entities are recognized. Two notable exceptions to this pattern were discussed in that review. The following discussion will elaborate on and update information in Crawford (1983), and it will be apparent that it is not possible to predict a *priori* with reasonable certainty genetic identities between infraspecific taxa, as is also the case with species.

Wheeler and Guries (1982) showed that mean genetic identities for populations of different subspecies of *Pinus contorta* (lodgepole pine) are comparable to values for the same subspecies, with all above 0.98. Wheeler et al. (1983) also studied varieties of *Pinus clausa* and found very little differentiation at isozyme loci, with a mean identity of over 0.99 for populations of var. *clausa* and var. *immuginata*. Conkle et al. (1988) found that the four subspecies of *Pinus brutia* have genetic identities higher than 0.90.

The lack of divergence among subspecific entities in *Pinus* is perhaps not surprising given the general lack of divergence among populations of the same species of conifers, that is, in conifers considerable isozyme variation is maintained within populations, and there is little subdivision among them. Such a lack may be caused by a combination of factors such as large population sizes, outcrossing, and the possibility for long-distance dispersal of pollen and seeds (Hamrick et al., 1981; Wheeler and Guries, 1982). Wheeler et al. (1983) also suggested that lack of genetic divergence among subspecies results from having been separated for a short period of time.

Doebley et al. (1984) examined the various subspecies and varieties of *Zea mays* (i.e., the cultigen and annual teosintes) for isozyme variation. Variation was found in genetic identities among the subspecific entities, with var. *huehuetenangensis* being the most distinct and exhibiting identity values ranging from 0.753 to 0.808 with other infraspecific taxa of *Z. mays*. When the different geographic races of var. *parviglumis* were compared with subsp. *mexicana*, lower identities (0.820–0.929) were found than were detected among geographic races within each infraspecific taxon, that is, 0.929 for the two races of var. *parviglumis*, and 0.899, 0.933, and 0.960 for the races of subsp. *mexicana*. Genetic identities of the cultivated *Z. mays* subsp. *mays* with the other subspecies vary from 0.802 to 0.975 (Doebley et al., 1984). In certain instances, the enzyme data correlate with morphology and other features in similarities among the subspecific taxa of *Z. mays*, but in other cases they do not correlate. Correlation exists with regard to var. *parviglumis* and subsp. *mexicana* because these taxa are separable by several morphological features and they grow at different elevations. They also exhibit lowered enzyme similarities (Doebley et al., 1984). The allozyme data for *Z. mays* var. *huehuetenangensis* are somewhat paradoxical when viewed against other information. This variety shows relatively low genetic identities compared with other subspecific taxa, but it is closest to var. *parviglumis* (Doebley et al., 1984). Morphological information and other types of molecular data indicate a close relationship between var. *huhuetenangensis* and the other subspecific taxa of *Z. mays*. Yet cytological data show it to be distinct from other *Z. mays*. Relationships have not yet been resolved because of the nonconcordant data sets.

McNeill and Jain (1983) examined allozyme divergence among subspecies and varieties of several species of *Limnanthes*. Genetic identities were calculated from estimated allelic frequencies for the taxa obtained by pooling all populations of each taxon. These workers found large differences in similarities among taxa in the three species *L. gracilis*, *L. alba*, and *L. floccosa*. The two varieties of *L. alba* show little differentiation, with an identity of 0.986. Comparisons of populations of the five subspecies of *L. floccosa* revealed a mean identity of 0.740 for the taxa, but the values range from 0.575 to 0.993. There is, however, concordance between morphological and genetic similarity for the subspecies. For example, *L. floccosa* subsp. *flocossa* and subsp. *bellingeriana* are most similar morphologically and are distinct from the other three subspecies. These two subspecies have a genetic identity of 0.993 and exhibit a value no higher than 0.762 with any other subspecies. By contrast, the other three taxa, subsp. *californica*, subsp. *grandiflora*, and subsp. *pumila*, show genetic identities among themselves ranging from 0.869 to 0.948

(McNeill and Jain, 1983). The concordance of morphological and electrophoretic data suggests that the subspecies of *L. floccosa* are diverging gradually in a number of genetically based features. Thus it seems reasonable to assume that the two groups of subspecies (as discussed above) have been separated for a longer period of time than have subspecies within each of the groups.

The two disjunct varieties of *L. gracilis* have a genetic identity of 0.760, which is perhaps surprisingly low, since they are treated as distinct only at the infraspecific level. However, hybrids between them have reduced fertility, and they are disjunct by 1200 km; in addition, they exhibit higher genetic identities with *L. alba* than with each other. McNeill and Jain (1983) indicated that these data demonstrate the need for additional studies of the varieties of *L. alba*, and they further suggested that "present taxonomic designations tend to obscure the evolutionary genetic relationships between these taxa." In section *Inflexae* of *Limnanthes*, enzyme electrophoresis has provided valuable supplemental data for inferring patterns of divergence and genetic relationships among subspecific entities. In addition, results of the enzyme studies should stimulate further research aimed at elucidating evolutionary relationships.

Wain (1982, 1983) carried out electrophoretic studies of the subspecies of *Helianthus debilis*. In the earlier investigation (Wain, 1982), the three subspecies occurring in Florida, subsp. *debilis*, subsp. *vestitus*, and subsp. *tardiflorus*, were examined electrophoretically. Subsp. *debilis* has a genetic identity (mean for all pair-wise comparisons of populations) of 0.886 and 0.902 with subsp. *vestitus* and *tardiflorus*, respectively. The latter two subspecies exhibit an identity of 0.985. *Helianthus debilis* subsp. *debilis* is more highly differentiated from the other two subspecies at isozyme loci than the two are from each other, paralleling the fact that subsp. *debilis* is the most morphologically distinct of the three subspecies (Heiser, 1956).

Wain (1983) examined two additional subspecies of *H. debilis* as well as additional plants treated either as two subspecies of *H. debilis* or as two subspecies of *H. praecox*. Synthetic F_1 hybrids between the so-called "debilis" and "praecox" breeding assemblages have very reduced pollen fertility and seed set relative to hybrids belonging to the same breeding assemblages. The mean genetic identity for taxa in the two different breeding groups is even somewhat higher than that found among subspecies within each assemblage (0.939 versus 0.923). Wain (1983) emphasized that genetic differentiation at isozyme genes may not parallel degree of reproductive isolation. We will see later that this is the case with a number of species as well.

Heywood and Levin (1984) detected very little divergence among the three varieties of *Gaillardia pulchella* at genes specifying isozymes, the values being 0.94 or higher. These authors observed that genetic identities based on isozymes may be of limited use in delimiting infraspecific taxa and suggested that morphological and other features often used to recognize infraspecific taxa may not be associated with complete barriers to gene flow. Gene flow will be retard divergence at isozyme loci.

Warwick and Aiken (1986) examined electrophoretic variation in annual wild rice, the genus *Zizania*. The five taxa had been classified either as one species, *Z. aquatica* (with five varieties), or as two species, *Z. aquatica* (with three varieties) and *Z. palustris* (with two varieties). Their results strongly supported the recognition of two species rather than one species and five varieties. The two groups were different, that is, monomorphic for alternative alleles or with the same alleles in very different frequencies, for four enzymes. These differences resulted in greatly lowered genetic identities for varieties of the different species as contrasted to the same species. For example, the ranges of identities for varieties within each of the two species are 0.86 to 0.99 and 0.86 to 0.98, whereas for varieties of the different species the values are in a very narrow range between 0.67 and 0.69. Other observations on these species indicated that while they overlap geographically there is no evidence that they grow sympatrically or that they hybridize in nature. The lack of additivity of alleles at isozyme loci supports the hypothesis that the species do not hybridize under natural conditions. The electrophoretic data were also useful for inferring the origin of the cultivated wild rice. A suggested origin was from var. *palustris*, with delay of disarticulation of spikelets being the critical development. The allozymes indicate, however, that the cultivar evolved from var. *interior* because the two share unique alleles at several loci.

Two studies of allozymes in the genus *Lens* have been carried out recently in different laboratories (Pinkas et al., 1985; Hoffman et al., 1986). The genus has been considered to consist of four or five distinct species, including the cultivated lentil, or of two species, *L. culinaris* and *L. nigrans*, with the former having three subspecies and the latter two subspecies. Both authors generally agree that the three subspecies assigned to *L. culinaris* exhibit relatively high genetic identities, although the values reported by Hoffman et al. (1986) are in general higher than those given by Pinkas et al. (1985). The range of mean values for pairwise comparisons of populations resulting from the former study was 0.80 to 0.90, whereas the latter authors reported a range of 0.65 to 0.86. It is of interest that both studies reported the same patterns of relative values for the same varieties. For example, they demonstrated clearly that subspecies *ervoides*, which is placed taxonomically in *L. nigrans*, is much more similar to other subspecies of *L. culinaris* than it is to subsp. *nigrans*. The electrophoretic data are concordant with recent cytogenetic information in suggesting that subsp. *ervoides* has closer affinities with *L. culinaris* than with *C. nigrans*.

Both studies addressed the question of the origin of the domesticated lentil, subsp. *culinaris*. Hoffman et al. (1986) concluded that their electrophoretic data alone could not distinguish between subsp. *orientalis* and subsp. *odemesis* as possible progenitors because they each had very similar genetic identities (0.87 and 0.80, respectively) with the cultivated lentil. By contrast, Pinkas et al. (1985) calculated an identity of 0.81 for subsp. *culinaris* and subsp. *orientalis*, whereas for the former and subsp. *odemesis* the value was only 0.65. From these results, they suggested that subsp. *orientalis* is the progenitor of the cultivated lentil. These differences ostensibly result primarily from the several different enzymes scored in the two studies rather than different results for the same enzymes, indicating that even in studies employing a large number of enzymes several differences in enzymes can affect the results.

Crawford and Bayer (1981) examined allozyme variation in *Coreopsis cyclocarpa* var. *cyclocarpa* and in var. *pinnatisecta*. The two taxa consist of suffrutescent diploid perennial plants that are disjunct by some 700 km in Mexico (Crawford, 1970). They are distinct morphologically, with var. *cyclocarpa* having simple linear leaves while those of var. *pinnatisecta* are pinnatisect; there are few other obvious differences between them. They both flower at the same time and occur in similar open grassy habitats of high plateau regions. When crossed, the two varieties produced fully fertile F_1 hybrids intermediate for the "key" feature of the leaves. The leaf character segregated in the fully fertile F_2 generation to provide some individuals indistinguishable from var. *cyclocarpa*, while other segregates were inseparable from var. *pinnatisecta*.

Mean genetic identities are 0.95 for populations of var. *pinnatisecta* and 0.98 for populations of var. *cyclocarpa*, whereas the mean value for populations of the two varieties is reduced to 0.75. Minimal genetic divergence has seemingly occurred for morphological features and ecological preferences for the two varieties of *C. cyclocarpa*. In addition, no chromosomal changes are apparent from the crossing studies, and there is no cross incompatibility between the taxa. The lowered identities, while perhaps a bit surprising at first glance, probably result from accumulated mutations at isozyme loci following geographic isolation in these long-lived perennials.

Crawford and Smith (1984) examined electrophoretic similarities among different varieties of another species of *Coreopsis, C. grandiflora*. This species is comprised of herbaceous perennials and three diploid, sometimes sympatric, varieties are recognized. Morphological differences are small but consistent, and the varieties are highly interfertile (E.B. Smith, 1973, 1976). Crawford and Smith (1984) detected no isozyme differentiation among the three diploid varieties of *C. grandiflora*; mean identity values for populations of different varieties (0.88 to 0.94) were essentially the same as populations of the same varieties (0.89 to 0.99). It was suggested that while small morphological differences have developed to produce the recognizable infraspecific taxa in *C. grandiflora*, sufficient time has not passed for detectable isozyme differentiation to occur. It is also possible that since the three diploid varieties have overlapping distributional ranges, occasional hybridization may be a factor preventing isozyme differentiation (Crawford and Smith, 1984).

Whalen (1979) found reduced isozyme similarities between infraspecific taxa in *Solanum heterodoxum* and *S. citrullifolium*, with identities of 0.83 and 0.71, respectively. The two varieties of *S. heterodoxum* are disjunct by several hundred miles, and the geographic ranges of the two varieties of *S. citrullifolium* barely meet at their extremes. Presumably, there is little chance for gene exchange between the taxa. The evolutionary history of the taxa may be tied to interspecific hybridization, so it is difficult to evaluate the meaning of the lowered genetic identities (Whalen, 1979). In addition, it is important to note that the values are based on seven gene loci, which is a rather low number.

Rieseberg et al. (1987) examined isozyme divergence among four varieties of *Allium douglasii*. They found that populations of varieties *columbianum* and *constrictum* showed a much higher mean identity to each other (0.93) than any other varieties did to each other (identities of 0.72 to 0.76). The four varieties are allopatric diploid perennials of rather limited distribution. Rieseberg et al. (1987) suggested that the low identities among certain varieties result from longer periods of geographic isolation with lack of gene flow, and that the high mean identity for populations of var. *columbianum* and var. *constrictum* indicate a recent divergence. Furthermore, analysis of variation in the two varieties suggests that var. *constrictum* is a recent derivative of var. *columbianum* because the former lacks 12 alleles present in the latter while containing only one unique allele. Variety *constrictum*, therefore, contains a limited extraction of the variation found in var. *columbianum*.

Allelic frequency data available for homosporous pteridophytes recognized as distinct at the varietal level indicate low genetic identities compared with flowering plants. Ranker (1987) estimated the extremely low identity value of 0.03 for two varieties of *Notholena candida*. In an extensive biosystematic study of the two subspecies (one woodland and the other on serpentine substrate) of the maidenhair fern *Adiantum pedatum*, Paris and Windham (1988) calculated a mean genetic identity of 0.495 for the two subspecies. This low identity, combined with other data, was used by Paris and Windham to argue for recognition of these entities as distinct species.

Several additional examples of lack of lowered identities between infraspecific entities are known. Levin (1977) demonstrated that subspecies of *Phlox drummondii*, which have weak barriers to gene exchange but are isolated ecogeographically, exhibit genetic identities comparable to populations of the same subspecies. Soltis (1981a,b) reported that the two highly interfertile sympatric varieties of the short-lived perennial *Sullivantia hapemanii* have high genetic similarity. Geographical races (varieties) of the weedy annual *Chenopodium incanum* are morphologically distinct yet populations of different varieties exhibit mean identities similar to values for populations of the same variety (Crawford, 1983). Wain et al. (1983) found the two varieties of *Cabomba caroliniana* (which are submersed aquatics) to be essentially identical electrophoretically. They suggested that morphological differences between the infraspecific taxa may result from phenotypic plasticity rather than genetic differentiation.

Domesticated plants are often treated as distinct from their putative wild relatives at the infraspecific level, and a number of electrophoretic studies have examined divergence between them and their wild progenitors. Examples from wild rice and lentils were discussed earlier. McLeod et al. (1983) found the wild *Capsicum baccatum* var. *baccatum* and its presumed cultivated derivative *C. baccatum* var. *pendulum* to be essentially identical electrophoretically (identity of 0.98). Wilson and Heiser (1979) used allelic data at one gene locus encoding leucine aminopeptidase (LAP) to investigate relationships between weed–domesticate pairs in the genus *Chenopodium* in Mexico and South America. The weed and domesticate in each area are fixed for the same allele, with pairs from the different geographic regions monomorphic for different alleles at the LAP locus. The allozyme data are concordant with information from crossing studies and morphology in suggesting that one species occurs in North and another in South America, with the domesticate and weed of each representing subspecies (Wilson and Heiser, 1979). More recent studies by Wilson (1988b,c), in which over 20 loci were examined, demonstrated identities of over 0.90 for the cultivated *Chenopodium quinoa* and the free-living populations, which are treated as distinct at the subspecific level.

In a study discussed earlier, Doebley et al. (1984) determined genetic divergence between the cultivated *Zea mays* subsp. *mays* and various other infraspecific taxa of *Z. mays* from which the domesticate presumably evolved. Genetic identities between the domesticate and the various wild forms range from 0.755 to 0.975. The similarities at isozyme loci perhaps appear rather high at first glance given the rather striking morphological differences between the domesticate and the wild forms.

SUMMARY

Mean genetic identities among conspecific populations are usually quite high, often above 0.90. This is particularly true for outcrossing plants where most populations contain the same high frequency alleles. In self-pollinating plants one of two situations normally is found for conspecific populations. In certain cases populations may be monomorphic for the same alleles at all (or nearly all) gene loci so that very high genetic identities occur. In other instances, populations may be fixed for different alleles at several loci to give lower mean identities than is normally found in outcrossers. Homosporous pteridophytes, like flowering plants, show a range of mean identity values for populations of the same taxon, but identities are often lower than in angiosperms. Populations of conifers show very high identities.

There is no one consistent pattern of electrophoretic divergence among infraspecific taxa (varieties and subspecies); most commonly they have the same high genetic identities that occur among populations not recognized as distinct taxonomically, indicating that the morphological differences used to distinguish infraspecific taxa often develop prior to divergence at isozyme gene loci. In wild–domesticate infraspecific taxa, the pattern may be particularly pronounced because humans have strongly selected for certain morphological features. This has brought about differences (in some cases quite pronounced) in external features in a relatively short period of time with no accompanying detectable mutations at loci specifying soluble enzymes. It should be pointed out that the actual genetic bases of morphological differences between infraspecific taxa are usually not known, but that conspicuous features may be based on one to several genes (Gottlieb, 1984a). Thus, while allozyme differences can be quantified and rather precise comparisons made, this is not the case with morphology, which must be kept in mind when comparing enzyme and morphological divergence between taxa.

Other studies have shown concordance between morphological and enzyme divergence: those taxa that appear most dissimilar also show lowered genetic identities. This pattern probably results from longer isolation times for the taxa, with progressively greater levels of divergence in morphology and isozymes reflecting increasing times since gene exchange last occurred between populations of the two taxa.

In rare instances, the level of divergence at isozyme loci appears large relative to morphological differences among the taxa. The major problem at present with attempting to interpret possible explanations is that most infraspecific taxa have not been studied in sufficient depth to provide a basis for inferring the genetic bases of differences in morphology or ecological preferences.

One fairly consistent pattern that has emerged is that wholly (or largely) allopatric infraspecific taxa may be well differentiated at isozyme loci, whereas sympatric or marginally sympatric taxa are less divergent. This pattern could reflect lack of gene flow between the geographically isolated taxa and/or the greater time of isolation for fully allopatric versus partially sympatric taxa. Lastly, discussions in this chapter have shown that considerable genetic divergence may occur between geographical races without the occurrence of speciation.

CHAPTER 8

Allelic Data in the Study of Species Relationships, Modes of Primary Speciation, and Phylogenetic Reconstruction

INTRODUCTION

The last chapter considered similarities among populations of the same taxa and infraspecific taxa; the present one focuses on allelic similarities (and divergence) among congeneric species, and how the data may be employed to ascertain relationships among species. The use of electrophoresis for inferring modes of plant speciation also will be evaluated. Lastly, the use of allelic data for constructing phylogenies will be presented briefly. Enzyme electrophoresis has found the widest application in plants at the species level, a fact reflected in the diverse kinds of studies included in this chapter.

In the first detailed review of electrophoretic data for congeneric plant species, Gottlieb (1977a) concluded that lowered mean genetic identities characterize populations of congeneric species (0.67) as compared with populations of the same species (ca 0.95). Updated reviews (Gottlieb, 1981a; Crawford, 1983, 1989) showed the same general conclusions, and additional results published since the last review have revealed a similar pattern. It is important to emphasize, however, that there is wide variation in mean genetic identities for populations of different congeneric species. Also, there is considerable variation *within* a genus, with some species very similar and others more divergent. This means that, given two species, it is not possible to predict a *priori* the mean value for populations of the two taxa. The varying patterns found among taxa will provide a framework for discussion in this chapter, and the different taxonomic, systematic, and evolutionary questions that may be addressed with the data will be considered.

TAXONOMIC APPLICATIONS

The simplest and most "straightforward" application of allozymes at the species level uses allelic data as additional characters for supporting taxonomic recognition of different populations (i.e., testing species concepts). While species cannot be recognized solely on the basis of electrophoretic differences, a consistent correlation between certain morphological features (small or cyptic as they may be in some instances) and the presence of particular alleles adds support to species recognition because it indicates that the taxa are not exchanging genes and thus are maintaining separate gene pools. Electrophoretic data may be particularly helpful for "problem" taxa that are very similar morphologically; several examples of this application will be considered.

Jeffries and Gottlieb (1982) examined electrophoretic variation at 30 gene loci in two species of *Salicornia*, *S. europaea* and *S. ramosissima*. The two are diploid, occur in salt marshes in Europe and England, and are very similar morphologically. These workers found the two species homozygous for the same alleles at 24 loci and monomorphic for alternative alleles at the remaining 6 gene loci. The enzyme data indicate that the two species do not exchange genes (are reproductively isolated); this information is particularly valuable because the taxa sometimes intergrade morphologically due to phenotypic plasticity and the differentiation of local populations (Jeffries and Gottlieb, 1982).

Enzyme electrophoresis has also been useful taxonomically in certain diploid members of *Chenopodium* occurring in the western United States. These weedy, annual, native plants exhibit marked phenotypic plasticity, often making taxonomic determinations difficult (Crawford, 1975, 1977). Allozymes were examined in three phenetically similar species, *C. atrovirens*, *C. desiccatum*, and *C. pratericola* (Reynolds and Crawford, 1980). These species were found to have genetic identities as high as populations of the same species (Crawford and Wilson, 1979; Crawford, 1983). By contrast, two species that are much more distinct phenetically, *C. hians* and *C. leptophyllum* (Crawford and Reynolds, 1974), show lowered identities (i.e., below 0.70) when compared with other diploid species (Crawford, 1983). *Chenopodium fremontii* and *C. incanum* had often been considered as doubtfully distinct (see discussion in Crawford, 1977), yet each exhibits several unique alleles, and they have a genetic identity below 0.50 (Crawford, 1979, 1983). The two species often appear similar morphologically when growing in similar habitats despite the fact that they are allozymically distinct. The similar morphologies must therefore result from phenotypic plasticity rather than hybridization; if hybridization were involved, it would be detectable electrophoretically. Lastly, allozyme data are concordant with other information in documenting that the previously recognized *C. incognitum* actually consists of two elements, one conspecific with *C. hians* and the other the same as *C. atrovirens* (Crawford and Wilson, 1979).

Walters (1988) has examined allozyme divergence among several additional diploid species of *Chenopodium* from western North America including *C. fremontii, C. neomexicanum, C. palmeri,* and *C. watsonii. Chenopodium fremontii* and *C. neomexicanum* have often been confused because of similarity in vegetative features, although features of the fruits have been used to place them in different subsections of the same section. The mean genetic identity for populations of the two species was calculated as 0.689, which indicates that they are clearly distinct electrophoretically. All other species of these weedy annuals had quite low identities, with the exception of *C. neomexicanum* and *C. palmeri* (mean identity of 0.974 as compared with 0.912 for different populations of *C. palmeri*). Walters (1988) suggested that the two entities are best treated as distinct at the varietal level rather than as separate species, presumably relying somewhat on the allozymic data in making this evaluation. Given that the species consist of morphologically distinct, as shown by the very thorough analyses included as part of the study, it is questionable how much weight should be assigned to the electrophoretic data in making a taxonomic decision.

The studies of Pinkas et al. (1985) and Hoffman et al. (1986) on *Lens* and Warwick and Aiken (1986) on *Zizania*, which were discussed in the last chapter, also have taxonomic implications because certain taxa that had previously been treated as distinct at the same taxonomic level were shown to exhibit very different levels of genetic identity. While electrophoretic data alone obviously cannot be used to make taxonomic decisions, they may be useful when employed with other information.

Nickrent et al. (1984) and Nickrent (1986) presented allozyme data for 19 taxa from the genus *Arceuthobium* (dwarf mistletoes), which are morphologically reduced parasitic flowering plants. A complete discussion of these interesting and provocative data is not possible here; rather, examples to illustrate the possible taxonomic (and evolutionary) implications of the results will suffice. The dwarf mistletoes, being very reduced morphologically, offer few characters for analysis; they have been subjected to various taxonomic treatments. The isozyme data do not support the placement of *A. douglasii* and *A. pusillum* together in the separate section *Minuta* because populations of these two species have lower identities (mean of 0.300, range 0.254–0.329) than they share with certain taxa in other sections. In fact, *A. divaricatum*, which has been placed in section *Campylopoda*, exhibits an identity of over 0.75 with *A. douglasii*. The populations of ten species in series *Campylopoda* exhibit high identities (above 0.80), and the conspecific populations in several instances do not group together, but rather are more similar to certain populations of other species.

In other instances, there are similarities between relationships inferred from morphology and isozymes. There are, however, more than a few cases in which the data do not correspond, and Nickrent et al. (1984) argued that morphological reduction and convergence in parasites such as *Arceuthobium* could serve to obscure taxonomic and phylogenetic relationships. They presented the reasonable view that all kinds of data should be considered in any taxonomic treatment of a group. It is apparent that isozyme data alone provided important insights into relationships in *Arceuthobium*.

APPLICATION TO STUDIES OF SYSTEMATIC AND PHYLETIC RELATIONSHIPS AMONG SPECIES

Divergence may occur at isozyme loci without speciation (Chapter 7). That is, divergence may take place among populations of geographic races viewed as members of the same species. The question that will be addressed now is whether speciation may occur without divergence at gene loci encoding soluble enzymes. The only way to determine this is to collect data from case studies, since a *priori* biological reasons for predicting whether isozyme divergence accompanies speciation do not appear.

Gottlieb (1973a) analyzed two species representing a progenitor–derivative pair with *Stephanomeria exigua* subsp. *coronaria* the progenitor and *S. malheurensis* the presumed recent derivative. The two species consist of annual diploid plants that are so similar morphologically they could well go unrecognized under casual observation in the field (see Gottleib 1973a, 1977b, 1978, 1979 for discussions). The presumed derivative species is known from only one locality in Oregon, where it grows together with its progenitor near the geographic limit of the latter. Gottlieb (1973a) demonstrated that (1) *S. malheurensis* is self-compatible and self-pollinating, which serves to restrict gene flow to the self-incompatible and outcrossing subsp. *coronaria*; (2) cross pollinations between the two species result in 50% reduction in seed set; and (3) the fertility of F_1 hybrids is reduced to about 25% because of structural differences in chromosomes. These factors permit the two species to grow together with hybridization occurring only in very rare instances; thus they are able to pass the "test of sympatry." Gottlieb (1973a, 1979) suggested that reproductive isolation may have developed rapidly, with a mutation at the self-incompatibility gene producing a plant of subsp. *coronaria* that was self-compatible and self-pollinating (such individuals are known in low frequency in the subspecies). Within these selfing plants, the chromosomal rearrangements could become homozygous quite rapidly. Pollen viability is increased markedly in F_2 hybrids as compared with F_1 plants, for example, from 25% to 60%, suggesting that few genetic differences may account for the isolation barriers (Gottlieb, 1973a).

Gottlieb (1973a) hypothesized that *S. malheurensis* was derived recently from subsp. *coronaria*. He noted that the former appears to encompass a limited amount of the variation found in the latter. Also, the putative recent derivative shows very few features not detected in its progenitor, something that would be expected because of insufficient time for genetic divergence since reproductive isolation. It should also be mentioned that none of the "novel" features displayed by *S. malheurensis* are of obvious adaptive value. In fact, one of them, lack of seed dormancy, is no doubt maladaptive because it does not prevent germination with the onset of the harsh winter in Oregon (Gottlieb 1973a, 1979).

The electrophoretic portion of Gottlieb's (1973a) study showed clearly that speciation had occurred without divergence at gene loci encoding soluble enzymes. The mean genetic identity of the two taxa is 0.94, with only one unique allele detected in *S. malheurensis*, whereas it lacks many alleles occurring in subsp. *coronaria*. Measures of variation at isozyme gene loci, such as proportion of polymorphic loci and observed mean heterozygosity, were likewise much lower for *S. malheurensis* than subsp. *coronaria*, the values being 0.12 versus 0.60 and 0.007 versus 0.159, respectively.

The study of *Stephanomeria* was significant for future electrophoretic investigations of plants for several reasons. First, it demonstrated that speciation can occur with essentially no divergence at gene loci encoding soluble enzymes. This is a most fortunate result for the plant systematist interested in employing enzyme electrophoresis for systematic or phyletic purposes, because if speciation were invariably accompanied by changes (mutations) at isozyme loci, the value of the data would be limited to taxonomic applications. The Gottlieb (1973a) study demonstrated concordance of the electrophoretic data with other biological information, suggesting that speciation was probably a recent and rapid event, and further that *S. coronaria* subsp. *exiqua* was the progenitor of *S. malheurensis* because of its greater morphological and allozymic variability. In fact, it was possible to quantify directly the genetic similarity (at least at isozyme loci) between the two taxa as well as measure accurately variation in each, procedures not feasible with other features such as morphological attributes.

Gottlieb (1974a) examined another annual diploid species pair electrophoretically, this time in the genus *Clarkia*. Harlan Lewis and his coworkers (Lewis, 1961, 1962, 1973; Lewis and Roberts, 1956), in a series of detailed investigations on *C. biloba* and *C. lingulata*, determined that the two species are highly intersterile because of structural differences in their chromosomes. Yet they are nearly indistinguishable morphologically; *C. lingulata* has entire petals as opposed to notched ones in *C. biloba*. In addition, *C. lingulata* is much less abundant and known from only two populations on the edge of the range of *C. biloba* (Lewis and Roberts, 1956). On the basis of a variety of data, Lewis developed the hypothesis that *C. lingulata* is a recent derivative of *C. biloba* (see Lewis, 1973 for a general discussion).

Gottlieb's (1974a) investigation of allozyme variation within and between the two species produced results similar to those for *Stephanomeria*. Populations of the two species are as similar at isozyme loci as populations of the same species (in the case of *C. lingulata*, only two populations can be compared). In addition, a lower mean heterozygosity was observed in *C. lingulata*, 8% as compared with 15% in *C. biloba*.

The results from *Stephanomeria* and *Clarkia* demonstrated that speciation can occur without detectable divergence at genes encoding soluble enzymes. In both genera, also, the presumed derivative species are less variable than their progenitors, indicating that the former represents a limited extraction of the allozymic variation found in the latter.

Subsequent to the studies in *Stephanomeria* and *Clarkia*, other examples of congeneric species with genetic identities comparable to conspecific populations were documented. In nearly all instances, the species are very similar morphologically and are thought to be examples of recent and rapid speciation. They include *Lycopersicon* (Rick et al., 1976); *Gaura* (Gottlieb and Pilz, 1976); *Coreopsis* (Crawford and Smith, 1982a,b); *Lasthenia* (Crawford et al., 1985); and *Camassia* (Ranker and Schnabel, 1986).

A study by Loveless and Hamrick (1988) examined allozymic variation within, and divergence between, two closely related species of perennial thistles, *Cirsium canescens* and *C. pitcheri*. The latter had been viewed as a derivative of the former species. *Cirsium canescens* occurs in the Great Plains, whereas *C. pitcheri* is found on the shorelines of the Great Lakes. The habitats in which *C. pitcheri* now occurs were probably available for its permanent occupation about 11,000 years ago. Using glacial history, Loveless and Hamrick (1988) concluded that the origin of *C. pitcheri* from *C. canescens* probably involved repeated colonizations and extinctions as migration occurred into the Great Lakes areas.

Electrophoretic data are of particular interest when viewed within the probable time and mode of origin of *C. pitcheri*. Loveless and Hamrick (1988) found much less variation in the proposed derivative species (*C. pitcheri*) than in the progenitor *C. canescens*. For example, observed heterozygosity was 0.018 in *C. pitcheri* and 0.174 in *C. canescens*. The mean genetic identity between populations of the two species is 0.778, which may seem rather low if only the value is considered. Loveless and Hamrick (1988) emphasized, however, that this low value results from an allele that is common in *C. canescens* but may not occur in *C. pitcheri*. In fact, no unique alleles were detected in *C. pitcheri*, and it is no more than a limited extraction of the genetic variation found in *C. canescens*. They suggested that migration, extinction, and drift, all of which were likely factors in the origin of *C. pitcheri*, could account for its reduced electrophoretic variation. The Loveless and Hamrick (1988) study is of particular interest because it interpreted the isozyme data from the perspective of the probable time and mode of origin of a species. It is also interesting that no novel alleles have appeared during the 11,000 years that *C. pitcheri* has presumably existed.

Another example of two species presumably related as a progenitor–derivative pair, namely *Clarkia franciscana* and *C. rubicunda*, was examined electrophoretically by Gottlieb (1973b). The pair appeared to be similar to the *Stephanomeria* and *Clarkia biloba–C. lingulata* cases discussed earlier (Lewis and Raven, 1958a,b), but electrophoretic study revealed the very low genetic identity of 0.28 for the two species (Gottlieb, 1973b). Also, Gottlieb (1974b) detected a gene duplication for alcohol dehydrogenase in *C. franciscana* whereas *C. rubicunda* lacks it (the topic of gene duplications will be considered in Chapter 10). Electrophoretic information may be used only to argue that the two species have not diverged from each other recently; They may not represent a progenitor–derivative situation, or they may be so related but the divergence may have been quite ancient.

Additional studies have demonstrated high genetic identities among congeneric species that are not extremely similar morphologically and ecologically. Also, reproductive isolation does not result (at least partially) from chromosomal restructuring, as is the case for other taxa discussed previously. Soltis (1985) examined allozymic variation within and similarity among four species in the genus *Heuchera—H. americana. H. parviflora, H. pubescens*, and *H. villosa*. These species consist of outcrossing, diploid, herbaceous perennial plants. They were the subject of a detailed biosystematic study by Wells (1979), and thus her data are available as a framework for viewing the electrophoretic results. Wells (1979) suggested that the four taxa represent two closely related species pairs: *H. americana–H. pubescens* and *H. parviflora–H. villosa*. This hypothesis is based on morphology, time of flowering, and fertility of hybrids. Wells (1979) indicated that the two species of the former pair are isolated by temperature factors; *H. pubescens* is a montane species and prefers cooler temperatures than *H. americana*. Where the species occur together, swarms of highly fertile hybrids are found. *Heuchera parviflora* and *H. villosa* are isolated from each other by ecological factors, substrate preferences in particular. Synthetic hybrids are highly fertile, but the two species apparently do not exchange genes under natural conditions. Synthetic hybrids between species belonging to different pairs show highly reduced fertility relative to crosses between members of the same pair (Wells, 1979).

Soltis (1985) found a mean genetic identity of 0.99 for populations of *H. americanum* and *H. pubescens*, and a value of 0.98 for *H. parviflora* and *H. villosa*, values characteristic of conspecific populations. Mean values for pair-wise comparisons of populations of species belonging to different pairs are lower and in the narrow range of 0.83 to 0.85. The high identities for each of these two pairs of species suggest recent divergence. Species of *Heuchera* differ from pairs of annual species discussed earlier because they are not isolated by strong internal barriers to gene flow. Soltis (1985) commented that because *H. americana* and *H. pubescens* are not completely isolated in nature and are similar morphologically, they perhaps should not be treated as distinct species; this seems a reasonable suggestion in the light of all data, including electrophoretic information. From an evolutionary perspective, it is possible that both pairs of taxa are in early stages of divergence, and, with time, additional barriers to gene flow may develop, just as they have between species of the different pairs. Alternatively, it is possible that under appropriate environmental conditions populations of the two taxa could merge into a single species.

The results of an electrophoretic study of *Sullivantia*, which, like *Heuchera*, is a member of the Saxifragaceae, showed comparably high levels of allozymic similarity (Soltis, 1981a). Plants of the two genera have many life history features in common (Soltis, 1985), and hopefully additional genera related to these two will be investigated electrophoretically to ascertain whether the same high levels of similarity are detected.

The genus *Tetramolopium* (Compositae) is composed of diploid suffrutescent perennials having a rather unusual disjunct distribution; it is entirely insular and restricted to New Guinea (25 species) and Hawaii (7 species) (Lowrey, 1986). Evidence suggests that the Hawaiian taxa may have evolved from a New Guinean ancestor within the past 20,000 years (Lowrey, 1986). The Hawaiian taxa are extremely diverse morphologically and ecologically, and have been treated as more than one genus by certain workers (see discussion in Lowrey, 1986; Lowrey and Crawford, 1985). Hawaiian species appear to be monophyletic because they have a diversity of sexual systems (unquestionably derived) unknown in their New Guinean counterparts. Also, all are diploid ($n = 9$) and are highly interfertile through the F_3 generation, but no interspecific gene exchange has been documented under natural conditions (Lowrey, 1986), ostensibly because of geographical and ecological factors. Population sizes are extremely small, with fewer than 100 individuals often present.

An electrophoretic study of Hawaiian *Tetramolopium* indicated that (1) variation within and among populations is extremely low; and (2) there is essentially no divergence among populations of different species, the mean genetic identity for all 19 populations being 0.95 with a range of 0.86 to 1.0 (Lowrey and Crawford, 1985). The results from enzyme electrophoresis are concordant with the Lowrey (1986) hypothesis that *Tetramolopium* was brought to Hawaii from New Guinea recently, probably by a single introduction. Following its introduction, *Tetramolopium* underwent rapid adaptive radiation into a variety of habitats in Hawaii, with morphological divergence accompanying the radiation. It is possible that few genetic changes underlie the seemingly large ecological and morphological differences among the taxa (see Hilu, 1984; Gottlieb, 1984a, for reviews). At any rate, the introduction of one or a few individuals from New Guinea probably resulted in very little initial variation at isozyme loci. Subsequent rapid adaptive radiation would provide little time for the accumulation of mutations at these genes.

The genus *Bidens* on the Hawaiian Islands is similar to *Tetramolopium* in that the plants are perennials and very diverse ecologically. In addition, the species produce fertile hybrids when crossed, but natural hybrids are rare because all species are largely allopatric (see discussions in Ganders and Nagata, 1984; Helenrum and Ganders, 1985). The Hawaiian members of *Bidens* differ from *Tetramolopium* in that they are hexaploid ($2n = 72$) instead of diploid. Ganders and Nagata (1984) suggested from a variety of evidence that the taxa of *Bidens* now on the Hawaiian Islands result from a single introduction.

Helenrum and Ganders (1985) conducted an electrophoretic study of 15 taxa of *Bidens* from Hawaii. Because the high ploidy of the plants results in gene duplication, assignment of alleles to particular loci is not a simple matter. These problems aside, very little divergence at isozyme loci has accrued during evolution and speciation in Hawaiian *Bidens*. Morphological and ecological divergence presumably has occurred rapidly via adaptive radiation, and very few mutations have accumulated at gene loci specifying soluble enzymes. In *Bidens*, these morphological and ecological changes have occurred at the hexaploid level.

A recent electrophoretic study of *Dendroseris*, a genus of Compositae endemic to the Juan Fernandez Islands, Chile, revealed high genetic identities among species (Crawford et al., 1987). The species are very different morphologically and occur in a wide variety of habitats (Stuessy et al., 1984). Despite this diversity, little divergence has occurred among species at isozyme loci.

Witter and Carr (1988) ascertained electrophoretic variation in the silversword alliance (Compositae–Madiinae), a monophyletic assemblage that is unquestionably one of the most diverse and fascinating groups of insular plants. Carr and Kyhos (1981, 1986) demonstrated that many species are interfertile, with some reduction in fertility caused by chromosomal differences. Witter and Carr (1988) found certain species of silverswords with much lower genetic identities (some pairs of populations below 0.50) than had previously been reported for any congeners on islands. In particular, species of *Dubautia* on the oldest island exhibit the lowest identities; these are also the species postulated from cytogenetic data to be the oldest in the genus (Carr and Kyhos, 1986). Species on younger islands have higher genetic identities. Witter and Carr's (1988) results implicate time as the important factor in divergence at isozyme loci. Their data demonstrate that divergence among congeners on islands may be comparable to that among continental species, given sufficient time.

Saidman and Vilardi (1987) studied allelic frequencies at 25 loci in the seven species comprising section *Algarobia* of the genus *Prosposis*. Theirs is one of the few electrophoretic studies of woody angiosperms; it is also of interest because relationships inferred from morphology are not concordant with electrophoretic information. The two species *P. alba* and *P. hassleri* are so distinct morphologically that they had been placed in different series within the section, yet their populations have a genetic identity of 0.99. A third species in the section, *P. caldenia*, had been placed in a series with four other species in a recent taxonomic treatment. Allelic frequency data suggest, however, that this species represents an isolated element in the section because its identities with other species are the lowest (0.757–0.825) of any species in the section. Saidman and Vilardi (1987) contrasted the electrophoretic data with other information available for the species and concluded that relationships inferred from allozymes and not morphology are concordant with other biological data. *Prosopis alba* and *P. hassleri* are very similar in phenolic compounds, and highly fertile hybrids are found in the field. These observations, despite the considerable morphological differences, suggest close evolutionary relationships between the species. *Prosopis caldenia* is more isolated reproductively from other species in the series than they are to each other, and this divergence corresponds to divergence at isozyme loci.

In the examples just discussed of high genetic identities among perennial plants, the species are, to a lesser or greater degree, distinct morphologically. By contrast, when annual species exhibit high identities, they are almost always very similar morphologically. Gottlieb et al. (1985) examined two species of *Layia* (*L. discoidea* and *L. glandulosa*) electrophoretically and morphologically. *Layia discoidea* was studied biosystematically in the 1940s by Clausen et al. (1941, 1947); even though it is morphologically very distinct, it forms vigorous, fully fertile hybrids with *L. glandulosa*. Clausen (1951) hypothesized that *L. discoidea* diverged from *L. glandulosa* in the distant past and represents an evolutionary relict. If this were the case, then we might expect low genetic identities between the two species at gene loci encoding soluble enzymes. Gottlieb et al. (1985), however, found *high* mean population identities (0.90) for *L. discoidea* and the two subspecies of *L. glandulosa*, which is nearly the same as mean values for the two subspecies of *L. glandulosa* (0.93) and for populations of *L. discoidea* (0.92). Although populational sampling was limited, the results demonstrated that in the geographic area where populations of both species occur they are no more divergent at isozyme loci than are populations of the same species. Gottlieb et al. (1985) concluded that the high similarity at isozyme loci is more concordant with *L. discoidea* being a recent derivative of *L. glandulosa* rather than the product of a more ancient divergence. Clausen believed that *L. discoidea* must have separated from *L. glandulosa* in the distant past because of its very different morphology, i.e., it must have taken considerable time for it to evolve its unusual features. Gottlieb and Ford (1987) studied the genetic and developmental bases of presence-absence of ray florets in synthetic hybrids between *L. glandulosa* and *L. disoidea*, and found that two gene loci control this feature. The isozyme data were useful for calling into question the previous hypothesis of the evolutionary relationship between these two species and additional data are concordant with electrophoretic information in suggesting that genomes of the two species are not highly divergent.

Sytsma and Schaal (1985a) examined allozymes of shrubby tropical plants in the *Lisianthius skinneri* complex occurring in central Panama. The populations are geographically isolated and occupy small areas. One species, *L. skinneri*, is widespread and occurs at lower elevations, whereas four species are more restricted in distribution and grow at higher altitudes. Three populations of *L. skinneri* were examined electrophoretically, but only one population of each of the other species was sampled (i.e., populational sampling was not extensive). Genetic identities for populations of the species ranged from 0.750 to 0.916 for the three populations of *L. skinneri* and 0.437 to 0.695 for populations of different species. Sytsma and Schaal (1985a) noted that these values are low both for the populations of *L. skinneri* and for populations of different species. The lowered identities occur because populations are monomorphic for alternative alleles at genes, suggesting that genetic drift and founder events have been factors in the evolution of the *L. skinneri* complex (Sytsma and Schaal, 1985a). In this regard, additional populational sampling of these species would be desirable. Electrophoretic studies of other tropical woody angiosperms are needed, and the results of Hamrick and Loveless (1986) demonstrate the feasibility of such investigations.

Congeneric species of ferns studied to date exhibit low genetic identities relative to values common in flowering plants. Haufler (1985) reported values for the majority of species comparisons in the genus *Bommeria* ranging from 0.100 to about 0.300. Werth et al. (1985a) ascertained genetic identities from less than 0.300 to slightly over 0.500 for the diploid species in the Appalachian *Asplenium* complex. Ranker (1987) measured mean values from less than 0.100 to 0.260 for various species pairs in the genus *Hemionitis*. For *Bommeria*, Haufler (1985) suggested that the low identities among the species could be caused either by extensive divergence from a common ancestor or convergence in morphological features so that the species are not as closely related as the morphology suggests. It may be that the low genetic identities calculated for species of ferns are a general reflection of their antiquity relative to species of flowering plants.

MODES OF PRIMARY SPECIATION

Earlier discussions in this and the previous chapter demonstrated that electrophoretic and morphological similarity may or may not go hand in hand. The same may be said with regard to isozymes and other biological attributes such as ecological preferences and interfertility. In a taxonomic sense, then, allozymes may not be "reliable" species-specific characters. Despite this "taxonomic" nonconcordance (and in fact because of it) between electrophoretic results and other data sets, enzyme electrophoresis has been highly useful for examining and testing modes of speciation in plants. The present discussion of this topic will be rather brief; a more detailed account was given by Crawford (1985). It is difficult to separate discussions of species relationships from modes of speciation; nonetheless, the distinctions appear useful for providing a sharper focus for the discussion.

Various classifications of modes of speciation have appeared, but for the present discussion certain aspects of the basic system of Grant (1981) will be followed. Grant's system will be simplified a bit, but hopefully its essence will remain. Speciation may involve evolutionary divergence, sometimes called primary speciation (discussed here), and hybrid speciation (discussed in the next chapter). Primary speciation in plants must involve, at the least, (1) divergence in some morphological feature or features so that the plants are distinguishable; and (2) the development of isolating barriers such that the characters separating the species are maintained. Both of these aspects should have a genetic and/or chromosomal basis, although the actual bases of these attributes often are not known.

The two major modes of divergent speciation may be roughly divided into quantum and geographical. Quantum speciation, as studied and developed by Harlan Lewis and co-workers (see Lewis, 1973 for a general discussion), is viewed as rapid. Reproductive isolation presumably develops in a very short time period. While these species are often intersterile because of differences in chromosome structure, it is not clear what factor(s) may have been important in their initial isolation (Carson, 1985). Species resulting from this process may be very similar (almost indistinguishable) morphologically. Grant's (1981) concept of quantum speciation is broader than the Lewis model and includes a variety of processes by which speciation could occur rapidly. Geographical speciation is viewed as a more gradual process in which divergence takes place in a variety of features; reproductive isolation develops as a by-product of this divergence. Thus there appear to be rapid (or abrupt) and gradual processes involved in speciation.

As discussed earlier, electrophoretic studies of nearly all species pairs thought to be examples of recent quantum speciation show high genetic similarity at isozyme loci. Such similarity would be expected if the isolating barriers arose both rapidly and recently because of insufficient time for the accumulation of mutations specifying unique polypeptides. Structural differences in chromosomes (rendering hybrids between plants with the two types sterile) and morphological similarity combined with high identities at isozyme loci are often indicative of recent quantum speciation. Crawford (1985) discussed several examples of electrophoretic studies of species pairs presumed to have originated by quantum speciation.

Electrophoretic studies of species in several different genera suggest that divergence may occur rapidly by other than the quantum mode as envisioned by Harlan Lewis. Several of these examples are discussed by Crawford (1985) and will be mentioned only briefly. Ornduff (1966) suggested that *Lasthenia maritima* originated from *L. minor*. The latter is self-incompatible and occurs in a wide variety of habitats in California, whereas the former is self-compatible and is almost totally restricted to seabird rocks and islands (where edaphic and other environmental conditions are harsh) from California north to British Columbia. Mutation at the incompatibility gene may have produced a selfing line within *L. minor*, and morphological and ecological differences between the taxa could have been fixed rapidly in the selfers, particularly if strongly selected. The two species are interfertile when crossed in cultivation but do not hybridize in nature (Ornduff, 1966). Populations of the two species are no more divergent at genes specifying soluble enzymes than are populations of each of the species (Crawford et al., 1985), and the data lend support to a recent rapid divergence for *L. maritima* from *L. minor*.

The several studies of congeners on oceanic islands indicate that rapid evolution and speciation via adaptive radiation can occur without divergence at isozyme loci, that is, speciation has been so rapid in adaptive features that there has been insufficient time for accumulation of mutations at isozyme genes.

It is critical to study species that have diverged quite recently; this point was emphasized by Leslie Gottlieb in his writings on plant speciation. Such an approach makes it more feasible to identify differences associated with the process of speciation as opposed to divergence in characters subsequent to speciation. Thus, the very low genetic identities between *Clarkia rubicunda* and *C. franciscana* alluded to earlier (Gottlieb, 1973b) may not be used as evidence against the quantum mode of speciation for these two taxa.

In geographical speciation, different selection pressures on allopatric populations (perhaps combined with drift) cause gradual divergence in a variety of morphological, ecological, and cytological features. Because this process occurs over a long period of time relative to quantum speciation, one might expect greater divergence at gene loci specifying isozymes. Furthermore, there should be some positive correlation between the level of allozyme divergence and degree of differences in other features.

Warwick and Gottlieb (1985) examined allozyme variation in six species belonging to the genus *Layia*, which had been studied biosystematically in some detail by Clausen et al. (1941) and discussed by Clausen (1951). The six diploid species ($n = 7$) fall into three groups on a variety of features, including fertility of hybrids. The groups include *L. jonesii*, *L. munzii*, and *L. leucopappa; L. fremontii* and *L. chrysanthemeides*; and *L. platyglossa*. Warwick and Gottlieb (1985) found mean genetic identities for populations of all species within each of the three groups to be 0.880, 0.824, and 0.901. By contrast, mean identities for populations of species belonging to different groups are 0.642, 0.576, and 0.576. Diagnostic alleles sometimes were found in one or two of the groups, that is, not all of the genetic divergence as measured by mean identity results from frequency differences. The electrophoretic results are concordant with the hypothesis that these species of *Layia* diverged gradually through time in a variety of features as they adapted to different habitats. The results are very different from those obtained with presumed examples of rapid speciation.

The study of McNeill and Jain (1983), discussed in the last chapter on infraspecific taxa, provided electrophoretic results suggestive of geographic speciation in *Limnanthes* sect. *Inflexae*. They found that, in general, mean genetic identities were highest among populations of the same taxon and were progressively lower in going from varieties or subspecies to species. This trend (some exceptions were noted) indicates progressive genetic divergence through time for different features. Also, a very high positive correlation was found between interfertility and genetic identity, suggesting that reproductive isolation has developed gradually as a result of genetic differences rather than abruptly by chromosome repatterning. Electrophoretic data are generally concordant with the hypothesis of geographical speciation in *Limnanthes* sect. *Inflexae*.

Kesseli and Jain (1984) studied allozyme variation in *Limnanthes* sect. *Reflexae* as part of a broadly based systematic and evolutionary study of the genus by Jain and collaborators. In most cases, they found a parallel between divergence in allozymes and morphology that suggested a geographic mode of speciation, as in section *Inflexae*, discussed earlier. This paper is also recommended reading because it provides a lucid discussion of the use and interpretation of electrophoretic studies for taxonomic and evolutionary purposes. Kesseli and Jain (1984) pointed out how electrophoretic data are properly interpreted in the light of other biological data for understanding relationships among entities and for inferring modes of speciation.

PHYLOGENETIC RECONSTRUCTION

In Chapters 1 and 6 general discussions of cladistic analyses of allelic data were presented; in this section a brief consideration of several examples will be given.

Whalen and Caruso (1983) conducted phenetic and cladistic studies of

Solanum sect. *Lasiocarpa* employing allozymes and morphology as separate data sets. Phenograms and cladograms constructed from each of the data sets were compared. For cladistic analysis of electrophoretic data, each allele was treated as a single character that was either present or absent in a species. The populations of *Solanum* studied by Whalen and Caruso (1983) are monomorphic at most loci examined, facilitating this type of character state coding. They did not attempt to polarize the character states, and stated that "We felt there was no reliable basis for polarizing the allozyme characters." As a result, unrooted Wagner networks were generated and then rooted at the midpoint of the longest path on each of the networks. The Wagner network is based on parsimony, and two most parsimonious very similar networks were produced; one of them is shown in Fig. 8.1.

When Whalen and Caruso (1983) compared cladograms based on morphology with ones generated from allozyme data, certain similarities were noted. Comparison of Figures 8.1 and 8.2 shows that *S. stramonifolium* and *S. sessiliflorum* are rather isolated regardless of which data are employed, but their positions are different with the two data sets. Also, *S. vestissimum* and *S. hyporhodium* appear very close together, and *S. candidum*, *S. quitoense*, *S. hirtum*, and *S. pseudolulo* all appear as a group (Figs. 8.1, 8.2). Whalen and Caruso (1983) then constructed a consensus tree, based on the method of Adams (1972), to portray groupings present in all analyses, that is, phenetic and cladistic analyses of morphological and allozyme data sets (Fig. 8.3). The tree shows clearly that both *S. stramonifolium* and *S. sessiliflorum* are quite isolated (at least with respect to species for which both morphological and enzyme data were available); one would expect the consensus tree to show this isolation given the positions of the two taxa on the cladograms generated from morphology and allozymes (cf. Figs. 8.1, 8.2). Although the four species *S. candidum*, *S. quitoense*, *S. hirtum*, and *S. pseudolulo* probably represent a monophyletic assemblage, the morphological and allozyme data do not resolve relationships among them (Fig. 8.3).

Whalen and Caruso (1983) used two measures of congruence, the consensus fork index developed by Colless (1980) and the consensus information index of Mickevich (1978) to compare quantitatively the cladograms produced from allozymes and morphology, the phenograms of the two data sets, and the cladograms and phenograms based on the same data sets. With each measure, the values may range from 0 to 1, with a value of 1 indicating an identical topology. The phenogram and

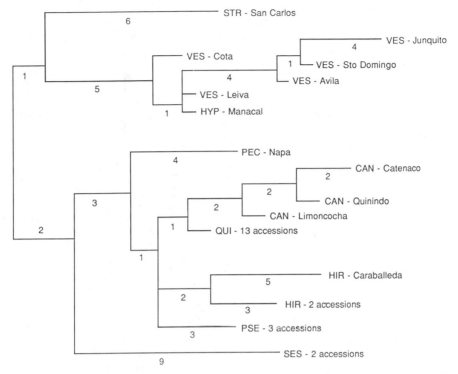

Figure 8.1 Rooted Wagner network for *Solanum* sect. *Lasiocarpa* constructed from allozyme data. Numbers refer to Manhattan distances. Species designations are *S. candidum*, CAN; *S. hirtum*, HIR; *S. hyporhodium*, HYP; *S. pectinatum*, PEC; *S. pseudolulo*, PSE; *S. quitoense*, QUI; *S. sessiliflorum*, SES; *S. stramonifolium*, STR. Names following the species designations indicate localities for seed sources; see Whalen and Caruso (1983) for exact localities. (Redrawn with permission from M.D. Whalen and E.E. Caruso, Syst. Bot. 8:369–380, 1983; American Society of Plant Taxonomists.)

cladogram based on allozyme data are more congruent than any of the other comparisons, suggesting that rates of allozyme divergence have been relatively uniform over time. By contrast, congruence between the phenogram and cladogram based on morphology is considerably lower; Whalen and Caruso (1983) attributed this finding to a less uniform rate of morphological evolution. While more studies employing these types of comparisons are needed, one note of caution raised in an earlier chapter should be interjected. Estimates of differences in morphological characters do not include any measure of the genetic bases of the character

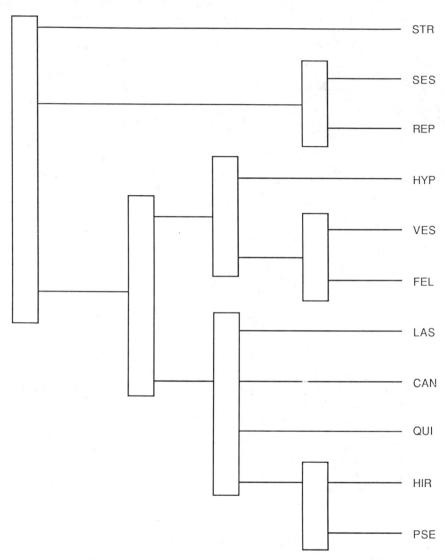

Figure 8.2 Cladogram for species of *Solanum* sect. *Lasiocarpa* using 20 morphological features. Abbreviations for the species are the same as in Fig. 8.1. Details of the analysis and construction of the cladogram are given in Whalen et al. (1981). (Redrawn with permission from M.D. Whalen and E.E. Caruso, Syst. Bot. **8**:369–380. 1983, American Society of Plant Taxonomists.)

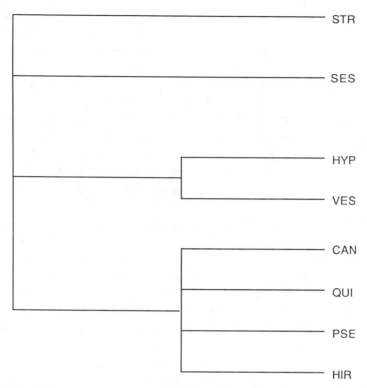

Figure 8.3 Adams consensus tree with clusters consistently present in all analyses of *Solanum* sect. *Lasiocarpa*. These include phenetic and cladistic analyses of both morphological and allozymic data. Species designations are the same as in Fig. 8.1. (Redrawn with permission from M.D. Whalen and E.E. Caruso, Syst. Bot. **8**:369–380. 1983, American Society of Plant Taxonomists.)

differences. Thus the relative precision with which different character states may be estimated in genetic terms differs considerably between allozymes and morphology. Sytsma and Schaal (1985a) used allozyme data to construct a phylogeny for five species in the genus *Lisianthius*, which is composed of woody shrubs occurring in central Panama. A number of morphological features unites members of the so-called *L. skinneri* complex, suggesting that the species comprise a monophyletic assemblage. The present discussion will be confined to phylogenetic aspects of the isozyme study; other facets of the investigation were considered earlier in this chapter.

Sytsma and Schaal (1985a) used two methods for constructing phylogenies from the allozyme data. One employed the algorithm of Fitch and Margoliash (1967), which is based on an overall measure of similarity or difference. The other was a Wagner network using the method of Nelson and VanHorn (1975), which constructs a network based on the fewest number of character state changes. Both result in unrooted networks. For the Fitch and Margoliash algorithm, genetic divergence values as computed by the method of Rogers (1972) were employed. Two networks were generated from this procedure because the divergence between LJ1 and LH1 on the one hand and LJ1–(LS4, LS6, LS7) on the other were nearly equal; the two networks differ only in the relative placement of these two populations (Fig. 8.4A,B). (LJ1, etc. are species designations.)

For constructing the Wagner networks, the 30 total alleles detected in the seven populations (or more correctly inferred to be present from electromorphs in the gels) were used as characters. Two character states were assigned, with an allele either present (at any frequency) or absent in a population. Sytsma and Schaal (1985a) presented the three networks necessitating the fewest steps to account for the distribution of alleles in the seven populations (Fig. 8.4D,E,F). Four additional, nearly identical networks were also produced from the data.

The unrooted networks produced by the two methods were each rooted by different methods. The networks produced by the methods of Fitch and Margoliash were not actually rooted; rather, a network based on DNA data rooted by the outgroup method was compared with them (Fig. 8.4C). The Wagner networks were rooted (converted to Wagner trees) by using a hypothetical ancestor assumed to have contained the allele at each gene that is most widely distributed in the group under study (Fig. 8.4D,E,F).

Once phylogenetic trees are constructed from the isozyme data, they may be compared with each other and with other phylogenies. In the case of *Lisianthius*, Sytsma and Schaal (1985a) compared trees generated from isozymes by the two methods and also compared them with the phylogeny generated from DNA data. The latter, referred to as providing "an unambiguous phylogenetic tree for the group," is taken as the standard against which the other phylogenies may be compared (Fig. 8.4C). One network constructed with the Fitch and Margoliash (1967) algorithm is very similar to the DNA-based phylogeny (Fig. 8.4A,C) while the other differs in several regards (Fig. 8.4B,C). The isozyme data do not detect that a western Panamanian population of *L. skinneri* (LS7) has diverged

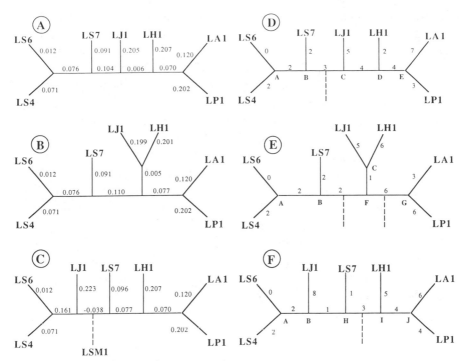

Figure 8.4 Estimates of relationships among species in the *Lisianthius skinneri* complex. Species designations are as follows: LS4, LS6, and LS7, different populations of *L. skinneri;* LA1, *L. aurantiarus*; LHI, *L. habuensis*; LJ1, *L. jefensis*; LP1, *L. peduncularis.* (*A, B*) Unrooted networks based on Rogers' divergence values for isozyme data. Numbers represent calculated divergence values for each branch on the diagrams. (*C*) Network based on divergence values of DNA (see discussion in text) with the root (the species *L. seemannii* used as the outgroup) shown by the dashed line. This tree is presented only for comparison with isozyme-based trees. (*D–F*) Unrooted Wagner networks constructed from presence–absence data for 30 isozymic alleles in populations. The algorithm of Nelson and Van Horn (1975) was employed to construct the shortest networks. Letters at nodes designate constructed hypothetical taxonomic units, and numbers along branches indicate the evolutionary steps that occur along each. Dashed lines show the hypothetical roots for each tree. Note that the topology of all trees is the same. (Redrawn from K.J. Sytsma and B.A. Schaal, Evolution, **39**:582–593 1985.)

as a separate lineage, yet the DNA data show this quite clearly (Fig. 8.4*C* versus *D,E,F*). Comparison of Wagner trees and the tree generated from DNA data reveals that one of the trees represents the same unrooted network as the DNA network, but the roots are different (Fig. 8.4*C,D*). Given that the tree based on DNA restriction site mutations was rooted using a "real" species as the outgroup, the root of the isozyme tree may be suspect. Perhaps an assumption made (i.e., the widespread allele is ancestral) in polarizing the characters was not valid in certain instances. The results of Sytsma and Schaal (1985a) show high concordance between phylogenies generated using different algorithms for the isozyme data, and between phylogenies produced by isozyme and DNA data. This concordance allows some confidence in the results, even though the small number of populations sampled per species combined with the fact that the populations are largely monomorphic at each locus could have affected results of the cladistic analyses.

Jorgensen (1986) used electrophoretic data to examine phylogenetic relationships in the genus *Hordeum* (barley). A Wagner network, which was not rooted, was constructed from Manhattan distances between any two taxa. Results of this rather extensive study indicated the presence of three groups, only one of which was recognized on the basis of morphology. Jorgensen (1986) suggested that the paucity of "good" morphological characters has precluded recognition of the three groups detected by isozymes. The electrophoretic data suggest that evolution and differentiation of *Hordeum* resulted from two centers of differentiation, one in Southwest Asia and another in South America. This study involved rather extensive sampling of populations and enzymes, and it produced insights into possible phylogenetic relationships not previously achieved with morphological features.

Several words of caution are necessary with regard to the use of allelic data for generating phylogenies. Witter (1986), in an electrophoretic study of the genera *Argyroxiphium, Dubautia,* and *Wilkesia* (the so-called silversword alliance, which is endemic to the Hawaiian Islands) pointed out that allelic polymorphisms at a locus can become fixed among different evolutionary lines so that the same allele may be shared by different lineages independent of phylogenetic relationships. It is necessary, therefore, to look at as many populations as possible. The probability of fixation of the same alleles in different lineages is greatest for plants with small population sizes, that is, when genetic drift is likely to be important, and it is also greater for selfers as compared with outcrossers.

The data set of Lowrey and Crawford (1985) indicates that the same alleles have become incorporated into certain populations in different sections of *Tetramolopium*. *Tetramolopium* is similar to the silversword alliance in having small population sizes on the Hawaiian Islands, and thus the probability of incorporation of alleles into different populations and ultimately into different lineages by drift alone seems quite high.

Potential problems in using allelic data for constructing phylogenies were discussed in Chapter 6. A brief review and elaboration of these points is in order. The locus may be used as the character and the alleles as character states. If two alleles occur at a locus, the ancestral and derived allele must be determined for the locus. The most common method is outgroup analysis (see Chapter 6), but this is feasible only if one of the alleles occurs in the chosen outgroup. If all taxa examined as possible outgroups lack both alleles, then it is very difficult (or impossible) to assign ancestral and derived alleles. If three or more alleles occur at a locus, then the problem becomes more complex because the proper transition series must be determined (see Chapter 6). If each allele is designated as the character, then presence of the ancestral allele (as determined by outgroup) will be treated as the ancestral condition, and absence of this allele (plus the presence of other alleles) will be viewed as the derived condition. Sampling might become a problem with this method because absence of low-frequency alleles could result from small sample size rather than true absence in a taxon.

SUMMARY

Many plant systematists study relationships among congeneric species, and the present chapter has considered the various kinds of problems that may be addressed at this level employing allelic data from allozymes.

At the most basic level, allozymic data may be employed as characters for distinguishing species or testing species concepts. The enzyme data may not "work" in every case, that is, plants judged to be different species will not always be distinct at isozyme loci, and, converely, infraspecific taxa may be allozymically divergent. In general, however, congeneric species exhibit lowered genetic identities relative to conspecific populations, thus often making isozymes of taxonomic value. Electrophoretic data may be particularly useful for delimitations in which phenotypic plasticity acts to obscure morphological boundaries between species.

Allozyme data have been useful for ascertaining evolutionary relationships among species and for inferring modes of speciation. In such studies, the isozyme information must be interpreted and evaluated in the light of biosystematic data such as fertility of and chromosome pairing in synthetic hybrids. Without this information, the value of isozyme data is greatly diminished. It is not fortuitous that many allozyme studies have been done on species that have also been subjected to extensive biological study; *Clarkia*, *Limnanthes*, and *Stephanomeria* are three genera representing outstanding examples of this situation.

The construction of phylogenies based solely on allelic data has not been attempted very often with plants. The studies discussed earlier show that phylogenetic trees constructed using specific algorithms and rooted according to certain assumptions are often in general agreement with phylogenies inferred from other data sets. However, in such studies strict adherence to specific rules and assumptions does not always produce phylogenies totally congruent with other lines of data. Nevertheless, if the potential problems (alluded to in this chapter and Chapter 6) are kept in mind, phylogenies constructed solely from isozyme data may be of interest (if not taken *too* seriously!) for comparison with other phylogenetic hypotheses and may bear on certain issues such as relative constancy of rates of evolution of allozymes and morphological features.

Plant taxonomists may be "put off" if data generated from a newer method are not concordant with taxonomies derived from morphology, that is, the data are not useful for strictly taxonomic purposes. Discussions in this chapter indicate that in almost all cases the reasons for taxonomic nonconcordance may be hypothesized from probable modes of speciation or evolutionary divergence for the plants. Thus taxonomic nonconcordance between isozymes and other data sets signals the need for additional biosystematic work to elucidate the biological situation.

CHAPTER 9

Allelic Data: Studies of Hybridization, Diploid Hybrid Speciation, and Polyploidy

INTRODUCTION

The previous chapter discussed the use of allelic data for studying genetic divergence; the present one will consider use of the data for examining two types of hybrid speciation, one at the diploid and the other at the polyploid level. Also, examples of the use of allelic data to study hybridization in plants will be presented. Documented cases of diploid hybrid speciation in plants are quite rare, whereas there are many known or suspected cases at the polyploid level. Discussion of polyploidy will also include the use of allelic data to assess whether a species represents an auto- or allopolyploid.

HYBRID SPECIATION AT THE DIPLOID LEVEL

It has been suggested that hybridization between congeneric diploid taxa (often species) followed by selection for one or more recombinant types can lead to the formation of new species (see Raven, 1976, 1980; Grant, 1981, for general discussions). Grant, in particular, has presented hypothetical arguments as well as carried out extensive experimental studies illustrating how new diploid species may result from hybridization (see Grant, 1981, for review).

Although it is generally accepted that hybridization may be an important evolutionary factor in plants (at least in some groups), there are few detailed studies of particular cases of species originating via the process.

The initial aspects of any such investigation must involve documentation of the hybrid origin of the plants under consideration. After this work has been done, detailed comparative studies of the biology of the plants and their putative parents, so important for understanding the factors and processes involved, can be initiated.

Plant systematists have employed a variety of techniques for documenting interspecific hybridization. Quantitative morphological characters have been most widely used, but there are problems and limitations with this approach (Gottlieb, 1972). Secondary chemistry (which has been especially popular the last 25 years) and karyotyping chromosomes have also been employed. The synthesis of interspecific hybrids is often considered the most definitive method for documenting the hybrid origin of plants from nature, but this approach simply demonstrates the possibility of hybridization under natural conditions rather than showing what actually occurs.

Using allozymes to document the hybrid origin of individual plants in natural populations offers certain advantages sometimes not available with the other commonly employed types of data (e.g., morphology and secondary chemistry). Allozymes allow one to determine the distribution of the products of individual gene loci without the complications of intermediacy and additivity (Gallez and Gottlieb, 1982). While both morphology and secondary chemistry may be useful in given instances, it is still not possible with these data to ascertain whether the products of alleles of specific loci in a putative hybrid do indeed represent the same products characteristic of the presumed parental species. By contrast, with allozymes it may be determined whether the products of the presumed parents are or are not present in the hybrids. Lastly, it is highly likely that the different allozymes detected by electrophoresis are not under strong selection (Gallez and Gottlieb, 1982); this fact is important if allozymes are to be used for detecting hybrid segregates.

Allozymic data will be useful for inferring the hybrid nature of a diploid species only if certain conditions exist with regard to the parental taxa. The parents must be extant and divergent at one to several loci (the more, the better) encoding the enzymes to be examined electrophoretically. The parental taxa cannot have diverged appreciably at the diagnostic gene loci after they were involved in the origin of the new species. Likewise, the hybrid species must have accumulated very few mutations at diagnostic loci subsequent to its origin (Crawford, 1983). From these considerations it is apparent that the situations most amenable to elec-

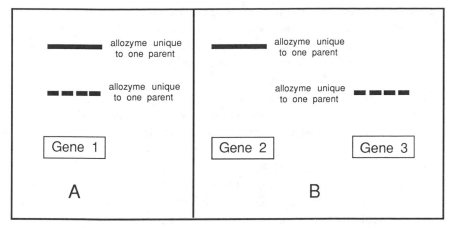

Figure 9.1 Types of enzyme additivity that could occur in plants of hybrid origin. (*A*) Parental-specific alleles of the same gene locus. (*B*) Parental-specific alleles of different loci.

trophoretic analysis of hybrid speciation involve two well-differentiated taxa, most likely species, which have produced the hybrid derivative relatively recently compared with their own origins (Crawford, 1983).

Consider next the types of divergence and additivity at isozyme loci that may be encountered in studies of hybrid speciation. Ideally, the parental species would have mutually exclusive alleles at several gene loci as opposed to having the same alleles in very different frequencies. If mutually exclusive alleles are detected at two or more loci in the parents, then two types of additivity could occur in their derivatives. Different alleles at one gene locus could be combined in individual plants (Fig. 9.1). Also, species-specific alleles could be detected among loci within a population of hybrid plants (Fig. 9.1). For example, one plant could have an allele specific to one parent at one locus and an allele specific to the other parent at another gene locus. A second plant might contain an allele specific to one of the parents at only one of these two loci, and so on. One would not expect to find additivity at every locus in every plant examined because the species is probably not composed of simple F_1 hybrids but rather different stabilized recombinant types. This means that electrophoretic documentation of diploid hybrid speciation should include examination of as many individuals from as many populations as possible. Also, as many loci as possible must be included in the study; one cannot reject the hypothesis of the hybrid origin of a species from lack of additivity at a few loci in one or several individuals. In summary, enzyme electrophoresis provides the best method readily available to the systematist for studying the hybrid origin of a species because one can look at numerous individual plants and detect different proteins encoded by alternative alleles of gene loci.

One of the few studies utilizing allozymes for studying the hybrid origin of a diploid species was carried out by Gallez and Gottlieb (1982). The three species are *Stephanomeria exigua, S. virgata*, and *S. diegensis*, with the latter thought to represent stabilized hybrid segregates of the former two (see Gottlieb, 1971, 1972 for discussion). The three species are annuals, have a chromosome number of $2n = 16$, and are obligate outcrossers. The hybrid nature of *S. diegensis* had been inferred from morphological, ecological, and other features (Gottlieb, 1971, 1972; Gallez and Gottlieb, 1982).

Gallez and Gottlieb (1982) studied allozymes in the three species as an additional test of the hybrid origin of *S. diegensis*. The products of 20 loci were examined electrophoretically; the two putative parental species are monomorphic for the same allele at eight loci and are polymorphic with the same high-frequency alleles at eight other loci. Although the two presumed parents are not divergent at 80% of the loci examined, the remaining four loci provided useful data. At one locus (encoding superoxide dismutase) *S. exigua* is monomorphic for an allele not detected in *S. virgata; S. diegensis* is monomorphic for the same allele found in *S. exigua*. At a second locus (encoding triosephosphate isomerase), *S. diegensis* is monomorphic for an allele found in high frequency in *S. exigua* but not detected in *S. virgata*. At the third locus, encoding aspartate–amino transferase, the same highest frequency allele was found in *S. diegensis* and *S. virgata*; this allele occurs in low frequency in *S. exigua*. Finally, at the fourth gene locus, encoding glutamate dehydrogenase, the allele common to *S. diegensis* and *S. virgata* was not detected in *S. exigua* (Gallez and Gottlieb, 1982). Thus, alleles at these four loci in *S. diegensis* represent a combination of those present in its putative parental species, and argue for its hybrid origin (Gallez and Gottlieb, 1982). Of the 31 alleles detected at polymorphic loci in *S. diegensis*, only one was unique to the species, that is, not found in either *S. exigua* or *S. virgata*. This unique allele was found in only one individual. The nearly complete lack of novel alleles in the hybrid derivative species suggests that it is of recent origin and that insufficient time has elapsed for the accumulation of mutations at isozyme loci (Gallez and

Gottlieb, 1982). This hypothesis is concordant with ecological data implicating a recent origin for *S. diegensis*; it occurs in pioneer and temporary habitats (Gottlieb, 1971).

Other examples of diploid hybrid speciation have been suggested in the literature and include *Lasthenia burkei* (Ornduff, 1969); *Delphinium gypsophyllum* (Lewis and Epling, 1959); *Potentilla glandulosa* subsp. *hansenii* (Clausen et al., 1940); and *Achillea rosea–alba* (Ehrendorfer, 1959). *Lasthenia burkei* is one of the best documented examples. Ornduff (1969) suggested that this species represents a stabilized hybrid derivative between *L. conjugens* and *L. fremontii*. One of the strongest arguments supporting the Ornduff hypothesis is that synthetic hybrids between the putative parental species are often indistinguishable morphologically from naturally occurring *L. burkei*. Recall that such an identical morphology is often viewed as the strongest evidence available that a naturally occurring plant is of a hybrid origin. Crawford and Ornduff (1989) examined the three species electrophoretically and found no support for the hybrid origin of *L. burkei*. Species-specific alleles in *L. conjugens* and *L. fremontii* are not combined in *L. burkei*. Also, *L. burkei* has the same number of unique alleles as the other two species. The distribution of alleles among the three species shows that *L. burkei* and *L. fremontii* each contain a different subset of the alleles detected in *L. conjugens*. Put another way, *L. conjugens* combines alleles found only in the other two species. The electrophoretic data suggest two alternative evolutionary hypotheses: either *L. conjugens* is the progenitor of the other two species, or it is a hybrid derivative of the other two species. Neither of these hypotheses appears concordant with morphology, and additional data are needed to resolve evolutionary relationships among the three species (Crawford and Ornduff, 1989). The important point is that enzyme electrophoresis provided strong evidence against the hypothesis of hybrid origin for *L. burkei*, a hypothesis for which there was considerable biosystematic support.

In addition to examining possible cases of hybridization one allele at a time, it is advisable to analyze alleleic frequency data for additional insight into hybridization. While, to our knowledge, such an analysis has not been done for suspected cases of diploid hybrid species, Doebley et al. (1988) used ordination techniques, including principal component analysis, to show the hybrid origin of a landrace of maize. They demonstrated that the present day corn belt or Midwestern dents are hybrid derivatives from Northern Flint and Southern Dent landraces.

D.E. Soltis and P.S. Soltis (1986a) used electrophoresis to document the occurrence of an intergeneric hybrid in the Saxifragaceae. A mixed population of plants containing *Conimitella williamsii, Mitella stauropetala*, and their presumed hybrids was examined electrophoretically. Four of the 18 gene loci served to distinguish the putative parental species, that is, they had no alleles in common at these loci. Plants that appeared morphologically intermediate showed additive banding patterns for these four diagnostic loci (Fig. 9.2 shows three of them). Two plants appeared more similar morphologically to *M. stauropetala*; they showed additive patterns at two of the diagnostic gene loci but were like *M. stauropetala* at two other loci. D.E. Soltis and P.S. Soltis (1986a) suggested that the allozymic and morphological data together indicate these plants may either be backcrosses to *M. stauropetala* or some sort of later generation hybrid segregates. Documentation of the occurrence of these hybrids lends support to the hypothesis that a number of genera in the Saxifragaceae may not be as divergent genetically as most genera of flowering plants (D.E. Soltis and P.S. Soltis, 1986a).

Lowrey and Crawford (1987) demonstrated electrophoretically that the widely cultivated *x Ruttyruspolia* in southern Africa is the result of hybridization between species in the two genera *Ruspolia* and *Ruttya*, both of which are members of the Acanthaceae.

ALLELIC DATA AND POLYPLOIDY

Polyploidy is common in flowering plants and is an important process in plant evolution (see W.H. Lewis, 1980 for a series of reviews). By classical definition, polyploids are plants containing three or more sets of chromosomes or genomes. Traditionally, two types of polyploids have been recognized, auto- and allopolyploids, although the distinction between them is not often clear-cut. A concise and lucid distillation of salient aspects of this complex situation was given by Soltis and Rieseberg (1986). In the extreme cases, new autopolyploids contain genomes that are identical chromosomally and very similar or identical genetically. By contrast, allopolyploids are plants with genomes that are highly differentiated chromosomally and genetically. It has generally been accepted that allopolyploidy is more common than autopolyploidy, although the reasons often given have recently been questioned (Levin, 1983). Also, because of the assumed rarity of natural autopolyploids,

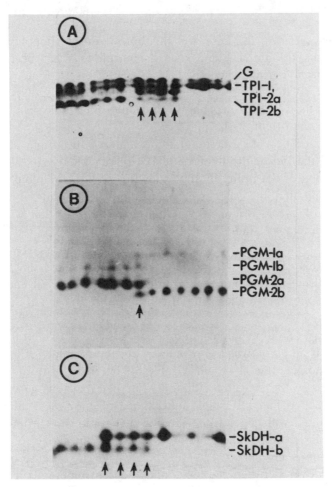

Figure 9.2 Allozyme patterns for plants of *Mitella stauropetala, Conimitella williamsii,* and their hybrids. Hybrids are designated by arrows, with extracts of *Mitella* to the left and *Conimitella* to the right. (*A*) Triosephosphate isomerase (TPI). G designates a ghost band of intermittent occurrence. In *Mitella*, TPI-1 and TPI-2a overlap in the gels. The three-banded heterozygote consisting of TPI-2a and TPI-2b may be seen in the hybrids. (*B*) Phosphoglucomutase (PGM). Additivity may be seen at both *Pgm-1* and *Pgm-2* in the hybrid plant (the enzyme is monomeric, so heterozygotes are two-banded). The two parental taxa have mutually exclusive alleles at *Pgm-1*. *Pgm-2a* is found very rarely in *Conimitella*; thus the two parental species do not exhibit mutually exclusive alleles at this locus. (*C*) Shikimate dehydrogenase (SDH). The two parental species exhibit alternative alleles, and the hybrids are heterozygous with two bands. The enzyme is a monomer. (With permission from D.E. and P.S. Soltis, Syst. Bot. **11**:293–297. 1986, American Society of Plant Taxonomists. Photographs courtesy of D.E. and P.S. Soltis.)

presenting a convincing case for the auto- versus allopolyploid of a species with "traditional" data such as morphology has been difficult. From a morphological perspective, the strongest cases for autopolyploidy come from isolated monotypic genera with diploid and polyploid plants because of the very low probability that any other biological entity was involved in the origin of the polyploid.

Electrophoretic data are useful for examining the origin of allopolyploids and the biochemical consequences of polyploidy. It follows from the discussion of the utility of allozymes for studies of diploid species of hybrid origin that the same approach should be useful for examining polyploids of hybrid origin. It was mentioned earlier that with diploid hybrid recombinants, individual plants would not be expected to display additivity at each gene locus having diagnostic alleles. By contrast, allotetraploids can show "fixed heterozygosity" at these diagnostic loci because pairing of homologous chromosomes ensures that each gamete receives the alternative alleles initially inherited from each of the parental taxa.

In contrast to allopolyploids, autopolyploids do not exhibit fixed heterozygosity. If an autotetraploid resulted from the doubling of an identical diploid genome, then it would exhibit no heterozygosity at those loci homozygous in the progenitor diploid. For heterozygous gene loci, one would expect to find tetrasomic inheritance of allozymes, although the ratios obtained will depend on whether segregation is chromosomal or chromatid in nature. If the gene locus encoding an allozyme is close to the centromere, chromosome segregation will occur because of the lack of crossing over between the locus and the centromere. If a tetraploid originated from within a single diploid species, then all alleles present in the polyploid should also be present in the diploid, assuming no mutation subsequent to its origin. Alleles unique to the diploid progenitor(s) could occur because they were present at the time the polyploid originated but were not incorporated into it, or they may have originated by mutation subsequent to the origin of the polyploid.

The study by Ownbey (1950) of allopolyploidy in *Tragopogon* is a classic investigation of the phenomenon in nondomesticated plants. Morphological and cytogenetic data documented the origin of the two allotetraploid species *T. mirus* and *T. miscellus*. The former species originated from the diploids *T. dubius* and *T. porrifolius*, whereas *T. miscellus* has *T. dubius* and *T. pratensis* as diploid progenitors. The diploid species are native to the Old World and were introduced into

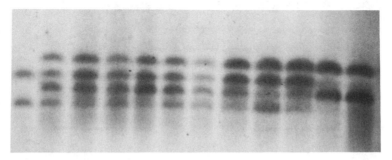

Figure 9.3 Banding patterns for leucine aminopeptidase (LAP) in plants of the two diploid species *Tragopogon dubius* and *T. pratensis*, and their allotetraploid derivative *T. miscellus*. Extract of *T. pratensis* is on the left and *T. dubius* on the right with all other extracts from plants of *T. miscellus*. Two gene loci are expressed for LAP in the two parental diploid species, and they have alternative alleles at each locus. The tetraploid plants show an additive banding pattern with the diagnostic polypeptides from each diploid species combined in *T. miscellus*. LAP is a monomeric enzyme. (Photograph courtesy of L.D. Gottlieb.)

western North America in the present century. This means that the two allotetraploid species are very young phylogenetically, certainly much younger than their three diploid progenitors.

Roose and Gottlieb (1976) carried out an electrophoretic study of the five species of *Tragopogon* (the three diploids and two tetraploids). The diploid species are monomorphic, or nearly so, for different alleles at about 40% of the 21 loci studied, that is, they are highly divergent. The electrophoretic data fully confirmed the Ownbey (1950) hypothesis on the origin of the two tetraploid species (not that it was in dire need of confirmation!). The tetraploids have additive patterns of polypeptides at each of the gene loci where their presumed progenitors are monomorphic for different alleles (see Fig. 9.3 as an example). As an aside, a study of flavonoid profiles in *Tragopogon* by Brehm and Ownbey (1965) provided less than conclusive evidence on the origin of the tetraploids because of the lack of species-specific flavonoid components in the diploids. Thus, isozymes are of considerably greater value than flavonoids as evidence for the origin of the tetraploid species in *Tragopogon*.

Roose and Gottlieb (1976) observed fixed heterozygosity in the tetraploids at those loci where different alleles were inherited from the diploids. For example, *T. mirus* has fixed heterozygosity at 9 of its 21 duplicated loci, and *T. miscellus* exhibits the same situation at 7 loci. As

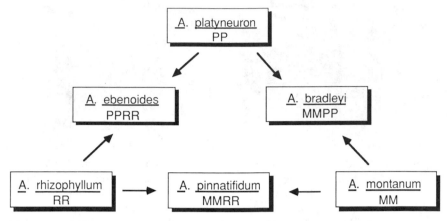

Figure 9.4 Appalachian *Asplenium* complex, showing the presumed origins (letters designate genomes) of the three allotetraploid species. (Redrawn with permission from C.R. Werth et al., Syst. Bot. **10**:184–192. 1985, American Society of Plant Taxonomists.)

discussed earlier, this is exactly what would be expected in an allotetraploid. For multimeric enzymes, the tetraploids exhibit novel heteromeric enzymes not detected in either of their diploid parents. The presence of these unique enzymes is readily accounted for because the genetic basis of the different allozymes in the diploids is known. Lastly, allozymes demonstrated that combining disparate diploid genomes results in the appearance of unique features (the novel heteromers) not found in either diploid. Also, the electrophoretic data documented the independent origins of several populations of *T. mirus*, a possibility that had been suggested previously on the basis of other information.

A classic study of reticulate evolution in plants is Wagner's (1954) investigation of the Appalachian *Asplenium* complex. Morphology, anatomy, chromosome number and meiotic pairing, and field observations showed that three diploid species of these ferns are the progenitors of three allotetraploid species (Fig. 9.4). The situation is a bit more complex than depicted in Fig. 9.4 because additional polyploid species are involved, but this illustration will suffice for the purposes of our discussion. Smith and Levin (1963), using chromatographic profiles of phenolic compounds (primarily flavonoids), demonstrated that each of the putative diploid parental species (*A. platyneuron, A. rhizophyllum*, and *A. montanum*) contains unique components. The allotetraploid species exhibit essentially all of the unique compounds from their diploid parents. Thus, secondary chemistry strongly supports Wagner's (1954) hypothesis.

Werth et al. (1985a,b) examined *Asplenium* electrophoretically and found the three diploid species (*A. platyneuron, A. montanum,* and *A. rhizophyllum*) (Fig. 9.4) highly divergent at isozyme loci, with genetic identities ranging from 0.269 to 0.516 for the three taxa. Furthermore, each diploid species is monomorphic for a unique allele at three gene loci, making these alleles valuable markers (Werth et al., 1985a). Even at polymorphic loci there is little overlap among the three species. The allotetraploid species, i.e. (*A. bradleyi, A. ebenoides,* and *A. pinnatifidum*) are heterozygous for the unique marker alleles from the diploid species Wagner (1954) considered to be their progenitors (Fig. 9.5). In addition, the allotetraploids are usually heterozygous for alleles at loci for which the presumed parental diploid species are highly divergent (though not necessarily monomorphic for different alleles).

Isozyme data documented the independent (or recurring) origins of two of the tetraploid species (Werth et al., 1985b). In particular, *Asplenium bradleyi* and *A. pinnatifidum* appear to have multiple origins because each species shows polymorphisms at the same loci that are polymorphic in their presumed diploid progenitors, that is, *A. montanum* × *A. platyneuron* and *A. montanum* × *A. rhizophyllum*, respectively.

The studies of Werth et al. (1985a,b) supported Wagner's (1954) hypothesis of the origin of the tetraploid species in much the same way as Roose and Gottlieb's (1976) investigation confirmed Ownbey's (1950) ideas on the origin of polyploids in *Tragopogon*. With *Tragopogon* there is no question that the polyploids are of recent origin, whereas the age of the polyploid ferns is much less certain. Werth et al. (1985a) indicated that while many generations have surely passed since the hybridization events, the genotypes of the allopolyploids must be very similar to those incorporated into the original hybrids. Such a similarity is so because the alleles present in the polyploids, with very rare exceptions, are electrophoretically indistinguishable from those in their diploid progenitors, indicating relative lack of divergence between the progenitor diploid taxa and the polyploids since the origins of the latter. The other electrophoretic data supporting a relatively "young" age for the polyploids is the lack (except for possibly two loci) of gene silencing, that is, loss of expression of duplicate loci, in the polyploids.

Murdy and Carter (1985) applied enzyme electrophoresis to test alternative hypotheses on the parents of the polyploid species *Talinum*

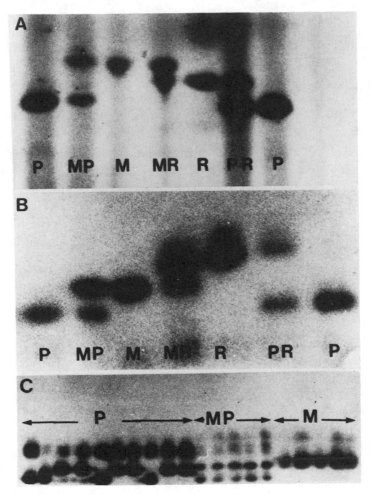

Figure 9.5 Electrophoretic banding patterns of diploid species of *Asplenium* and their allopolyploid derivatives. M, *A. montanum*; P, *A. platyneuron*; R, *A. rhizophyllum*; MP, *A. bradleyi*; MR, *A. pinnatifidum*; PR, *A. ebenoides*. (*A*) Leucine aminopeptidase (LAP), a monomeric enzyme. Additivity in the form of two-banded heterozygotes are seen in the allotetraploid plants. (*B*) Skikimate dehydrogenase (SDH), a monomer. The same type of additivity seen for LAP is apparent for this enzyme. (*C*) Phosphoglucomutase (PGM), a monomeric enzyme. Only *A. montanum* and *A. platyneuron* and their allotetraploid derivative *A. bradleyi* are shown. The faster migrating (toward top) bands, designated as PGM-1, are faintly stained, but additivity may be seen in plants of *A. bradleyi*. Variation in the slower, more densely staining bands is more complex, and the original publication by Werth et al., 1985 should be consulted. (With permission from C.R. Werth et al., Syst. Bot. **10**:184–192. 1985, American Society of Plant Taxonomists. Photographs courtesy of C.R. Werth.)

WAGNER'S HYPOTHESIS:

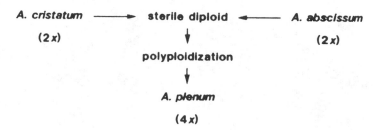

MORZENTI'S HYPOTHESIS:

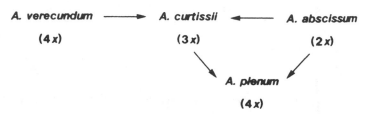

Figure 9.6 Diagram of the two alternative hypotheses for the origin of the tetraploid fern species *Asplenium plenum*. See text for additional discussion. (With permission from G.J. Gastony, Am. Jo. Bot. **73**:1563–1569. 1986, Botanical Society of America. Figure courtesy of G.J. Gastony.)

teretifolium. One parental diploid species is unquestionably *T. parviflorum*, whereas the other parent, on the basis of morphology and other attributes, could be either *T. calycinum* or *T. mengesii*. While geographic distribution suggests that *T. mengesii* is more likely the other parent, *T. calycinum* could not be dismissed on available data. Electrophoretic data suggested that *T. parviflorum* and *T. mengesii* are the parents of the tetraploid *T. teretifolium*, and that *T. calycinum* is not one of the parents. Although the diploid species are not highly divergent at isozyme loci (as in *Tragopogon* and *Asplenium*), the tetraploid is fixed for alleles unique to each of the two diploid species at two loci. Also, at several genes the tetraploid lacks alleles unique to the "alternative" diploid parent *T. calycinum*. Electrophoretic data provide strong evidence in support of one of two alternative hypotheses for the percentage of *Talinum teretifolium* (Murdy and Carter, 1985).

Gastony (1986) used electrophoretic data to test two alternative hypotheses for the origin of the tetraploid fern species *Asplenium plenum*; the two competing hypotheses are shown in Fig. 9.6. It may also be

Figure 9.7 Fronds from species of *Asplenium*. Left to right, *A. cristatum, A. abscissum, A. plenum, A. curtissii*, and *A. verecundum*. (With permission from G.J. Gastony, Am. Jo. Bot. **73**:1563–1569. 1986, Botanical Society of America. Figure courtesy of G.J. Gastony.)

seen by looking at the fronds of the species (Fig. 9.7) that *A. plenum* appears intermediate morphologically in both hypotheses. The Wagner hypothesis appears more plausible because it involves only crosses between two sexual species followed by chromosome doubling. The Morzenti hypothesis requires that an unreduced spore from the triploid *A. curtissii* function to produce gametes and that backcrossing occur to *A. abscissum* to produce *A. plenum*.

Gastony's (1986) electrophoretic data rejected the more parsimonious Wagner hypothesis and supported the Morzenti hypothesis. In particular, *A. plenum* does not contain alleles specific to *A. cristatum* (Fig. 9.8), in contrast to what would be expected with the Wagner hypothesis. On the other hand, *A plenum* does contain alleles specific to *A. verecundum* (and to *A. curtissii*), and thus the Morzenti hypothesis is supported by the electrophoretic data (Fig. 9.8). The study by Gastony (1986) is particularly interesting because it provided definitive data for choosing between alternative hypotheses for the origin of a polyploid species rather than confirming data for a widely accepted hypothesis.

While studying diploid subspecies of *Adiantum pedatum*, the maidenhair fern, Paris and Windham (1988) discovered a tetraploid cytotype of the species. Electrophoretic study of tetraploid plants indicated fixed heterozygosity at several loci. These heterozygous patterns were additive for alternative alleles characteristic of two diploid subspecies of *Adiantum*. The electrophoretic data were particularly useful for documenting the origin of the tetraploid plants because the two diploid subspecies are highly divergent at isozyme loci (Paris and Windham, 1988).

Many examples in which isozymes have proved useful for examining the ancestry of polyploids include ferns. A recent review by Werth (1989) provides additional examples from ferns as well as presenting general comments on using allelic data for inferring the ancestry of polyploids. Werth (1989) argues that if unique alleles are not present in the putative diploid parents, then the frequencies of alleles in the diploids can give likelihood estimates of which species contributed an allele to the polyploid. Thus, if an allele that occurs in a polyploid is found in one putative diploid species in a frequency of 0.100 and in a second possible species in a frequency of 0.900, then it is more likely that the second species contributed the allele. While such estimates are not absolutedly conclusive, they are useful for suggesting what probably occurred.

In certain instances, electrophoretic data have been suggestive but not definitive in elucidating the parentage of allopolyploids. Walters (1987) examined the putative diploid progenitors of the widely distributed tetraploid species *Chenopodium berlandieri* and found that it contained alleles at several loci that were not detected in any of the diploids. Walters (1987) suggested several possibilities explanations, including extinction of one of the diploid progenitors and divergence of the tetraploids subsequent to their origin.

Bryan and Soltis (1987) studied allozymes in diploid, triploid, and tetraploid cytotypes of *Polypodium virginianum*. Although the diploid cytotypes were very similar electrophoretically to the triploid and tetraploid plants, the latter two differed from the former in that they had additional alleles. This finding suggests at the least that another species was involved in the origin of tetraploid *P. virginianum*. Preliminary data given by Bryan and Soltis (1987) implicate *P. amorphum* as the other diploid species. Since this species is nondivergent from diploid *P. virginianum* at all the same loci that fail to distinguish the different cytotypes of *P. virginianum*, the electrophoretic data do not provide evidence against *P. amorphum* as the other parent. Also, *P. amorphum* has two of the same alleles at two loci as are found in the triploid and tetraploid

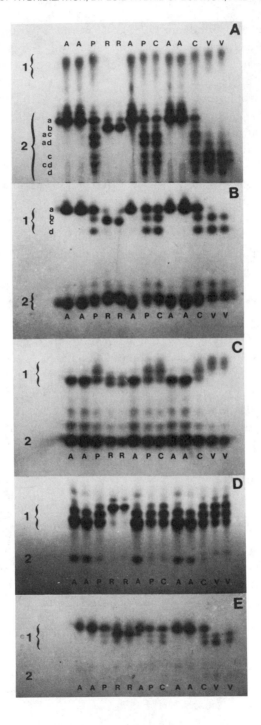

cytotypes (Bryan and Soltis, 1987). Because the putative parental taxa are so similar at isozyme loci, the electrophoretic data provide suggestive but not compelling data on the origin of the polyploids.

Holsinger and Gottlieb (1988) used enzyme electrophoresis to study the origin of the tetraploid species *Clarkia gracilis*, which is composed of four subspecies. Lewis and Lewis (1955) suggested that *C. amoena* and *C. lassenensis* were the parents, while a previous hypothesis had been that *C. arcuata* was one of the parents. Allelic data support the hypothesis that *C. amoena* (in particular, subspecies *huntiana*) is one of the parental diploids because it shares many of the same alleles at all loci examined. The evidence does not indicate that either *C. lassenensis* or *C. arcuata* are ancestral to *C. gracilis* because alleles unique to or common in the diploid species are lacking from the tetraploid. Holsinger and Gottlieb (1988) suggested that the second diploid ancestor of *C. gracilis* is extinct.

Bayer and Crawford (1986) employed enzyme electrophoresis in an attempt to confirm the parentage of two widely distributed polyploid largely agamospermous species of *Antennaria*, *A. parlinii* and *A. neodioica*. Morphological information (Bayer, 1985a,b) implicated a total of five sexual diploid species in the parentage of the two polyploids, with *A. parlinii* containing the genomes of *A. plantaginifolia*, *A. racemosa*, and *A. solitaria*. Diploid genomes thought to be present in *A. neodioica* are *A. neglecta*, *A. plantaginifolia*, *A. racemosa*, and *A. virginica*. The two polyploid species are thought to have in common the genomes of *A. plantaginifolia* and *A. racemosa* while *A. solitaria* is unique to *A. parlinii*, and *A. neglecta* and *A. virginica* are present only in *A. neodioica*. Dis-

---

Figure 9.8 Electrophoretic banding patterns for five variable enzymes in *Asplenium*. A, *A. abscissum*; P, *A. plenum*; R, *A. cristatum*; C, *A. curtissii*; V, *A. verecundum*. (*A*) PGI. Note that A,R, and V have species-specific bands at PGI-2. The pattern of C contains bands specific to A and V, and P is a summation of bands from A and C. (*B*) PGM. Species-specific patterns occur for A,R, and V, with those of C a summation of A and V, and the banding pattern of P an addition of A and C. (*C*) TPI. TPI-1 with fixed multibanded patterns except in A. Distinct patterns occur in A,R, and V, with C the addition of A and V, and P the summation of A and C. (*D*) 6PGD. 6PGD-1 consists of a single band (with lighter "ghosts") in A and R while in V it is a three-banded heterozygote. C has the sum of patterns of A and V while P adds patterns of A and C. (*E*) LAP. LAP-1. A,R, and V have specific bands; C combines the bands A and V while P adds bands of A and C. (With permission from G.J. Gastony, Am. Jo. Bot. **73**:1563–1569. 1986, Botanical Society of America. Figure courtesy of G.J. Gastony.)

tribution of alleles among the diploid and polyploid species provides support for the hypotheses on genomic constitution of the polyploids. One allele unique to *A. racemosa* and one unique to *A. solitaria* occur together in *A. parlinii*. The third putative diploid progenitor of *A. parlinii*, *A. plantaginifolia*, has no unique alleles relative to the other four diploid species. Thus, the presence of all three diploid genomes in *A. parlinii* could not be verified electrophoretically. Allelic distributions were of more limited use for documenting the parentage of *A. neodioica*. Two alleles unique to *A. virginica* occur in *A. neodioica*, which would be expected. An allele unique to *A. racemosa* was not detected in *A. neodioica* even though it is thought to have contributed a genome to *A. neodioica*. Since neither *A. neglecta* nor *A. plantaginifolia* contain diagnostic alleles, it was not possible to document their genomes in *A. neodioica* (Bayer and Crawford, 1986). Lack of divergence at isozyme loci in the five diploid species precluded the use of allelic distribution as a robust independent test for the parentage of the two polyploid species of *Antennaria*. In no instance, however, were the electrophoretic data nonconcordant with previous hypotheses on genomic constitutions of the polyploids.

The previously discussed studies represent examples of the use of enzyme electrophoresis for documenting naturally occurring allopolyploids in plants. Additional similar cases come from such genera as *Gossypium* (Cherry et al., 1972), *Nicotiana* (Smith et al., 1970; Reddy and Garber, 1971; Sheen, 1972), *Triticum* (Hart, 1970, 1979; Jaaska, 1978; Torres and Hart, 1976), and *Stephanomeria* (Gottlieb, 1973c).

Enzyme electrophoresis may also be employed to examine suspected instances of autopolyploidy. If a polyploid originated from the doubling of genetically identical genomes, then it should have no unique alleles relative to the diploid. Also, if a polyploid resulted from genome multiplication within a single diploid species, it should likewise exhibit the same alleles that occur in the diploid. All alleles present in the diploid members of a species, however, probably will not be incorporated into the polyploid. An autopolyploid should differ from an allopolyploid electrophoretically in that the former would not exhibit the fixed heterozygosity characteristic of the latter. Instead, for example, tetrasomic inheritance would be expected for some of the enzymes in the autotetraploid.

Epes and Soltis (1984) studied diploid, triploid, and tetraploid cytotypes of the species *Galax urceolata* of the family Diapensiaceae. This species has been recognized as one of the best examples of autopolyploidy (see Stebbins, 1980 for discussion), the primary reason being its isolated taxonomic status (it is a monospecific genus), which effectively removes any other taxa as potential contributors to the polyploids. Also, the polyploids are morphologically very similar to the diploids, and the flavonoid chemistry of the cytotypes is identical (Nesom, 1983; Soltis et al., 1983a). Epes and Soltis (1984) found the same alleles expressed in the triploid and tetraploid plants as in the diploids, that is, no "novel" alleles were detected in the polyploids. In addition, there was no evidence of fixed heterozygosity in polyploid *Galax*.

Coreopsis grandiflora consists of four recognized taxonomic varieties; three of them are diploid ($n = 13$), whereas the restricted Texas endemic var. *longipes* is hexaploid (Smith, 1976). No tetraploid cytotypes have been reported, and var. *longipes* has been viewed as originating from within diploid *C. grandiflora* because of its morphological similarity to the diploids (Smith, 1976). Crawford and Smith (1984) detected no novel alleles in the hexaploids. No evidence of fixed heterozygosity was found, as many as three alleles of certain gene loci were expressed in individuals, and there was evidence of apparent dosage effects with differential staining of bands.

An extensive electrophoretic study of autopolyploidy in *Tolmiea menziesii* (Saxifragaceae) was carried out by Soltis and Rieseberg (1986). The genus contains only this one species, and within it both diploid and tetraploid cytotypes occur. Diploid plants occupy a more southern distribution than the tetraploids. A variety of other data, including secondary chemistry (Soltis and Bohm, 1986; P.S. Soltis and D.E. Soltis, 1986) and karyotypes (Soltis, 1984) implicated autopolyploidy in *T. menziesii*. Soltis and Rieseberg (1986) found that the diploid cytotypes contain only the minimum conserved number of isozymes typical of diploid plants (Gottlieb, 1982). Isozyme multiplicity is common in the tetraploids, with an individual plant sometimes exhibiting three or four alleles at a locus (Fig. 9.9). Differential staining intensities of different allozymes within a plant suggest unbalanced heterozygotes (Fig. 9.9). There is genetic evidence of tetrasomic inheritance of allozymes in the tetraploid plants (D.E. Soltis and P.S. Soltis, 1988a). The two cytotypes are very similar in the allozymes expressed, with two allozymes found only in the diploids and five unique to the tetraploids. These observations could result from mutations in the two cytotypes since the origin of the tetraploids and/or sampling error (Soltis and Rieseberg, 1986).

Hauber (1986) studied diploid populations of two subspecies of *Haplopappus spinulosus* and tetraploid plants thought to be hybrids between the two subspecies. In addition, synthetic hybrids at both the diploid and

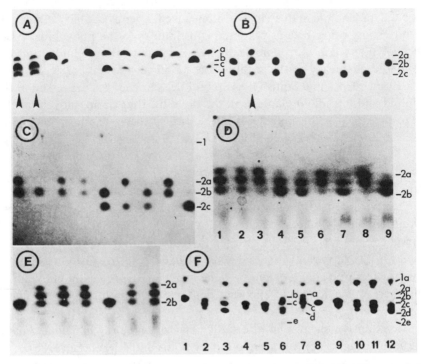

Figure 9.9 Electrophoretic patterns in starch gels of diploid and tetraploid cytotypes of *Tolmiea menziesii*. Gels C and E are of diploid plants while gels A, B, D, and F are from extracts of the tetraploids. (*A*) SDH, a monomeric enzyme. The arrowheads indicate two plants with three allozymes (not found in diploid plants). Differential staining intensities of bands in individual plants suggest dosage effects of different alleles. (*B*) PGM-2, a monomer, in tetraploid plants. One plant (at arrowhead) exhibits three allozymes. Differential staining is evident for two of the plants. PGM-1 is not detectable in most plants, but it occurs above PGM-2. (*C*) PGM-2 in diploid *Tolmiea*. No individual plant exhibits more than two allozymes, and no unbalanced heterozygotes are present. PGM-1 is barely perceptible. (*D*) TPI-2 (a dimer) in tetraploid plants of *Tolmiea*. Plants 1 and 2 are balanced heterozygotes, with two each of alleles 2a and 2b. The other plants consist of unbalanced heterozygotes (dosage effects) in which presumably either allele 2a or 2b is present three times and the alternative allele only once. (*E*) TPI-2 in diploid *Tolmiea*. Heterozygotes are balanced with the intermediate band staining most heavily. (*F*) PGI, a dimeric enzyme, in tetraploid plants. Individuals 1 and 8 show a single allozyme. Plants 3, 4, 5, 9, and 12 are unbalanced heterozygotes, as seen by the differential staining. Individuals 6 and 7 exhibit more than two allozymes. (With permission from D.E. Soltis and L.H. Rieseberg, Am. Jo. Bot. **73**:310–318. 1986, Botanical Society of America. Figures courtesy of D.E. Soltis and L.H. Rieseberg.)

tetraploid levels were examined. Meiotic configuration frequencies in the tetraploids indicated they are autoploids. In addition, the gene locus for cytosolic phosphoglucose isomerase exhibited tetrasomic inheritance in the progeny of controlled crosses.

Genetic studies demonstrated tetrasomic inheritance of allozymes in tetraploid alfalfa (*Medicago sativa* and *M. falcata*) and disomic inheritance in diploid alfalfa (Quiros, 1982, 1983). These data are in agreement with other information in documenting that the plants are autotetraploids. Extensive genetic investigations of diploid and tetraploid potatoes (*Solanum tuberosum*) have likewise documented their autotetraploid nature (Quiros and McHale, 1985).

Doebley et al. (1984) found no fixed heterozygosity for any isozymes in the tetraploid *Zea perennis*, which is in agreement with the view that it is an autopolyploid derived from *Z. diploperennis*.

Gastony (1988) has reevaluated previous hypotheses on the origin of polyploid agamospermous taxa in the fern genus *Pellaea*. A widely distributed agamospermous triploid species, *Pellaea atropurpurea*, had been implicated in the parentage of two tetraploid agamospermous varieties of *Pellaea glabella*. Gastony (1988) demonstrated that the agamospermous tetraploids contain none of the diaganostic alleles of *P. atropurpurea*. In fact, the polyploid *P. glabella* contains a subset of the allelic variation found in diploid *P. glabella*, and Gastony (1988) concluded that the tetraploids are auto- and not allopolyploids.

Allozymes also allow genetic insights into tetraploid (or higher ploidy level) plants that are not available with certain other methods such as secondary chemistry. For example, the evidence for autopolyploidy from flavonoid chemistry is whether the compounds are identical (or nearly identical) in diploid and polyploid plants. With allozymes, in addition to measuring genetic similarity, it is possible to assess such factors as the number of alleles expressed in different individuals of diploids and polyploids, apparent dosage effects, and levels of genetic divergence between the cytotypes.

SUMMARY

Allelic data generated by enzyme electrophoresis can provide compelling information on the hybrid origin of a diploid species because the products of alleles of individual gene loci specific to the presumed parental taxa either *are* or *are not* combined in the putative hybrid species. The ability to include a number of loci and to examine a large number of plants electrophoretically is also important because segregation and recombination make it very unlikely that alleles diagnostic for each parent will be found at every locus in every individual.

Allopolyploids are characterized electrophoretically by the presence of alleles unique to each of the parental diploid species and by fixed heterozygosity at these gene loci. The latter results from the pairing of homologous chromosomes, thus ensuring that each gamete receives an allele originally inherited from each parent. For dimeric and tetrameric enzymes, novel heteromers will occur in the polyploids if the polypeptide subunits unique to each of the parents are able to associate and produce an active enzyme.

Allozyme data are useful for examining the origin of polyploids. If a polyploid originated within a species, then all (or nearly all) alleles found in it should also be detected in the diploid members of the same species, assuming no electrophoretically detectable mutations in either diploid or polyploid plants subsequent to the origin of the latter. In a new or "raw" autotetraploid, for example, tetrasomic inheritance of allozymes would be expected, and fixed heterozygosity should not be observed.

Enzyme electrophoresis is less effective for inferring the ancestry of polyploids in those instances in which the diploid progenitors are not divergent at isozyme loci or in which the polyploids have diverged subsequent to their origin. Thus, if the polyploids are of recent origin relative to their diploid progenitors, then alleleic data will be of value for inferring ancestry of the polyploids.

Given the common occurrence of polyploidy in plants, it is surprising that so few electrophoretic studies have been done on them. Clearly, allelic data generated from electrophoresis can be valuable for elucidating the origin and evolution of polyploids.

CHAPTER 10

Isozyme Number and Plant Phylogeny

INTRODUCTION

Previous chapters dealt with the value of allelic data generated from enzyme electrophoresis for plant systematics. The present chapter changes focus and considers how the number of isozymes may be used for systematic and phylogenetic purposes. The general conceptual framework for such studies in plants was produced in large measure by the investigations of L.D. Gottlieb and collaborators, which will become apparent during discussions of isozyme number in this chapter.

As indicated in Chapter 5, minimum isozyme number is highly conserved in diploid plants for many of the enzymes routinely examined in electrophoretic studies (Gottlieb, 1982). In plants, the enzymes with conserved isozyme numbers are nearly always ones with known natural substrates, which means they can be assayed for in gels by employing the same substrates they use in vivo. It was mentioned also in Chapter 5 that the minimal conserved number of isozymes is related to subcellular localization of the enzymes, because the enzyme must be present in any subcellular compartment where its catalytic function is needed.

Any increase above the minimal number of isozymes typical of diploid plants is of potential value for inferring systematic and/or phylogenetic relationships. This statement is true because an increase in isozyme number *in diploids* very likely results from gene duplication, and the presence of the duplicated gene in species of the same genus (or possibly closely related genera) may result from a single event in a common ancestor of the taxa. If this is the case, then it follows that taxa possessing the duplication represent a monophyletic assemblage. In most instances (exceptions will be noted below), presence of the duplication in a group of plants probably represents the derived condition.

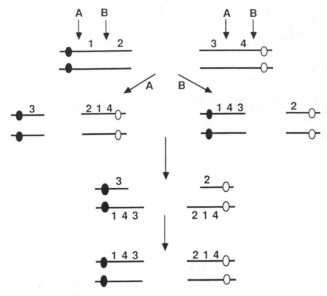

Figure 10.1 A possible mechanism for generating gene duplications in diploid plants. Two partially overlapping reciprocal translocations occur in two different plant lines. The translocations are designated A and B. Crosses between plants with the different translocations produce individuals heterozygous chromosomally. Either self-pollinations or crosses between individuals heterozygous for the same chromosomal arrangements will yield one progeny class with the duplicated segments; these segments will be on nonhomologous chromosomes, and thus they will exhibit independent assortment. (Redrawn with permission with L.D. Gottlieb, Genetics **86**:289–307. 1977, Genetics Society of America.)

Gottlieb (1983) discussed the use of isozyme number for phylogenetic purposes and pointed out that several mechanisms may produce gene duplications. Those resulting from unequal crossing over, for example, may occur repeatedly and may be of less significance in considering isozyme number and phylogeny, but there could be exceptions. Another proposed mechanism by which gene duplications can occur in plants involves a series of rather rare, independent events taking place in proper sequence. The process, as redrawn from Gottlieb (1977b), is shown in Fig. 10.1. Two partially overlapping reciprocal translocations must occur in different plants, and a cross between them yields the heterozygous condition. Either selfing or crosses between the same heterozygous types will produce four progeny classes, one of which carries a duplicated segment of chromosome. If the gene of interest is on this segment, then it will also be duplicated (Gottlieb, 1983). The two genes will not be linked, that is, they will assort independently because they are on nonhomologous chromosomes. Gottlieb (1983) suggested that annual plants may be particularly amenable to this type of duplication because different congeneric species of annuals often exhibit different chromosome rearrangements such as translocations. Change in number of loci for particular enzymes could result from aneuploidy because the loci are on the extra or missing chromosome.

GENE DUPLICATIONS AND THEIR USE IN PHYLOGENY

The gene duplication for cytosolic phosphoglucose isomerase (PGI) has been studied both genetically and from a phylogenetic perspective in the genus *Clarkia* (Gottlieb, 1977c, 1983; Gottlieb and Weeden, 1979). Harlan Lewis and collaborators (see Lewis, 1962, 1973; Lewis and Lewis, 1955; Lewis and Roberts, 1956; Lewis and Raven, 1958a,b, as examples) studied speciation in *Clarkia* and produced a hypothesis of phylogenetic relationships for the genus (Fig. 10.2). Gottlieb and Weeden (1979) viewed the presence of the duplicated PGI gene within the framework of Lewis' phylogeny for *Clarkia* (Fig. 10.2). Certain diploid species of *Clarkia* contain only one gene for cytosolic PGI, which is the "normal" condition in plants, whereas other species have two structural genes (i.e., exhibit a gene duplication) (Fig. 10.3). Distribution of the duplication is of taxonomic and phylogenetic interest because it occurs in all diploid species (for one notable exception, see below) of sections viewed as morphologically advanced by Lewis and Lewis (1955) (Fig. 10.2). By contrast, species in the less advanced sections lack the duplication. However, the advanced sections are not closely related in this hypothesized phylogeny, and if the phylogeny were correct, it would mean that the duplication must have originated several times (Fig. 10.2). Gottlieb (1983) reported that Lewis, in a more recent appraisal of the phylogeny of *Clarkia* taking into account isozyme number for PGI, has made certain modifications from the earlier hypothesis so that only one duplication event is required (Fig. 10.2).

The only exception to homogeneity of isozyme number within a section involves *C. rostrata* of section *Peripetasma*. This species expresses only one isozyme for cytosolic PGI, whereas other members of the section have two gene loci. At least three feasible alternative explanations

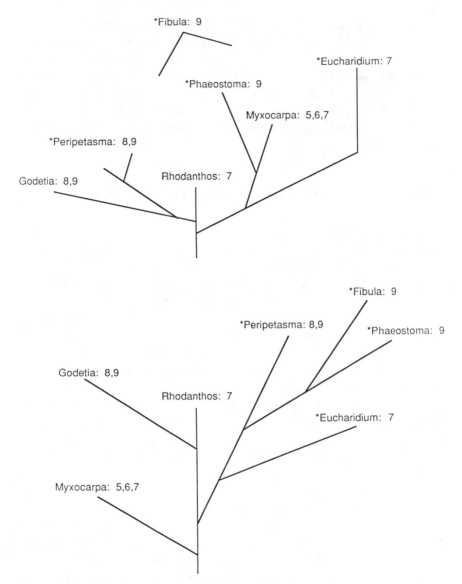

Figure 10.2 Two hypotheses of phylogenetic relationships among the seven sections of the genus *Clarkia*. (Top) Modified from Lewis and Lewis (1955). (Bottom) Modified from H. Lewis (1980). Asterisks denote sections with the duplicated gene locus for cytosolic PGI; numbers following sections refer to gametic chromosome numbers. (Redrawn with permission from U. Jensen and D.E. Fairbrothers, "Proteins and Nucleic Acids in Plant Systematics" 1983, Springer-Verlag, Berlin Heidelberg.)

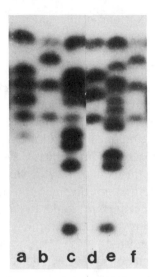

Figure 10.3 Banding patterns for PGI in species of *Clarkia* gene locus with a duplicate gene locus for cytosolic PGI. The most anodally migrating form (at top) occurs in the plastids and is not of concern with regard to the duplication. Polypeptides encoded by different loci associate to form active heterodimers. Three-banded patterns are produced by plants homozygous at each locus (b, d, and f); six-banded phenotypes result from plants homozygous at one locus and heterozygous at the other (a); ten-banded patterns are produced by plants heterozygous at both gene loci (c and e). Comigration of different enzymes in gels prevents detection of all bands in the six- and ten-banded patterns. (From N.F. Weeden and L.D. Gottlieb. Evolution **33**:1024–1039. 1979. Photographs courtesy of L.D. Gottlieb.)

are possible. First, *C. rostrata* may not express the additional isozyme because it has never had it. If this were the case, then this species would be of considerable phylogenetic interest; Gottlieb and Weeden (1979) discussed a possible scenario. Specifically, *C. rostrata* has been viewed as morphologically similar to *C. lewisii* of the nonduplicate section *Rhodanthos*. This hypothesis suggests that these two species, one duplicate and one presumably nonduplicate, may be closely related phylogenetically to the *Clarkia* in which the duplication originated. If this were the case, the two species must have been separated genetically for a long period of time because they have no PGI alleles in common. Also, the disjunct distribution of the two species indicates that their extant populations probably are relicts (Gottlieb and Weeden, 1979).

Another possibility is that one of the genes for cytosolic PGI has been silenced (either by mutation or deletion) and is no longer encoding an active enzyme. Such a silencing apparently has occurred for another enzyme in *Clarkia* (see later discussion) and cannot be ruled out for PGI in *C. rostrata*. It is not possible, using enzyme electrophoresis, to demonstrate the presence of a structural gene that, because of mutation, no longer encodes an active enzyme (i.e., a silenced gene) in a plant. It is possible to do so using techniques from molecular biology, although this is not a practical approach for most plant systematists working with isozymes. If a probe for the gene for cytosolic PGI could be obtained, and if the noncoding gene had not diverged too much from the one being expressed, then the silenced gene could be detected. If deletion had caused the silencing, then it would be extremely difficult to show that the duplicate gene had ever existed.

The last possible explanation is that two genes are present for cytosolic PGI but that they encode enzymes with overlapping mobilities in gels. The extensive electrophoretic analyses carried out by Gottlieb and Weeden (1979) effectively eliminated this possibility. Restriction site analyses of chloroplast DNA by Sytsma and Gottlieb (1986b) showed clearly that *C. rostrata* is properly placed in section *Peripetasma* and that it is not closely related phylogenetically to *C. lewisii* of section *Rhodanthos*. These results, to be discussed in Chapter 17, negated the hypothesis that *C. rostrata* never possessed the PGI duplication.

Isozyme number for cytosolic PGI is also of value in assessing the relationships of two other genera to *Clarkia*. One question is whether to include section *Eucharidium*, consisting of two species that are very distinct morphologically from other members of *Clarkia*, in the genus or to recognize it at the generic level. Despite their morphological divergence from "good" *Clarkia*, both species have the duplication, which provides evidence for their inclusion in *Clarkia* (Gottlieb and Weeden, 1979; Gottlieb, 1983). *Heterogaura heterandra* has been viewed as closely related to *Clarkia*, although this monotypic genus exhibits a variety of unusual morphological features. *Heterogaura* has the PGI duplication (Gottlieb, personal communication), which argues for its inclusion in *Clarkia*. Restriction site analyses of chloroplast DNA provide strong support for the notion that *Heterogaura* originated within section *Peripetasma* of *Clarkia*, which likewise has the PGI duplication (Sytsma and Gottlieb, 1986a) (see Chapters 16 and 17).

Pichersky and Gottlieb (1983) determined the number of gene loci encoding cytosolic and plastid isozymes of triose phosphate isomerase (TPI) in *Clarkia* and other genera of the *Onagraceae*. They found that both genes are duplicated in *Clarkia*, giving two isozymes each for cytosolic and plastid TPI. Genetic data presented by Pichersky and Gottlieb (1983) showing an independent assortment of the allozymes encoded by different gene loci suggested an independent origin for each of the duplications.

The taxonomic distribution of the two duplications differs. Two gene loci for cytosolic TPI have been detected in all but two genera of the Onagraceae examined (Gottlieb, 1983). Interestingly, the additional isozyme is present in the taxa viewed as primitive members of the family (Raven, 1979). Lack of the duplication in two genera from different tribes has been viewed as probably an independent loss of "extra" isozymes (Gottlieb, 1983), which seems a reasonable interpretation. The duplication for plastid TPI has a much narrower taxonomic distribution than the cytosolic one, and it is seemingly restricted to *Clarkia* (Pichersky and Gottlieb, 1983; Gottlieb, 1983). The differing distributions of the two gene duplications suggest that, while they must both be rather old, the one specifying cytosolic TPI originated considerably earlier than the one encoding the extra plastid isozyme (Gottlieb, 1983). It also follows that both of the TPI duplications must have preceded the one for cytosolic PGI discussed earlier.

Odrzykoski and Gottlieb (1984) examined numbers of isozymes of 6-phosphogluconate dehydrogenase (6PGD) in *Clarkia*. In a situation similar to that found for TPI, they detected an extra isozyme for both the plastid and cytosolic forms of 6PGD throughout most of the genus (Fig. 10.4). Four species lack one or both of the duplications, and all of them belong to section *Peripetasma*, which is viewed as advanced or specialized within the genus (see Fig. 10.2, and discussion in Odrzykoski and Gottlieb, 1984). Two species, *C. rostrata* and *C. epilobioides*, express only one isozyme for both plastid and cytosolic 6PGD, whereas *C. cylindrica* and *C. lewisii* lack the duplication for the cytosolic form (Odrzykoski and Gottlieb, 1984). Since section *Peripetasma* is viewed as specialized (Fig. 10.2), the interpretation of Odrzykoski and Gottlieb (1984) that the lack of duplicated 6PGD loci in this section results from loss of expression seems reasonable. Three species, *C. cylindrica*, *C. lewisii*, and *C. rostrata*, are assigned to subsection *Peripetasma*, whereas *C. epilobioides* is placed in its own subsection (*Micranthae*) within *Peripetasma* (Lewis and Lewis, 1955). Because the three species of subsection *Peripetasma* appear to be

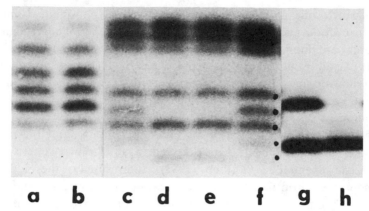

Figure 10.4 Electrophoretic banding patterns for 6PGD in diploid species of *Clarkia*. The more anodal bands (toward top) are cytosolic, while the slower migrating forms are in the plastids. (*a,b*) Both plants have the same three-banded patterns for plastid and cytosolic forms; they have duplicate loci for both forms of 6PGD. (*c,−f*) All plants with the same three-banded pattern for cytosolic 6PGD. Plant *c* has one three-banded pattern for the plastid form, while plants *d* and *e* have an alternative pattern with a different slow form. Plant *f* is heterozygous for the two alleles found in plant *c* versus plants *d* and *e*, giving a six-banded pattern with the bands indicated by dots. Comigration of two enzyme forms make up the central band and thus only five rather than six bands are resolved. (*g,h*) Plants lacking both duplications for 6PGD and thus containing only two isozymes of 6PGD. Lane *h* contains enzymes from isolated chloroplasts. (With permission from I.J. Odrzykoski and L.D. Gottlieb. Syst. Bot. **9**:479–489. 1984, American Society of Plant Taxonomists. Photographs courtesy of L.D. Gottlieb.)

closely related (Davis, 1970), Odrzykoski and Gottlieb (1984) suggested that the absence of the duplicated locus for cytosolic 6PGD probably results from its loss in an ancestor common to all of them. They further suggested that at a later time *C. rostrata* lost the duplication for plastid 6PGD.

The case of *C. epilobioides*, which like *C. rostrata* lacks both duplications, is not clear. The species is now assigned to its own subsection, but Lewis and Lewis (1955) viewed it as equally closely related to subsection *Peripetasma* and subsection *Lautiflorae*. Odrzykoski and Gottlieb (1984) argued that lack of the duplication of cytosolic 6PGD in *C. epilobioides* places it closer to subsection *Peripetasma* (keep in mind that subsection *Lautiflorae* has both duplications). They suggested that lack of the extra isozyme for plastid 6PGD may not indicate a close relationship with *C.*

rostrata, the critical question being whether the losses result from a single or independent events. In Chapter 17, data from chloroplast DNA will be considered relative to these phylogenetic questions (Sytsma and Gottlieb, 1986b), and comparison of different data sets will be presented, compared, and contrasted. Suffice it to say that restriction site analyses clearly group together these four species lacking one or both gene duplications. Furthermore, the DNA data demonstrated that the two species lacking one duplication (*C. lewisii* and *C. cylindrica*) group together, as do the two (*C. epilobioides* and *C. rostrata*) lacking both duplications. Clearly, silencing of the duplicate loci for the cytosolic isozyme occurred in an ancestor common to all four species, whereas loss of the extra isozyme in the plastid took place at a later time in an ancestor common to *C. epilobioides* and *C. rostrata*.

P.S. Soltis et al. (1987) detected duplicate genes for both plastid and cytosolic phosphoglucomutase (PGM) in *Clarkia*. All but two species have the duplication for plastid PGM, and the taxonomic distribution of these species suggests independent loss of duplicate gene expression. The extra cytosolic isozyme occurs in all members of sections *Godetia* and *Myxocarpa* and one species in section *Rhodanthos* (Fig. 10.2). This distribution could be explained either by common ancestory or by the independent origin of the duplication in the species in section *Rhodanthus* and an ancestor common to the other two sections (i.e., *Godetia* and *Myxocarpa*).

Crawford et al. (1990) determined that two loci encode plastid PGI in species in 9 of the 11 sections of North American *Coreopsis*. The two sections lacking the additional isozyme are *Electra* and *Anathysana*. E.B. Smith (1975) originally viewed these two sections (especially *Electra*) as primitive elements in North American *Coreopsis*. Since one plastid PGI isozyme is the common condition in flowering plants (Gottlieb, 1982, 1983), one could interpret isozyme number as concordant with the original phylogenetic hypothesis because the primitive members of *Coreopsis* would then have the ancestral condition while the duplication present in other sections would represent the derived condition.

Subsequent studies (Crawford and Smith, 1983a,b; Jansen et al., 1987) suggest that sections *Electra* and *Anathysana* represent a terminal lineage of North American *Coreopsis* that is not ancestral to other taxa. Additional surveys for the number of plastid PGI isozymes in genera closely related to *Coreopsis* (i.e, *Bidens*, *Coreocarpus*, *Cosmos*, and *Thelesperma*) revealed two plastid PGI isozymes. This finding indicates that the ancestral condition for North American *Coreopsis* is two isozymes and that the single plastid PGI in sections *Electra* and *Anathysana* represents the derived condition. The joint absence of the additional isozyme in sections *Electra* and *Anathysana* suggests that the loss occurred in an ancestor common to the two sections, which in turn indicates that the two sections represent a monophyletic assemblage within North American *Coreopsis*. Data from morphology, secondary chemistry, and chloroplast DNA restriction site analyses are concordant with this hypothesis (Crawford and Smith, 1983a,b; Crawford et al., unpublished data).

Additional examples of gene duplications of potential phylogenetic significance include the recent report of an extra isozyme for PGM in *Layia* (Warwick and Gottlieb, 1985) and related genera of the subtribe Madiinae of the Heliantheae (Gottlieb, 1987). This duplication appears to occur in the whole subtribe. Additionally, a duplication (perhaps the same as in *Layia*?) for cytosolic PGM is clearly present in several genera of the subtribe Coreopsidineae and appears to be much more widely distributed in the Asteraceae (Crawford, unpublished data). The taxonomic distribution and possible phylogenetic significance of this duplication need to be determined. Warwick and Gottlieb (1985) and Gottlieb (1987) detected two isozymes for cytosolic isocitrate dehydrogenase (IDH) in species of *Layia* with a chromosome number of $n = 7$, whereas species with $n = 8$ have three isozymes.

Tanksley and Kuehn (1985) documented the duplication of a gene encoding the cytosolic form of 6PGD in tomato (*Lycopersicon*; Solanaceae). They ascertained that the subunits (polypeptides) specified by the different loci differ in molecular weight, the magnitude of which corresponds to 15 amino acid residues. This finding led Tanksley and Kuehn (1985) to suggest that the duplication does not represent a recent event, and that it is probably found more widely than in *Lycopersicon esculentum* alone. Indeed, they reported the duplication in three other genera of the Solanaceae, and additional survey work would be desirable to ascertain its taxonomic distribution.

Lumaret (1986) determined that two loci for cytosolic PGI in *Dactylis glomerata* are tightly linked, indicating that they may represent a tandem duplication resulting from unequal crossing over. While *Dactylis* is monospecific, it is widespread and variable, and the duplication occurs in all diploid subspecies. Since closely related genera do not have the duplication, and since it is found throughout *Dactylis*, it means that the duplication originated early in the evolution of the genus (Lumaret, 1986).

Gene duplications have also been detected as polymorphisms in populations and species. For example, Ennos (1986) documented a single individual in a population of *Cynosurus cristatus* (Gramineae) with a duplication for alcohol dehydrogenase (ADH). A polymorphism for a duplication for ADH occurs in *Stephanomeria exigua* (Roose and Gottlieb, 1980). Gottlieb (1974b) documented an extra isozyme for ADH in *Clarkia franciscana*, a highly restricted species known from only one population growing on serpentine soil in California. Given the apparent absence of the duplication in all other species of *Clarkia*, there appears to be little question that it originated either with or subsequent to the differentiation of this single species. These observations document that duplications arise within populations, as one would expect. Rieseberg and Soltis (1987b) found that two subspecies of *Helianthus debilis* have one isozyme for cytosolic PGM, whereas three other subspecies have two cytosolic PGM isozymes.

In discussing isozyme number and plant systematics, Gottlieb (1983) concluded that the ultimate value of gene duplications for inferring taxonomic and evolutionary relationships remains to be determined. While their phylogenetic utility in the genus *Clarkia* has been demonstrated in a series of studies, at the same time several caveats are in order. The taxonomic distribution of gene duplications for different enzymes varies within *Clarkia* and other genera related to it, reflecting the relative times of origin of particular duplications in the evolutionary history of the genus. For example, the duplication for cytosolic PGI undoubtedly occurred subsequent to the origin of *Clarkia*, whereas the TPI duplications must represent events that took place earlier because all species of *Clarkia* have them. Additionally, the extra gene for cytosolic TPI is apparently older than the one for the plastid form because the former occurs in several related genera whereas the latter is restricted to *Clarkia*. The duplicate gene for ADH in *C. franciscana* results from a more recent event because of its restriction to this one species. More recent duplications would be polymorphic within populations or species. It is apparent, then, that the taxonomic level at which extra isozymes may be most informative for phylogenetic inferences will vary with the group, and the enzyme.

What represents the ancestral and derived condition for gene number must be evaluated critically in each case. In many instances, consideration of the outgroup (in cladistic terminology) will be a powerful (indeed perhaps the only) method for making this choice. In most cases the conserved number characteristic of the vast majority of diploid plants will represent the primitive character state. However, apparent exceptions such as the loss of extra isozymes for 6PGD in *Clarkia* and for plastid PGI in *Coreopsis* have been and no doubt will continue to be found.

It is apparent that gene duplications in diploid flowering plants are more common than they were once thought to be. As more investigators become "tuned in" to looking for extra isozymes, it is likely that more and more will be reported. It will be interesting also to see whether duplicate loci encoding soluble enzymes will prove to be more common in plants with certain attributes and/or more common for certain enzymes than others. For example, duplications are quite common in *Clarkia*, a genus well known for the occurrence of chromosome repatterning during the course of evolution. The question remains as to whether the extensive electrophoretic surveys of *Clarkia* by Gottlieb and collaborators account for the large number of duplications reported for *Clarkia* or whether the common occurrence of chromosome translocations generate more duplications than occur in genera without extensive chromosome repatterning. Only additional studies in a variety of genera will provide the answer.

ISOZYME NUMBER AS EVIDENCE FOR THE DIPLOID OR POLYPLOID NATURE OF PLANTS

As indicated earlier, increase in isozyme number could be caused by duplications at the diploid level or by polyploidy. Gottlieb (1981b) used gene number as evidence for diploidy or polyploidy. The plants in question belong to the tribe Astereae of the Compositae, in which chromosome numbers range from $n = 2$ to $n = 9$, with $n = 4$ or 5 being very common. Raven et al. (1960) argued that all plants are diploid with aneuploid reduction accounting for the lower chromosome numbers. An opposing view (Turner et al., 1961; Turner and Horne, 1964) suggested that the plants with $n = 9$ are allotetraploids originating from hybridization between plants with $n = 4$ and $n = 5$ chromosomes. Although chromosome number per se often indicates whether plants are diploid or polyploid, the number alone was not adequate with regard to the Astereae. Stucky and Jackson (1975) found that species with $n = 9$ actually have less nuclear DNA than those with $n = 4$ or 5, which argues against the former being simple allotetraploids. Arguments based on other data were inconclusive. Gottlieb (1981b) stressed that the essence of polyploidy is not chromosome number but rather the number of genomes present. Once this premise is accepted, then the alternative hypotheses can be tested in a more explicit manner.

If the $n = 9$ plants are recent simple allotetraploids, then they should exhibit increased numbers of isozymes relative to the plants with lower numbers (i.e., the diploids), as was discussed earlier for *Tragopogon* and *Asplenium*. If, however, all plants are diploids they should have the same or nearly the same number of isozymes because there is no evidence that aneuploid reduction results in loss of structural gene loci (at least those coding soluble isozymes) below the minimum conserved number (Roose and Gottlieb, 1978; Crawford and Smith, 1982a).

The isozyme data provided support for the hypothesis that the $n = 9$ plants are diploids (and thus that lower numbers result from aneuploid reduction) because they have the same number of isozymes as the plants with the lowest numbers (Gottlieb, 1981b). It might be argued that (1) the $n = 9$ plants are allopolyploids that originated from hybridization between diploids not differentiated at any gene loci specifying soluble enzymes (thus the duplicate genes would be encoding the same polypeptides as in an autoploid); or (2) they could be ancient allopolyploids in which all duplicate loci have been silenced. These explanations cannot be dismissed, although several considerations make them less likely than the view that the plants are all diploids. Species with $n = 4$ or 5 are highly divergent from each other at isozyme loci; thus if species with $n = 9$ are allotetraploids one would expect to see isozyme multiplicity. The question of whether duplicate genes have been silenced in plants with $n = 9$ chromosomes is more difficult to address. While morphological similarity among the species suggests a recent rather than ancient origin and thus makes wholesale silencing of duplicate genes unlikely, the evidence is by no means conclusive.

Chase and Olmstead (1988) surveyed the subtribe Oncidiinae of the Orchidaceae for isozyme number. A remarkable range of chromosome numbers ($n = 5-30$) occurs in this group, and it has been debated whether the low numbers represent the ancestral or base number and the high numbers are polyploids, or whether the plants with the highest numbers are diploid and the lower numbers are derived by aneuploid reduction (see discussion in Chase and Olmstead, 1988). The same number of isozymes was detected in plants regardless of chromosome number, which indicates that the species with the highest chromosome numbers are not polyploid, that is, lack of isozyme multiplicity in the plants with high numbers suggests that they have the same number of genomes as

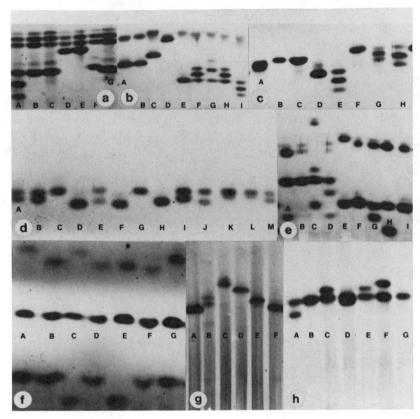

Figure 10.5 Electrophoretic patterns in species of *Bommeria*. Letters beneath bands refer to particular populations of given species and will be mentioned only where of interest. (*a*) Triosephosphate isomerase (TPI). The more anodal area of activity (at top) is composed of a consistent three-banded pattern, suggesting a gene duplication. The slower migrating bands consist of one- or three-banded patterns in individual plants. (*b*) Phosphoglucose isomerase (PGI). Only the lower area of activity will be considered. All plants exhibit either single-banded phenotypes or different three-banded heterozygous patterns. (*c*) Isocitrate dehydrogoenase (IDH). One isozyme for this dimeric enzyme stained darkly enough for visualization on gels. Plant E is the triploid apomict *B. pedata*, and it exhibits a fixed three-banded heterozygous pattern. The single-banded (homozygous) and three-banded (heterozygous) patterns found in other species are shown. (*d*) Shikimate dehydrogenase (SDH). Plants K and L are the triploid apomict *B. pedata* and have two bands that run very close together, that is, plants are fixed heterozygotes for this monomeric enzyme. Other species exhibit one or more allozymes, with no evidence of more than one gene locus being expressed in any plant. (*e*) Phosphoglucomutase (PGM). A monomeric enzyme. Two isozymes are

plants with low chromosome numbers. Isozyme number supports the hypothesis of an aneuploid reduction number in the subtribe Oncidiinae of the Orchidaceae (Chase and Olmstead, 1988).

A series of isozyme studies of homosporous pteridophytes (Gastony and Darrow, 1983; Haufler and Soltis, 1986; Haufler, 1985, 1987; Soltis, 1986; D.E. Soltis and P.S. Soltis, 1988b; P.S. Soltis and D.E. Soltis, 1988b; Wolf et al., 1987) has provided interesting phylogenetic information. The results show that these pteridophytes exhibit the "minimum" conserved number of isozymes typical of diploid flowering plants despite having high chromosome numbers, that is, plants with chromosome numbers so high that they would be viewed as polyploids if they were flowering plants behave as diploids isozymically. For example, in the fern genus *Bommerica*, Haufler (1985) showed that sexual species with a number of $n = 30$ have the typical diploid number of isozymes (Fig. 10.5). A triploid apomictic species, *B. pedata*, does show fixed heterozygosity at nearly half the loci examined, which demonstrates that the type of gene duplication typical of allopolyploid plants is present in this species (Fig. 10.5).

Haufler and Soltis (1986) reviewed both previously published and unpublished data on isozyme numbers in homosporous ferns. The tabulated results for over ten genera of ferns indicate that, with rare exceptions, all plants viewed as "diploid" (having the lowest known number for the taxon) show only the typical diploid isozyme number. These gametic

---

resolved in most species, but only the lower (more cathodol) forms will be considered. The apomict triploid *B. pedata* is represented by plant C and shows three fixed allozymes. Plant D is a representative of *B. ehrenbergiana*, and it likewise displays three bands, suggesting a duplicated locus. Individuals of all other species exhibit one or two bands, which would be expected for plants with one gene locus. (*f*) PGM from individual haploid gametophytes of *B. ehrenbergiana* from a single sporophyte (see *e*, plant D). Three loci are expressed in these haploid individuals, with segregation at genes specifying the most anodal and cathodal isozymes. The expression of three enzymes in the gametophyte could occur only if there were a gene duplication. (*g*) Leucine aminopeptidase (LAP), a monomeric enzyme. All species except the triploid *B. pedata* exhibit one or two allozymes and show no evidence of gene duplication. Plant C, a member of *B. pedata*, shows a fixed heterozygous pattern; the two bands migrate very near to each other, but both bands may be seen with close scrutiny. (*h*) LAP, showing variation within a single population of *B. hispida*. Both single-banded homozygotes and two-banded heterozygotes are present. (With permission from C.H. Haufler, Syst. Bot. **10**:92–104. 1985, American Society of Plant Taxonomists. Photograph courtesy of C.H. Haufler.)

numbers range from 29 to over 40. By contrast, polyploid species, that is, those species having a multiple of the lowest number for the genus, do exhibit isozyme multiplicity and fixed heterozygosity in all the genera examined.

In addition to ferns, homosporous pteridophytes have been surveyed for isozyme number. They include *Equisetum* (Soltis, 1986); *Lycopodium, Diphasiastrum*, and *Hyperzia* (D.E. Sotis and P.S. Soltis, 1988b); and *Psilotum nudum* (P.S. Soltis and D.E. Soltis, 1988a). Soltis (1986) examined three species of *Equisetum*; despite a chromosome number of $2n = 216$, all species had the minimal conserved number of isozymes typical of diploid plants. There was no evidence for the multiplicity of isozymes expected in polyploids. D.E. Soltis and P.S. Soltis (1988b) found that three genera of lycopods (*Lycopodium, Diphasiastrum*, and *Hyperzia*) with high base chromosome numbers have the number of isozymes typical of diploid seed plants for 12 of the 13 enzyme systems studied. P.S. Soltis and D.E. Soltis (1988a) found that *P. nudum*, with a chromosome number of $2n = 104$, has the number of isozymes typical of diploid plants for the 15 enzymes examined.

Two possible alternative hypotheses for explaining observed isozyme numbers in homosporous pteridophytes have been proposed (Haufler and Soltis, 1986). Isozymically, diploid pteridophytes with high chromosome numbers may be the product of ancient, repeated occurrences of allopolyploidy. According to this model, following each episode of polyploidy in fern lineages there was a silencing of duplicated loci and chromosomal diploidization to produce "diploids" (Haufler, 1987) (Fig. 10.6). Also, the progenitor diploids in any one cycle became extinct. The same process could then occur repeatedly. The alternative hypothesis is that the high chromosome numbers in pteridophytes are the original diploid condition (Haufler and Soltis, 1986). Presumably, for plants with very high basic numbers such as *Equisetum* ($2n = 216$), several cycles of polyploidy with complete silencing of duplicate loci must have occurred if one accepts the hypothesis that these plants are ancient polyploids. Available data do not allow a clear choice between the two hypotheses, but several lines of investigation were suggested to evaluate the two hypotheses (Haufler and Soltis, 1986).

The series of studies discussed in this chapter provide clear-cut distinctions between diploid and polyploid plants in isozyme number; species with multiples of a base chromosome number show extensive duplications relative to those with the base number. There are other instances in

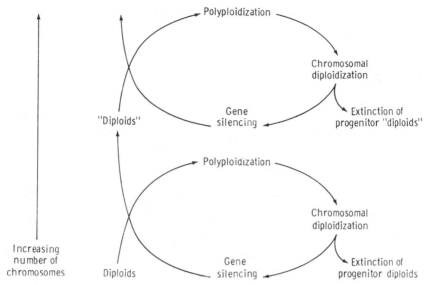

Figure 10.6 Diagram of model for explaining the evolution of high chromosome numbers and genetic diploids in homosporous pteridophytes. Essential processes are polyploidization, chromosomal diploidization, silencing of duplicate gene loci, and extinction of progenitor diploids. (With permission from C.H. Haufler, Am. Jo. Bot. **74**:953–966 1987, Botanical Society of America. Courtesy of C.H. Haufler.)

which isozyme number per se is not definitive for inferring ploidy level, since gene duplications can occur at the diploid level, and ostensibly silencing of some duplicate loci can take place in polyploids. This means that convergence on a certain percentage of duplicated isozyme loci could occur via two different mechanisms. For example, in the genus *Clarkia*, at least 25% of the loci examined are duplicated in some or all species (Gottlieb, 1986). Nearly 25% of the loci in diploid *Coreopsis* appear to be duplicated as well (Crawford, unpublished data).

Certain angiosperm genera that have been surveyed for isozyme number appear to be of polyploid origin. The lowest known chromosome number in the genus *Antennaria* is $2n = 28$, and since the closely related genus *Gnaphalium* has species with $2n = 14$, it appears that *Antennaria* is of polyploid origin (Bayer, 1988). The percentage of duplicate loci detected in *Antennaria* is lower than in *Clarkia* or *Coreopsis*. Elisens and Crawford (1988) found only one duplication in the genus *Mabrya* despite rather compelling evidence that it is of polyploid origin. Each situation must be examined critically when using isozyme number as evidence of ploidy level.

SUMMARY

Minimal isozyme number is conserved in diploid plants, and any increase at the diploid level almost certainly results from gene duplication. A duplication is of potential phylogenetic interest because its presence may delineate a monophyletic assemblage of taxa. While the higher isozyme number will probably represent the derived condition in most instances, each case must be evaluated with all available data because silencing of duplicated loci may have occurred (Wilson et al., 1983). The taxonomic level at which duplicated gene loci will prove useful phylogenetically varies both with the plant group and the enzyme. This variation reflects the time in the evolutionary history of a lineage when a particular duplication occurred.

The frequency of duplications in diploid plants remains as open question. It is not known whether chromosomal repatterning, which is common in annual plants, is responsible for generating extra isozymes or whether duplications are equally abundant in herbaceous and woody perennial plants. The ultimate phylogenetic value of gene duplications in diploid plants remains to be elucidated. Now that workers are aware of the "expected" number of isozymes, perhaps more duplications will be detected and studied with the same detail as in *Clarkia*.

Electrophoretic studies of the homosporous pteridophytes have documented that, despite high chromosome numbers in these plants, they are genetically diploid with respect to isozymes. These results raise interesting questions as to whether the high basic chromosome numbers in different genera are the result of repeated episodes of polyploidy and gene silencing or whether they represent the original diploid numbers.

While isozyme number may provide useful information on whether plants are diploids or polyploids, the data may not be definitive in some cases, since isozyme number can increase in true diploids via duplications while it seems likely that it can decrease via gene silencing in polyploids. Plants with one-fourth to one-third of their isozyme loci duplicated could be diploids with extensive duplications, or polyploids with gene silencing.

CHAPTER 11

Amino Acid Sequences and Angiosperm Phylogeny

INTRODUCTION

Amino acid sequencing, first performed on animals for phylogenetic purposes over two decades ago (Fitch and Margoliash, 1967), helped usher in "hard core" macromolecular approaches to phylogeny. Hard core refers here to the complete or partial sequencing of molecules rather than comparative electrophoretic studies or other methods whereby particular properties of the molecules are employed as the data. Shortly after the first papers on protein sequencing in animals appeared, Donald Boulter and his collaborators began publishing amino acid sequences of cytochrome c in plants along with discussions of the possible phylogenetic significance of their results (see Boulter, 1972, 1973a,b as early examples, and Giannasi and Crawford, 1986, for more comprehensive references).

The phylogenetic and systematic implications of Boulter's work were discussed and debated at greater length than other kinds of chemical data previously employed, despite the widespread recognition that the number of taxa sampled was much too small for meaningful interpretations of the evolution of flowering plants. Rather, interest in the sequence data stemmed from its potential for assessing phylogeny in angiosperms at much higher taxonomic levels than was possible with other previously available chemical information. For the first time, an independent means for constructing phylogenies for part or all of the flowering plants was seemingly available, and these phylogenies could then be compared with those systems developed from more "conventional" data.

Another curious aspect of amino acid sequence data is that they were of interest to many plant systematists even though the information could never be applied to the small group of plants (genera or groups of

171

congeneric species) on which they were working. There were several reasons why the data were not applicable at lower levels (such as among congeners), including the time and expense of doing the research, the problems with data analysis, and the likely inapplicability of the data at these levels because sufficient variation was not present.

Since protein sequencing was among the newest macromolecular approaches at the time (important publications on enzyme electrophoresis began appearing at about the same time), and since the initial data were of such interest, one might expect to find expanded research in this area at the present time. Such is not the case, however, and amino acid sequencing as a means of generating phylogenetic data for plants (and other living organisms) is employed with decreasing frequency. The primary reason is the greater technical ease of sequencing DNA, if the coding gene can be isolated. Thus, amino acid sequencing is giving way to nucleic acid sequencing (a topic considered in Chapter 13).

Sequencing of amino acids was appealing phylogenetically because it represented the best available method for inferring the sequence of nucleic acids in the structural gene encoding the protein. Thus, the statement of Harborne and Turner (1984, p. 406) that sequencing of DNA is "an alternative method of sequencing proteins which may become more widely adopted in the future" seems misplaced in its emphasis; for purely phylogenetic purposes, sequencing of proteins has served as an alternative method for inferring nucleic acid sequences. Also, not everyone would agree with Harborne and Turner (1984, p. 456) that once a protein has been sequenced there is no point in sequencing its coding portion of DNA. It is not true that "clearly the message is the same whether it is read in the DNA code, the RNA code or in the amino acids in the protein." Earlier considerations of the basic processes of protein synthesis mentioned information loss in going from DNA to protein. Because of both technological advances and considerations of information content, amino acid sequencing is used with decreasing frequency in phylogenetic studies less than two decades after it was first employed.

The present chapter is in some respects an historical discussion of amino acid sequencing as a phylogenetic tool for the plant systematist. This is not to say that recent publications on the topic will not be considered and that these papers do not represent valuable contributions, but is rather to emphasize that the impact and importance of amino acid sequences should be considered more in historical perspective than in future contributions toward plant phylogeny. The discussion will include the strengths and weaknesses of the approach and the impact of the data on plant phylogeny.

OVERVIEW OF ADVANTAGES AND LIMITATIONS FOR PHYLOGENETIC STUDIES

The publication by Fitch and Margoliash (1967) over two decades ago was historic partly because it demonstrated high concordance between phylogenetic hypotheses for animals based on amino acid sequences of the protein cytochrome c and hypotheses generated from "most all other data." Possible implications of their results were profound for those who took the time to ponder them because the information contained in this one molecule agreed with phylogenetic hypotheses developed from data gathered from many disciplines over decades of time. The optimist of the time could imagine the possibility of unraveling the phylogeny of a group of organisms by simply sequencing one or several proteins.

For studying the phylogeny of plants, particularly in the angiosperms, the value of amino acid sequence data seemed to offer even greater potential than for animals, given the problems of convergence of morphological features in plants, the meager fossil record, and other factors complicating the understanding of their evolutionary history.

The potential advantages of using the primary structure of a protein versus morphology for phylogenetic purposes are apparent. Comparisons of the same (or homologous) characters across the angiosperms can be made with proteins, whereas such comparisons are more difficult with many morphological features because of parallel or convergent evolution. Such features as the inferior ovary and bilaterally symmetrical flowers have developed independently in flowering plants; this probably applies to many other characters, but the extent of the problem is not known.

The use of amino acid sequence data allows one to assess the differences (or similarities) between taxa in a quantitative manner, that is, it may be determined that species A differs from species B at a certain number of sites for a given protein molecule. From this data, the similarity between the two species for this protein may be calculated. This method cannot be used with the same precision for morphological features or secondary chemistry. Methods are available for quantifying phenetic similarities in morphological characters and in secondary chemistry, but these similarities do not represent a precise quantification of *genetic* similarities.

With morphological characters, it is often difficult to compare features across large taxonomic groups because of the presence–absence problem. How does one compare, for example, the absence of petals in one group with its absence in another group? If petals are present in one group and absent in another, comparisons are likewise difficult. The problem is seemingly circumvented with protein sequencing because proper choice of molecules for study ensures that they are present in all plants under consideration.

Heywood (1973) (and others) have pointed out that the age (i.e., time of origin) of a group of plants may be determined with reasonable certainty only from an adequate fossil record. Needless to say, the flowering plants are notorious for their poor fossil remains despite the continuing efforts of paleobotanists, who have produced significant new discoveries. Attempts to infer time of origin of extant plants from morphological features are generally regarded as difficult at best and futile at worst, the reason being the uneven rate at which external features may change through time, that is, one cannot infer with any degree of accuracy the relative times of divergence of taxa by their measured phenetic differences. Also, relative phenetic divergence may not represent relative genetic divergence in a precise manner.

An important question is whether a given protein undergoes a fairly constant rate of substitution through time in different organisms. One hypothesis is that for one protein, change at a constant rate occurs because many of the amino acid substitutions are largely neutral and the rate of change will depend on mutation rate. The phenomenon of constant rate of change for a protein has been called the molecular evolutionary clock, and the selective equivalence of different amino acid substitutions is generally known as the neutral theory of molecular evolution (both of these terms may be applied to nucleic acids as well as proteins). If a molecular clock exists and if it is reasonably accurate, then its value for inferring the origin of plant groups is obvious. The existence of the clock, or how well it keeps time, is a matter of continuing debate (see Kimura, 1968, 1979, 1983, 1984; Thorpe, 1982; Wilson et al., 1977; Harborne and Turner, 1984; Nei, 1987 versus Boulter et al., 1979; Cronquist, 1976; Fitch and Langley, 1976; Margoliash et al., 1976; Romero-Herrera et al., 1979, for contrasting views). Nei (1987, pp. 53–59) and Hartl and Clark (1989, pp. 358–367) presented very useful discussions of the topic. A series of papers treating various aspects of the molecular clock appeared recently (Dover, 1987; Jukes, 1987; Kimura, 1987; Ohta, 1987; Preparata and Saccone, 1987; Zuckerkandl, 1987).

Despite the impressive list of seeming advantages of amino acid sequencing for constructing phylogenies, several weaknesses and problems are also associated with it, concerned either with data generation or analysis and interpretation of information.

Harborne and Turner (1984, p. 396) gave a list of desirable attributes of a protein to be sequenced. The protein must show some variation in amino acid sequence among the taxa to be examined. It might be added that, by the same token, sequence variation cannot be too great. The size of the enzyme molecule must be relatively small (preferably around 100 amino acids or fewer) so that sequencing is feasible without huge investments of time and effort. The enzyme should occur in "reasonable" amounts in plants and at the same time be relatively easy to isolate and purify. Lastly, the enzyme must represent a homologous protein in the organisms to be compared. That is, the enzyme must be encoded by genes having a common origin that are related by descent.

Despite selection for these features in an enzyme for sequencing, Scogin (in Fairbrothers et al., 1975) emphasized that obtaining adequate enzyme often requires beginning with large amounts of plant material. For example, the respiratory enzyme cytochrome *c* requires huge quantities of rapidly germinating seeds at the beginning for sufficient amounts of enzyme to be isolated. Harborne and Turner (1984, p. 396) noted that the yield of cytochrome *c* is up to 100 mg per kg in some vertebrates, whereas seedlings yield 1 to 4 mg per kg. The low yield of an important enzyme like cytochrome *c* limits the plants that may be examined, and many of the species sequenced are crop or horticultural plants. Plants from phylogenetically interesting groups such as members of the presumably primitive Magnoliales have been excluded from analysis because of these limitations. This is not a problem with nucleic acids if the gene of interest can be cloned, or amplified by some other method.

The time required for complete sequencing of a protein is greater than gathering data from secondary chemistry and electrophoresis. Indeed, more time is required to sequence proteins than nucleic acids, and the differential between the two becomes greater with time. Even with improved methods and technology such as the use of sequencers, there are still problems with sequencing larger proteins. Thus, limitations exist for both the taxa and the enzymes that may be investigated.

As mentioned earlier, one of the seeming advantages of the method when amino acid sequencing was initiated was that it provided an independent source of data for constructing a phylogeny. As Scogin (1981) emphasized, the method is free of earlier biases, and is "objective" in the sense that data and methodology are spelled out in great detail. Scogin (1981) correctly (and cleverly) noted that "it may be wrong but it is reproducible!" Harborne and Turner (1984, p. 400) (and others before them) have said that phylogenies generated from a single protein are not really phylogenies of the organisms, but rather are phylogenetic hypotheses of the molecules contained within the plants. Therefore one would not expect a perfect "match" between phylogenies generated rather intuitively from many other kinds of data and those produced from information in one molecule using explicit methods and assumptions. In reality, for the plant systematist a phylogeny produced from amino acid sequence data cannot be divorced completely from phylogenies for organisms generated from other data. Rather, the phylogenies of the molecules will invariably be used for comparisons with other phylogenetic hypotheses for the organisms.

A potential problem with the use of amino acid sequences concerns the existence of heterogeneity within individual plants, that is, two or more gene loci encode different "forms" of the protein. Such heterogeneity arises from gene duplications, and if present it may cause problems in analysis and interpretation of results. This phenomenon, which has not received the attention it probably deserves, was discussed by Harborne and Turner (1984, pp. 397, 402, 405). Present evidence indicates little heterogeneity in the protein cytochrome c, with possible detection limited to one species (Ramshaw, 1982). Haslett et al. (1977) demonstrated plastocyanin polymorphisms in four species of Asteraceae, but did not determine whether they result from variation among plants or the presence of more than one type of protein within individual plants. These workers also cited unpublished data showing polymorphism for plastocyanin from a single plant of *Malva*. Lastly, two different forms of the protein ferrodoxin have been reported from a number of different species.

Problems associated with analysis of amino acid sequence data and methods for constructing trees from the data are considerable, but a detailed consideration is beyond the purpose and intent of the present discussion. The reader is referred to Boulter (1980), Boulter et al. (1979), Cronquist (1976), Scogin (1981), Wilson et al. (1977), Giannasi and Crawford (1986) and Chapter 1 of this book for more general discussions and overviews. Publications dealing in depth with methodology are too numerous to mention, but the reviews by Peacock (1981), Nei (1987, Chapter 11), and especially Felsenstein (1988) should prove useful to the nonspecialist.

The basic assumption in using amino acid sequence data is that identity (or similarity) of sequences in the same protein of different species of plants results from common ancestry. If similarity is caused by convergence resulting from chance and/or selection for particular amino acids at specific sites in the molecule, then there are obvious problems with phylogenetic interpretation. The critical question is ascertaining the magnitude of the problem, and estimates place the percentage of convergent substitutions in the range of 30 to 40% for cytochrome c and plastocyanin (Boulter, 1980). These values represent a significant factor, particularly when the identity of sequences being compared falls below 25% (see Harborne and Turner, 1984 pp. 401; and Doolittle, 1981). Doolittle (1981) presented an extensive discussion of the problem of deciding whether similarities of proteins in different organisms result from convergence or common ancestry. Phylogenies produced from amino acid sequences may suffer problems of convergence similar to those generated from morphology; a basic difference is that it is possible to gain a better estimate of the magnitude of the problem with amino acids as compared with morphology.

CYTOCHROME c: ANGIOSPERM ORIGIN AND PHYLOGENY

As discussed by Harborne and Turner (1984, p. 395), cytochrome c, a molecule consisting of slightly more than 100 amino acids that functions as a respiratory enzyme, holds a significant place in plant chemosystematics. The success of Fitch and Margoliash (1967) in producing from cytochrome c sequences a phylogeny for animals that corresponded very well with generally accepted relationships no doubt was an incentive to gather sequence information for plants. In 1972, the first phylogenies generated from sequence data from plants were published. Boulter et al. (1972) produced a tree of relationships for 14 species of flowering plants and the gymnosperm *Ginkgo biloba*. Ramshaw et al. (1972) employed the cytochrome c data to discuss the time of origin of the flowering plants and times of divergence of various groups. The latter paper will be considered first.

Ramshaw et al. (1972) used the divergence between cytochromes c from different flowering plants as a basis for inferring the time of their origin. To do this, they assumed the same rate of evolution of the enzyme in plants as in animals. In animals, the fossil record allows the clock to be set for cytochrome c because there is good agreement between the times of divergence of lineages in the fossil record and the level of sequence differences in their cytochromes c. Ramshaw et al. (1972) used the value of one amino acid substitution about every 20 million years as the "unit evolutionary period" for plants, and they stated quite clearly that this value was based on data from animals. They also mentioned that this assumption may not be valid, but that given the homology of the enzyme in plants and animals it seems reasonable. It must be added that, given the paucity of the fossil record of plants, there appears to have been little choice in the matter! When these calculations were employed for estimating when the flowering plants appeared, a time of around 400 million years ago was obtained. This value is several geological periods (i.e., over 250 million years) earlier than the occurrence of flowering plants in the fossil record. Wilson et al. (1977) suggested that the unit evolutionary period for the cytochromes c of mammals is 15 million years. If this value is applied to flowering plants, then the difference between the fossil record and the calculations from sequence data is narrowed somewhat, but a wide discrepancy still exists. Also, it should be remembered that no sequence information is available for dicots with the presumably primitive monosulcate pollen; these data could widen the gap further.

The differences between the fossil record and cytochrome c data as indicators of when the angiosperms appeared were discussed by Cronquist (1976) and Turner (1977). These two workers viewed the situation from quite different perspectives. Cronquist (1976) felt that since pollen is well preserved in the fossil record, its occurrence should provide a rather accurate indication of when flowering plants first appeared. Turner (1977), while recognizing the potential problems with assumptions about sequences, argued that the molecular data are preferable to morphological information because an imperfect clock (the former) is preferable to no clock (the latter). As discussed earlier (Giannasi and Crawford, 1986), the matter ultimately comes down to choosing between the sparse protein data and uncertainties about the molecular clock on the one hand and an incomplete fossil record with the considerable difficulties of attempting to estimate divergence times from the comparative morphol-

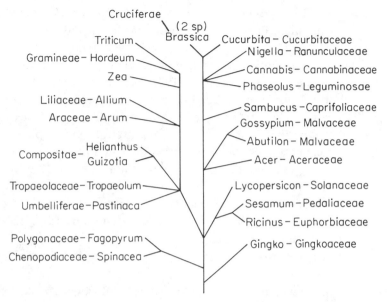

Figure 11.1   Phylogenetic tree of selected species of flowering plants constructed from amino acid sequences of cytochrome *c*. (Redrawn and modified from Boulter, 1974 and Scogin, 1981 and used with permission as published in D.E. Giannasi and D.J. Crawford, Evol. Biol. **20**:25–248. 1986, Plenum Publishing Co.)

ogy of extant plants on the other hand.

The discussion of divergence times of particular groups of angiosperms will be integrated into a consideration of phylogenetic relationships among the groups as shown in trees produced from cytochrome *c* sequence data. Figure 11.1 represents a tree redrawn with modifications from several sources, including primarily Boulter (1974) and Scogin (1981). It will serve as the point of reference for discussion. The early split of the Caryophyllidae (one representative each from the Chenopodiaceae and Polygonaceae), estimated at 300 to 400 million years ago by Ramshaw et al. (1972) based on 20 million years as the unit evolutionary period, contrasts with the presence of the first known fossil pollen for the group some 50 to 60 million years ago. Cronquist (1976) and Turner (1977) read this situation in the same way as each did for angiosperm origins, that is, the former sided with the pollen record and the latter with the molecular interpretation. Also, Cronquist (1976) did not like the Caryophyllidae branching off the main line prior to the divergence of the monocots and dicots (Fig. 11.1). These points aside, it is well recognized that the Caryophyllidae represents a distinct monophyletic assemblage of plants with many unifying features (Mabry, 1977; Rodman et al., 1984), and thus it would be surprising indeed if the two representatives of the subclass did not come out together. In this regard, Scogin (1981) presented the view that the several distinctive features of the group argue for its antiquity. Giannasi and Crawford (1986) indicated that while this may be true, the degree of phenetic divergence may not be a valid indicator of the time of divergence of the group because one does not know without an adequate fossil record when distinctive features first appeared. Related to this problem is the fact that one cannot infer antiquity from phenetic divergence in morphological features and/or secondary chemistry because there is no evidence that anything approaching an evolutionary clock exists for either of them.

Another interesting feature of cytochrome c phylogeny is the early split of the Compositae from the dicots; indeed, they diverge from the "mainline" dicots at about the same point (and time) as the monocots (Fig. 11.1) As might be expected, the same arguments for and against these phylogenetic hypotheses were made by Cronquist (1976) and Turner (1977), with the latter arguing that the Compositae may be more ancient and isolated than generally considered, while the former is not willing to accept this hypothesis on the basis of information from a single molecule. It is reassuring to note that representatives of the Compositae come out together, as do the three grasses (Fig. 11.1).

It was mentioned earlier that cytochrome c data are not available for representatives of woody, presumably primitive plants having monosulcate pollen and belonging to the Magnoliidae (Cronquist, 1981). As also indicated earlier, cytochrome c occurs in small amounts in plant tissues (as opposed to animals), which necessitates using large quantities of germinating seedlings to obtain sufficient quantities. This obviously is not practical for woody plants growing at great distances from the laboratory. These difficulties were the apparent reasons for Boulter switching from sequencing cytochrome c to plastocyanin (see comment by Martin and Jennings, 1983). The result has been that cytochrome c data are not available for some "critical" angiosperms, and, as emphasized by Harborne and Turner (1984, p. 408), the information is also lacking for ferns and gymnosperms.

PLASTOCYANIN AND PLANT SYSTEMATICS

Harborne and Turner (1984, pp. 402–404) and Giannasi and Crawford (1986) provided the most recent discussions of the use of sequences from the protein plastocyanin. This is a copper protein functioning in electron transport in the chloroplast. Harborne and Turner (1984, pp. 402) indicated that one of the reasons the Boulter laboratory switched to sequencing of plastocyanin is its higher concentration in plant tissue. Also, the molecule is small (99 amino acids) and is amenable to automatic sequencing.

Of the some 65 species in which plastocyanin has been examined, the entire molecule has been sequenced for 13 species, whereas for the remainder only the 40 N-terminal residues were determined. The rationale for partial sequencing is that most of the variation has been detected in the first 40 residues, and also it speeds up the automatic sequencing considerably.

Amino acid sequences of plastocyanin have not proved extremely useful for constructing phylogenies. Among the problems are the high mutation rates (about twice the rate as for cytochrome *c*) and the occasional heterogeneity within individuals (Haslett et al., 1977). The latter phenomenon suggests the presence of more than one gene encoding the enzyme, which could complicate the use of the protein for phylogenetic purposes.

A number of studies have employed plastocyanin sequences for generating phylogenies. Boulter et al. (1978) used partial sequences to construct a tree for 22 members of the Compositae. Boulter et al. (1979) produced a tree for ten families of angiosperms using partial sequences. Grund et al. (1981) used the 40 N-terminal sequences from five species of the Ranunculaceae to infer relationships among the representatives of the family as well as their relationships to other families. The reader is referred to Harborne and Turner (1984, pp. 402–404) and to Giannasi and Crawford (1986) for more extensive discussions of the results of these studies; only the most salient aspects will be given here. The results of Boulter et al. (1978) produced a phylogeny for the Compositae in which species belonging to the same tribe in general occurred together. Certain aspects of the phylogeny are difficult to reconcile with accepted relationships. As examples, two species of *Senecio* occur widely separated on the tree, as do different members of the Cichorieae.

The tree generated by Boulter et al. (1979) for ten families of angiosperms is concordant with what would be expected from other data. That is, the different members of such families as the Apiaceae, Asteraceae, Brassicaceae, and Caprifoliaceae all occur together. Certain peculiarities such as the presence of two species of the Leguminosae on different main branches of the tree and the occurrence of seven members of the Leguminosae between two species of the Rosaceae are anomalous. Certain results of Grund et al. (1981) for the Ranunculaceae and other families are likewise difficult to reconcile with accepted phylogenetic relationships. The original paper as well as Giannasi and Crawford (1986) may be consulted for further details.

The conclusion that must be drawn from the plastocyanin sequence studies is that the molecule is of little utility at higher taxonomic levels, such as comparisons of families. In fact, even within families the results sometimes appear inconsistent with established relationships. The high mutation rates for the molecule undoubtedly cause convergence to be a factor.

SMALL SUBUNIT OF RUBISCO: USES IN CONSTRUCTING PHYLOGENIES

In a recent series of papers Martin and collaborators reported the sequences of the 40 N-terminal residues of the small subunit of rubisco and constructed phylogenies from these data when they were combined with previously available sequences from other proteins (Martin and Jennings 1983; Martin et al., 1983, 1985, 1986; Martin and Dowd, 1984a,b,c, 1986a,b). The first paper (Martin and Jennings, 1983) provided an overview of the methods employed in isolating and sequencing the small subunit. In addition, this paper furnished a concise yet valuable discussion of previous work on cytochrome c and plastocyanin, indicating that the difficulty in purifying the former protein from plants and the rather unsatisfying results obtained with the latter are the primary reasons for lack of continued sequencing of their amino acids.

Martin and Jennings (1983) spelled out the rationale for choosing the small subunit of rubisco and for sequencing only the 40 N-terminal sequences. The abundance of the protein in plant material makes it possible to obtain sufficient quantities for sequencing from 100 g of leaves. The large subunit is not as desirable because it is more difficult to purify, and preliminary results showed there is not sufficient variation in the molecule to be useful for comparative studies. Martin and Jennings (1983) suggested that the total sequencing of the molecules would slow the survey work, and indicated that the additional data would not be worth the extra effort. The complete sequences available at that time suggested that the level of variation in the remaining 80 or so residues is comparable to that detected in the first 40 amino acids, that is, it does not appear that any bias is introduced by looking only at the first 40 amino acid residues. In all their papers, Martin and collaborators stressed the limited taxonomic sampling within families and even attached the prefix pro- to each family to denote this.

Martin et al. (1983) constructed phylogenetic trees for selected families of angiosperms employing sequence data from the small subunit of rubisco. In addition, they used published sequences of cytochromes *c* and plastocyanins together with the small subunit data to construct consensus trees. Trees were constructed according to maximum parsimony. The minimal length tree for eight families based on the small subunit sequences is shown in Fig. 11.2*a*. Some of these results are unexpected, particularly the placement of the Chenopodiaceae and Poaceae. When data from cytochrome *c* were used, two minimal length trees were produced (Fig. 11.2*b,c*) differing from each other in several respects as well as differing from the rubisco-based tree. In the tree constructed from all sequences, as opposed to the rubisco tree, the single monocot is distinct from the dicots (cf Figs. 11.2*a* and 11.3). Still, there are certain peculiarities about the tree: the Apiaceae and Chenopodiaceae are grouped together, and the Fabaceae and Brassicaceae occur together (Fig. 11.3).

Martin and Dowd (1984a) determined sequences for three species each of Ranunculaceae and Malvaceae and used the data (together with available sequences from cytochrome *c* and plastocyanin) to generate trees. The two minimal length trees employing all three proteins are shown in Fig. 11.4. In both trees, the Brassicaceae and Fabaceae occur together, as do the Apiaceae and Chenopodiaceae. Equally puzzling are the positions of the Asteraceae and Ranunculaceae.

Martin and Dowd (1984c) incorporated small subunit sequence data from members of the Magnoliaceae and Polygonaceae as well as the gymnosperm genus *Metasequoia* into their trees. Also included were sequences from the Fagaceae and Proteaceae reported earlier (Martin and Dowd, 1984b). The tree shown in Fig. 11.5 was constructed from sequence data from two or three proteins, depending on the taxon. The addition of these families altered the arrangements of several families

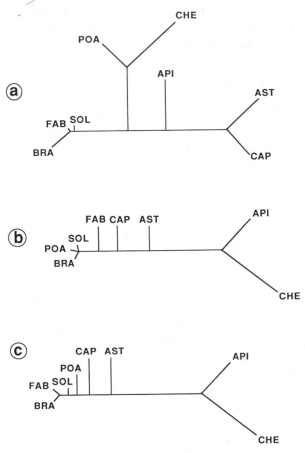

Figure 11.2 Minimal phylogenetic Steiner trees for eight representatives of eight families of flowering plants. API Apiaceae; BRA Brassicaceae; CAP Caprifoliaceae; SOL Solanaceae; POA Poaceae; CHE Chenopodiaceae; FAB Fabaceae; AST, Asteraceae. (*a*) Tree based only on sequences of small unit of rubisco. (*b,c*) Two minimal trees based on sequences of cytochrome *c*. (Redrawn with permission from P.G. Martin et al., Austr. Jo. Bot. **31**:411–419. 1983, CSIR.)

compared with the tree shown in Fig. 11.4. For example, the Chenopodiaceae and Polygonaceae now occur together rather than the Apiaceae being near the Chenopodiaceae, which is much more congruent with prevalent concepts of relationships. The Apiaceae now occupies a rather isolated position. The close grouping of the Brassicaceae and Fabaceae remains; the same is true for the Asteraceae and Ranunculaceae. These

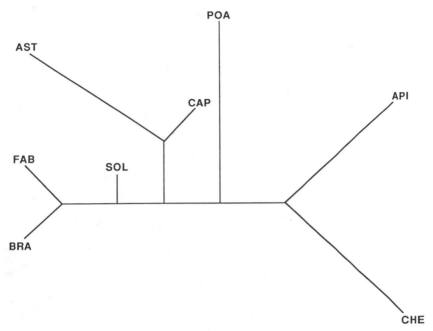

Figure 11.3   A phylogenetic tree representing the preferred minimal tree constructed from the analysis of combined sequence data from the three proteins
cytochrome *c*, plastocyanin, and the small subunit of rubisco. This tree represents
(1) one of four minimal trees produced from the combined data of the three
proteins; (2) the consensus tree produced from the 10 shortest trees from the
combined data for the three proteins; and (3) the minimal tree from combined
small subunit and cytochrome *c* analyses. Familial abbreviations are the same as
in Fig. 11.2. (Redrawn with permission from P.G. Martin et al., Australian Jour.
Bot. 31:411–419. 1983, CSIR.)

arrangements seem rather puzzling compared with present concepts of
phylogenetic relationships for these families. There is uncertainty about
the relative positions of the Caprifoliaceae, Fagaceae, Magnoliaceae,
Malvaceae, and Proteaceae based on available sequence data (Fig. 11.5).
Note also the uncertain position of the Poaceae.

   Martin et al. (1985) combined data from ferredoxin, 5S-ribosomal-
RNA, and the three previously mentioned proteins to construct phylogenetic trees for 11 species of flowering plants. The results did not differ
fundamentally from those shown in Fig. 11.5. Regardless of which combination of sequence data was used and which families were included in
the anlysis, the Chenopodiaceae and Polygonaceae still occur together,

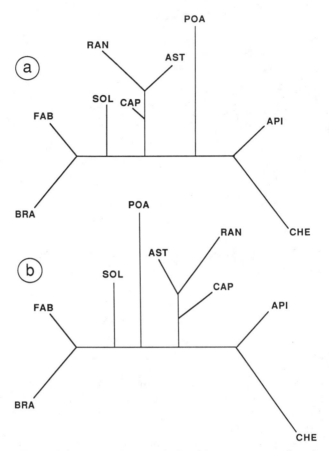

Figure 11.4   Two minimal length trees derived from sequence data for the three proteins cytochrome *c*, plastocyanin, and small subunit of rubisco. Trees *a* and *b* differ in the position of POA. RAN Ranunculaceae. Otherwise familial designations are the same as in Fig. 11.2. (Redrawn with permission from P.G. Martin and J.M. Dowd, Aust. Jo. Bot. 32:283–290. 1984, CIRS.)

the Brassicaceae and Fabaceae likewise are near each other on the tree, the Apiaceae remains an isolated element, and the Solanaceae retains its same basic position. The relative positions of the Asteraceae, Caprifoliaceae, Malvaceae, and Ranunculaceae varied depending on the families included and whether weighted or unweighted data were used. The latter procedure weighted nuclcotide sites based on the ratio of observed to expected incompatibilities for each position.

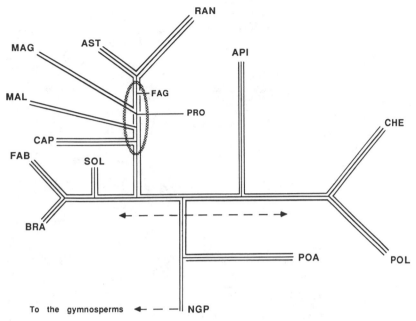

Figure 11.5   A composite tree for the familial nodes for 14 families of flowering plants. FAG Fagaceae; MAG Magnoliaceae; MAL Malvaceae; PRO Proteaceae. Other family designations are the same as in Fig. 11.2. Sequences for three proteins (cytochrome *c*, plastocyanin, and the small subunit of rubisco) were available for all families except FAG and PRO (only rubisco), MAG (rubisco and plastocyanin), and MAL (cytochrome *c* and rubisco). NGP refers to the node for *Ginkgo*. The horizontal dashed line with the arrows indicates that the POA could join anywhere along the limits of the line. The line surrounding the nodes for five families denotes that the order in which the families join in this region is uncertain. (Redrawn with permission from P.G. Martin and J.M. Dowd, Aust. Jo. Bot. **32**:301–309. 1984, CSIR.)

Martin and Dowd (1986a) reported sequence data for the small subunit from species of monocots and gymnosperms and included them with previously reported sequences to construct a tree (Fig. 11.6). For the monocots, the results agree with the general large groupings. The lines surrounding the Poaceae and the Liliaceae denote that it is not possible to resolve the branching patterns with certainty. The gymnosperms included in the studies represent each of the major groups, with *Ephedra* of the Gnetales being closest to the dicots.

Martin and Dowd (1986b) studied the small subunit of rubisco in

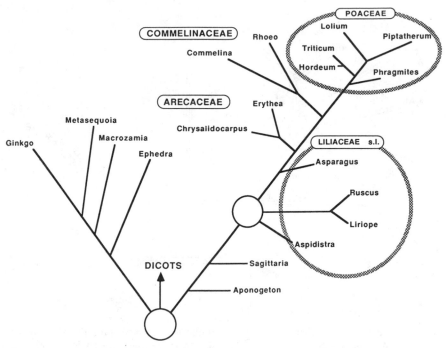

Figure 11.6 Phylogenetic tree for monocots and gymnosperms based on sequences of the small subunit of rubisco. The two circles in the tree indicate that the order of joining of internodes is not certain within the bounds of the circles. One of these circles includes the gymnosperms, dicots, and monocots. Lines surrounding Liliaceae and Poaceae denote that it is not possible to resolve branching patterns within each family. (Redrawn with permission from P.G. Martin and J.M. Dowd, Taxon, 35:469–475. 1986, International Association for Plant Taxonomy.)

several families of the Myrtales together with representatives of the Thymelaeaceae, Euphorbiales, and Malvales. The authors commented that one problem they faced was lack of a clearly defined phylogeny against which they could compare their phylogenies generated from sequence data. One result of the study was that the Thymelaeaceae grouped with families of the Myrtales. This is in agreement with the views of Cronquist (1984), as compared with Dahlgren and Thorne (1984), who considered it to be nearer the Euphorbiaceae. It should be mentioned, however, that representatives of the Euphorbiales occurred on the same side branch of the tree as the Myrtales. The Onagraceae, considered by most workers as clearly belonging in the Myrtales, come off the main branch of the tree at a slightly different place than other "good" families of the Myrtales, which is a bit unexpected.

During the past several years Martin and collaborators have produced sequence data for the small subunit of rubisco from a variety of flowering plants as well as several gymnosperms. From these data and previously published information for other molecules, phylogenetic trees have been produced. In all publications, Martin and coworkers have carefully pointed out the limited taxonomic sampling that has been done as well as the problems associated with inferring phylogenetic relationships from a single molecule. Thus, the papers are refreshing in their candor and discussions of the strengths and limitations of their data. One must ask whether additional phylogenetic insights have been gained from these most recent efforts. The answer seems to be a qualified "yes." Martin and collaborators emphasized that the small subunit data by themselves are not reliable for producing trees but rather they must be combined with sequences from other proteins for more stable phylogenies. This raises the question of the availability of sequences from other proteins (primarily plastocyanin and cytochrome c) to be used along with the newly generated rubisco data. That is, if the trees are most reliable when data from three proteins are combined, then sequences must be generated for plastocyanin and cytochrome c from the same species for which the small subunit sequences are being generated. It seems the availability of these data will be a limiting factor as additional species are surveyed for rubisco sequences.

Bremer (1988) subjected data sets of sequences used by Martin and collaborators to further cladistic analyses. In particular, Bremer (1988) was concerned with calculating the consistency indices for the most parsimonious cladograms for two data sets. The values of 0.583 and 0.689 indicate extensive homoplasy (i.e., paralleleisms and reversals) in the data. When cladograms with one to several steps more than the most parsimonious tree are used to construct strict consensus trees, there is little resolution of monophyletic groups. Put another way, one or two changes in sequences in species can change the topology of the cladograms rather drastically. Bremer (1988) concluded quite correctly that more information is needed, whether from sequencing more proteins or longer parts of the same proteins, before greater confidence can be placed in the phylogenetic hypotheses generated from protein sequences.

SUMMARY

Amino acid sequencing has not had the impact on plant systematics and evolution that its strong advocates suggested it would 20 years ago (which is not unusual for the development of a new method or approach). In the case of amino acid sequencing, problems in gathering and analysis of data were two of the biggest limitations. Also, very few laboratories have actually done sequencing with a phylogenetic motive in mind. As a result, data are available for very few proteins (cytochrome c, plastocyanin, and the small subunit of rubisco being the major ones), and the taxonomic sampling for these three molecules is rather limited. The method may be a victim of technology in the sense that emphasis is changing from amino acid sequencing to nucleotide sequencing for generating phylogenies.

It would be misleading to evaluate the impact of amino acid sequence data on plant systematics only in the context of phylogenetic hypotheses generated from the data. As discussed previously (Giannasi and Crawford, 1986), these data stimulated plant taxonomists to think more carefully and rigorously about angiosperm phylogeny. Taxonomists had to come to grips with such questions as the evaluation of different data sets with regard to rates of evolution and times of origin of different taxa. Other issues raised as a direct consequence or as a by-product of sequencing studies were comparisons of homologous features across higher taxonomic levels. Donald Boulter and his collaborators must be given credit for raising these issues. While Boulter's sequence data may not have influenced the actual construction of phylogenies to the extent originally envisioned, his contributions to plant systematics have been substantial. Lastly, basic questions raised by the cytochrome c data such as the ages and relative times of splitting of various lineages of flowering plants remain unanswered. Perhaps data from nucleic acid sequences and fossils will eventually throw additional light on these issues.

CHAPTER 12

Systematic Serology

INTRODUCTION

This chapter, on systematic serology, is the final one dealing primarily with the use of proteins in systematic studies. As should be evident by the conclusion, contributions to modern serology have come from a relatively small (and dedicated) group of researchers, perhaps because live animals must be used, or because the method is theoretically and technically complex.

The first section describes the antigen–antibody reaction that forms the basis of systematic serology, the second deals with the various methods of measuring the serological reaction and evaluating the data, the third discusses representative systematic studies employing serology, and the final section of the chapter provides an evaluation of the strengths and weaknesses of systematic serology, the contributions made by the approach, and its future prospects.

THE SEROLOGICAL REACTION

As Fairbrothers (1977) indicated, systematic serology of plants originated from or is an application of immunology; he also suggested that it is not desirable to use the terms serology and immunology interchangably, since the word immunology suggests that immunity is somehow involved. This is a valid point and should be kept in mind, although certain principles of immunology must be considered when discussing the reactions involved in systematic serology.

The immune response in mammals involves initiation of the production of proteins called immunoglobulins when certain foreign substances, normally proteins, are introduced into the animal. These foreign substances are called antigens, and the immunoglobulins are termed antibodies. The molecular structure and genetics of immunoglobulins are beyond the scope of the present discussion. Lewin (1987, Chapter 32) presented a readable and rather comprehensive discussion of these topics.

The utility of the serological reaction for the plant systematist results from the specificity of interactions between antigens and antibodies. This specificity resides on certain parts of the molecules called determinants (epitopes) when occurring on the antigens and antideterminants (paratopes) when present on the antibodies. A determinant is normally composed of 5 to 10 amino acids, with each plant protein (antigen) having perhaps 5 or 6 up to 20 or more determinants (Fairbrothers, 1977, 1983; Jensen, 1973). Evidence indicates clearly that amino acid substitutions in regions of the antigen molecules other than the determinants can affect the binding of antigens and thus change the affinity or even specificity of the reaction.

The specificity of the antigen–antibody response represents a way of probing for differences in proteins (antigens) between different taxa. When foreign proteins (i.e., antigens) are injected into an animal, its immune system responds by producing antibodies specific to those antigens. These antibodies are then reacted with antigens from other taxa. The degree of this reaction (see discussion of methods below) is thus an indication of antigen similarity between the taxa.

When antigens are reacted with the antibodies raised aginst the same antigens, the process is referred to variously as the reference, the self, or the homolgous reaction; it should give the strongest reaction and is normally taken as the standard. Reactions between antigens and antibodies raised to other antigens are called cross or heterolgous reactions.

METHODS FOR MEASURING THE REACTION AND INTERPRETING THE RESULTS

Because the data for systematic serology are represented by the reaction between antigen and antibody, appropriate methods for measuring the extent or degree of this reaction are of vital concern. Also, any factors affecting the reactions, such as the animals (usually rabbits) employed and the number and timing of the antigen injections, must be considered and cvaluatcd. This scction will discuss various aspects of methodology and data interpretation in systematic serology.

Proteins extracted from seeds, pollen, or other tissues are most frequently used as the antigens in systematic serology. The protein extract, which may be either a crude mixture of proteins or else protein(s) purified to various degrees, is injected into the animal, usually a rabbit. Individual animals respond differently to the injection of antigens; therefore inbred animals are used to decrease genetic variation among rabbits. The number of injections and their timing will alter the amounts and types of antibodies produced by the animals (Fairbrothers, 1977). For different purposes it appears desirable to have either long or short injection series for maximum benefit from the serological reaction. Fairbrothers (1977) explained that this procedure is not experimental manipulation, but rather it uses the method to fullest advantage. Long-term injection courses may increase not only the amounts of antibodies produced but also the numbers of different antibodies, thus providing additional characters for comparative purposes. On the other hand, long-term injection courses may have several undesirable attributes. Reactions may not increase with continued injections because the animals lose sensitivity and become immune-tolerant to some antigens. The continued increase in numbers of different antibodies with additional injections may in fact reduce the discriminating capacity of the antiserum. Thus, it is important to state the number and timing of injections when presenting results for a given study.

Since the antigen–antibody reaction provides the basic data for plant systematic serology, the appropriate method for measuring this reaction is an important consideration, or, more specifically, what are the advantages and disadvantages of the various methods? A number of the techniques most commonly employed in plant systematic serology will be considered.

One method used for measuring the antigen–antibody reaction involves allowing it to take place in a liquid and then measuring the turbidity resulting from the formation of insoluble antigen–antibody complexes. With this precipitin reaction, it is common practice to employ a fixed amount of antiserum (antibodies) and then react this with increasing concentrations of antigen (Fig. 12.1). If the amount of precipitate (the turbidity) is plotted against increasing concentrations of antigens, a curve results in which turbidity increases with increasing antigen concentration but eventually declines with excess antigen. The point at which there is neither excess antigen nor excess antibody is referred to as the equivalence point, but it may not correspond to the point of greatest precipita-

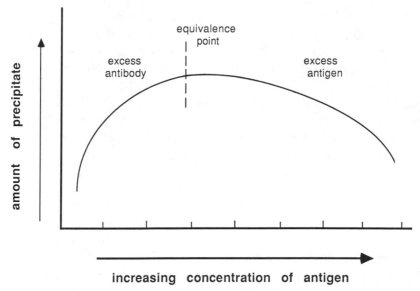

Figure 12.1 A precipitin curve (or Boyden curve) showing the amount of pre-cipitate as the antigen concentration is increased. The equivalence point is defined as the condition in which neither antigen nor antibody occurs in the supernatant fraction after the reaction. This point usually does not coincide with maximum precipitation.

tion; the latter usually occurs with slight excess of antigen (Fig. 12.1) The reason that highest precipitation occurs at optimal proportions of antigens and antibodies may be appreciated by reference to Fig. 12.2. With excess antibody there are so few antigen molecules by comparison that very little lattice formation can occur, and thus there is little precipitate formed. By contrast, an excess of antigen saturates all antibody binding sites with different antigen molecules and prevents formation of the lattice (Fig. 12.2).

In plant systematic serology, the precipitin curve produced by dilution of antigen concentration is called the Boyden curve (Fig. 12.1). The greatest reaction will occur (or *should occur*) with the self (homologous or reference) reaction, that is, when antigens are reacted with antibodies raised against themselves. This reaction is used as the standard, and the cross (or heterologous) reactions will be less, as measured by the area under the Boyden curve. The percent of the area under the curve in the heterologous reaction as compared with the self reaction is called the percent serological correspondence.

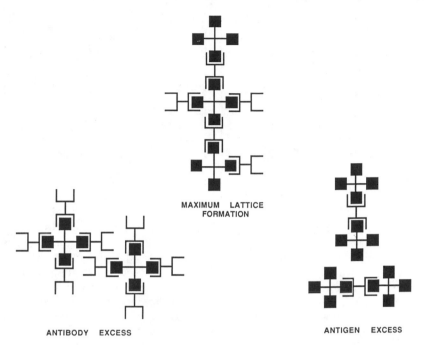

MAXIMUM LATTICE
FORMATION

ANTIBODY EXCESS

ANTIGEN EXCESS

Figure 12.2 Diagrams illustrating antigen–antibody formation with roughly equal amounts of antigen and antibody, and with an excess of each. Solid squares represent antigenic determinants and the three-sided open figures denote the corresponding antideterminants. It may be seen why maximum lattice formation (precipitate) occurs when there is not an excess of either antigen or antibody.

Measuring the precipitin reaction in liquid (the Boyden procedure) can provide useful systematic data provided relative placement series and valid comparisons can be made. This is possible only if one individual animal or a mixture of antisera against several animals are the sources of antibodies. One limitation of the technique is the rather large amounts of antiserum needed in order to obtain a complete curve. Another shortcoming with regard to its use in serotaxonomy is that the precipitin reaction does not provide an indication of how many different kinds of antigen and antibody systems are reacting because these are not distinguished in any way.

The use of gel media for the precipitin reaction provides several advantages over liquid media. Gels permit the separation of different antigen–antibody reactions, which in turn reveals the number of precipitin lines as a measure of the serological reactivity between taxa. The most commonly used technique for measuring the precipitin reaction in sys-

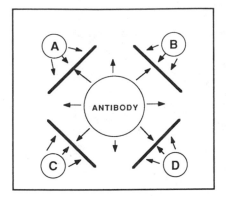 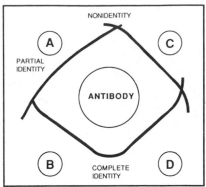

Figure 12.3 Diagrams showing the double-diffusion method in gels to measure the precipitin reaction. (Left) Antibody is placed in the central well, and antigens from four different plants (A, B, C, and D) are placed in the peripheral wells. The proteins will migrate in the gels as indicated by the arrows. (Right) Three patterns may be formed in a double-diffusion gel when an antibody is reacted with antigens. Fusion of a precipitin line indicates that the antigens of two plants share the same determinants with the antibody. For example, plants B and D share the same determinants relative to the antibody employed. Two antigens may share completely different determinants with the antibody, that is, the antigens have no determinants in common with the antibody. This reaction produces precipitin lines that cross in the gel without interacting. Plants A and C, and C and D, show this pattern. Two antigens may share certain determinants with the antibody, but one of the antigens may have additional determinants in common with the antibody, that is, one antigen has more determinants in common with the antibody than does the other. The pattern produced by this situation is a spur, and the spur points toward the well with the antigen having fewer determinants in common with the antibody. Plants A and B have certain determinants in common with the antibody, but A shares more determinants with the antibody than does B. See text for further explanation.

tematic studies of plants is double diffusion in gels, often referred to as the Ouchterlony technique. Usually antigens and antibodies are placed in wells cut in agar gel plates, with the central well containing antibodies and the peripheral wells different protein extracts (antigens) (Fig. 12.3). The antigens and antibodies will then diffuse out from the wells. The rate of diffusion depends on concentration, molecular size, and other factors. If a diffusing antigen and antibody meet in the gel and react, then they will form a precipitate, the so-called precipitin line or arc (Fig. 12.3). In systematic serology, antibodies raised against antigens from a particular species are usually placed in the central well, and antigens from other species are put in the different peripheral wells (Fig. 12.3). With regard to two given systems and one antibody system, three possible patterns may form in the gel where the precipitin lines meet. If two proteins contain identical antigenic determinants, then a reaction of identity occurs, and one observes two fused lines (Fig. 12.3). This line develops because all antigen and antibody molecules can accumulate in a common precipitin line; thus after they meet they remain fused. At the other extreme, the antibody mixture may share some determinants with two proteins, but the two antigens have no determinants in common. This results in a reaction of nonidentity; the two lines form independently and thus cross in the gel without any interaction (Fig. 12.3). A third pattern is produced when two antigens share some determinants that react equally with some antibodies but other determinants are specific to one of the antigens. The pattern produced has both a connected curve, reflecting shared determinants, and also a line or spur beyond the curve that results from determinants not shared by one of the two antigens (Fig. 12.3) The spur always points toward the well with the antigen that lacks one or more determinants present in the antibody system. The drawings in Fig. 12.3 show only single precipitin lines in the gel, but when protein mixtures are used, as is often the case in plant systematics, several to many lines will be visible (Fig. 12.4). Dyes can be used to enhance the visibility of the lines for observation and photography.

The numbers and patterns of precipitin lines in the gels represent data for systematic purposes. One source of data is the number of lines, which are sometimes referred to as immunoprecipitating systems or IPS (see Piechura and Fairbrothers, 1983, as an example). Comparisons of the number of IPS in the cross or heterologous reactions as compared with the self or homologous reaction provide useful data. Also, whether antigens in adjacent wells form smooth curves, crossed lines (double spurs), or single spurs is an important source of information. The density of a particular reaction (i.e., density of the precipitin line) is also a factor to consider. Dense, sharp lines indicate that both antigens and antibodies are plentiful and have a high affinity for each other; diffuse lines indicate lower affinity of the antibodies for the antigenic determinant sites.

Immunoelectrophoresis (IEP) is commonly employed in serotaxonomy; it offers the advantage of combining the separation of molecules electrophoretically with the recognition of these molecules by the precipitin reaction (Fig. 12.5). IEP also allows the resolution of complex mixtures of antigens. A mixture of proteins serving as the antigens is placed

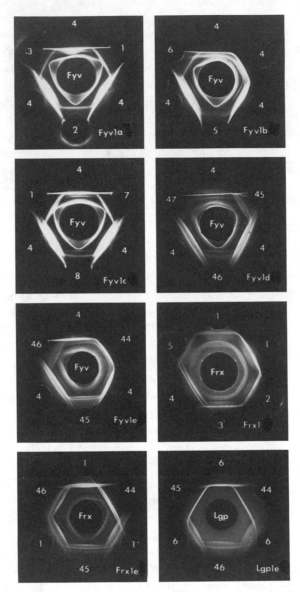

Figure 12.4 Photographs illustrating the precipitin reaction with double diffusion in gels. Gels 1–5 have the same antiserum, as do gels 6 and 7, and 8. The peripheral wells contain antigens from different species. Patterns of complete fusion of precipitin lines, spur formation, and crossing of lines may be seen. (With permission from J.E. Piechura and D.E. Fairbrothers, Am. Jo. Bot. **70**:780–789. 1983, Botanical Society of America. Photograph courtesy of J.E. Piechura and D.E. Fairbrothers.)

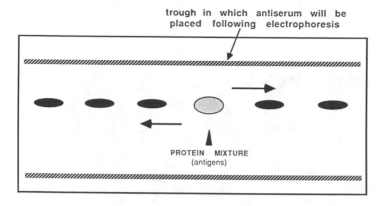

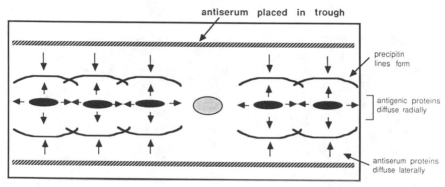

Figure 12.5 Diagrams of immunoelectrophoresis for visualizing the precipitin reaction. (Top) Proteins (antigens) are placed in the gel and then separated electrophoretically. (Bottom) Antiserum is placed in the troughs. The antigenic proteins diffuse radially, while the antiserum proteins diffuse laterally and the precipitin lines form as arcs.

into a well cut in a gel, and they are subjected to electrophoresis. Immediately following electrophoresis, antiserum is placed in the troughs (Fig. 12.5). The antigenic molecules diffuse radially while the antibodies diffuse laterally; where reactive molecules meet in the gel precipitin, lines or arcs are formed (Figs. 12.5, 12.6). The primary data generated by immunoelectrophoresis are the numbers of lines (arcs) formed between the antiserum and particular antigens. If lines of similar mobility, width, and so on are formed when two antigens are reacted with the same antibody system, then it is usually assumed that the antigens are the same. Much better evidence for identity of the precipitin lines would

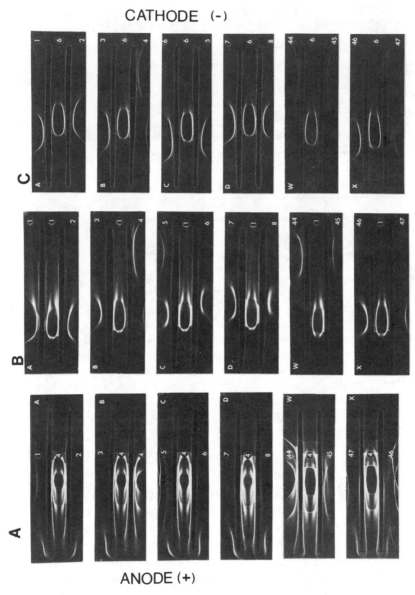

CATHODE (−)

ANODE (+)

Figure 12.6 Patterns of precipitin lines produced by immunoelectrophoresis. In each column of gels, A, B, and C proteins from the same species are used as the antigens and are separated electrophoretically. In a gel, antiserum from a different species is placed in the two troughs. The numbers denote the sources of the antigens and antibodies. (With permission from J.E. Piechura and D.E. Fairbrothers, Am. Jo. Bot. **70**:780–789. 1983, Botanical Society of America. Photograph courtesy of J.E. Piechura and D.E. Fairbrothers.)

come by allowing the lines to form beside each other and see whether the lines join or cross, as was discussed for double diffusion (Fairbrothers, 1967). If the antigens of two taxa produce the same number of lines when reacted to the same antiserum, this does not necessarily mean they share the same antigens with each other, but rather that they each share the same number with the species providing the antiserum (Fairbrothers, 1967).

Radial immunodiffusion (RI) is a double-diffusion method that differs from the Ouchterlony technique by having the antiserum incorporated into the gel prior to well formation. Wells are filled with protein extracts of a known standard concentration, and then circular precipitation patterns are observed and the diameters of the outer precipitin rings measured. Each antigen diffuses into the gel and in the process precipitates with antibody molecules in the gel. The precipitate forms a ring in the gel, but as antigens continue to diffuse from the well, they accumulate until antigen excess causes the precipitate to dissolve; it is then redeposited farther out from the well. Thus the ring migrates out from the well, and the area of the circle within the ring or halo is proportional to the original concentration of antigen in the well. Each measurement is compared with the reference extract. Because mixed antiserum (antiserum produced to the total antigen content of the extract) is used, the technique can be employed *only* as a general screening for detecting serological similarities among a large number of taxa. This qualitative use of the technique for systematic purposes contrasts with the original employment of radial immunodiffusion for quantifying specific antigens. Thus the method represents a valuable first screening or sorting technique for numerous taxa (Petersen and Fairbrothers, 1983).

Rocket immunoelectrophoresis (RIE) derives its name from the rocket-shaped precipitin bands that appear after electrophoresis of proteins in a gel containing an antiserum. Although originally used for the quantification of a single protein, it has proved valuable for screening proteins (antigens) from different taxa. RIE provides an approximation of the degree of serological correspondence in comparative studies of species, genera, and families. The height of the "rockets" that develop in cross reactions compared with the reference reaction provides the measure of serological correspondence. The method has proved to be more sensitive (in ability to detect reactions) than several other techniques for measuring the precipitin reaction in systematic studies (Lee, 1977).

Microcomplement fixation is a method that may be employed to measure the antigen–antibody reaction. It will be considered only briefly here; the reader is referred to Champion et al. (1974) for details of the procedure and to Giannasi and Crawford (1986) for a general discussion. A basic advantage of the method is that only small quantities of protein are used. Proteins referred to as complement are bound when the reaction takes place; thus the amount of complement "used" during antigen–antibody reactions is a measure of the extent of the reaction. A known amount of complement is added, and then the amount bound during the reaction is ascertained by determining the amount of free (unbound) complement available to lyse red blood cells following the precipitin reaction.

The technique of radioimmunoassay (RIA) has been little used in plant systematic serology, but it appears to have potential for such studies. The basic principle of the method is simple and takes advantage of the fact that radioactively labelled antigens or antibodies may be used to measure very precisely the amount of antibody (or antigen) reacting with a given antigen (or antibody). One method will be described very briefly as an example; the reader is referred to Tsu and Herzenberg (1980) for more detailed descriptions of methodology. An antigen solution of a given species is coated on a solid surface. Next, unlabelled antisera from another species of systematic interest (or the same species in the reference reaction) are added and allowed to react with the antigen. Next, a labelled antibody is added, and, following an appropriate incubation period, the amount of bound labelled antibody is measured. The amount of labelled antibody that is bound is proportional to the amount of unlabelled antibody from the species of interest that was bound. The major advantage of RIA for systematic studies is its sensitivity, meaning that smaller amounts of protein are required for the reactions and that precise quantitative measures of the antigen–antibody reaction can be made. The major limitations are the expense and the problems associated with handling radioactive materials.

A method similar to RIA for measuring the antigen–antibody reaction is enzyme immunoassay, sometimes referred to as enzyme-linked immunosorbent assay and abbreviated ELISA. An enzyme (instead of a radioactive label) is linked to an antibody. Enzymes that produce a colored product when they convert the substrate are generally used, commonly alkaline phosphatase and horseradish peroxidase. Antibody with the enzyme is bound to antigen and substrate added. The amount of substrate conversion is a measure of how much antibody is bound to antigen. Two advantages of ELISA over RIA are lower cost and greater safety.

An important methodological aspect of systematic serology is pre-saturation or absorption. Rather extensive discussions of this topic were presented by Lester et al. (1983) and Lester (1984), and only a general treatment based on these papers will be given. The procedure involves taking antiserum raised against antigens of one species (A) and reacting it with antigens from a second species (B). This process will precipitate (and thus remove) all precipitating systems that species A and B have in common. The absorbed antiserum can then be reacted with antigens from other species. The procedure is useful for a variety of reasons. Reaction of the absorbed antiserum with other antigens reduces the number of precipitin lines and eases interpretation of gel patterns. Absorption re-moves antibodies that will react with determinants common to all antigen systems included in the study, and thus it causes no loss in resolving power (Lester, 1984). Fairbrothers (1977) discussed the value of pre-saturation with regard to understanding the formerly designated "antisys-tematic" or "asystematic" reactions. These "unexpected" cross reactions occur among distantly related species, and it is now known that they result from determinants that are widely distributed in the plant kingdom. Rubisco (see Chapter 4) is an example of a universally distributed protein in flowering plants in which reacting determinants occur in very distantly related plants (Fairbrothers, 1977).

INTERPRETATION AND USE OF SEROLOGICAL DATA FOR SYSTEMATIC PURPOSES

Once the antigen–antibody reaction has occurred and been measured appropriately, the data must be interpreted and applied to systematic problems. There is no simplistic answer for the best way to perform these tasks. The method by which the reaction is measured, that is, turbidity, double diffusion, and so on, will determine the type of data generated. An exhaustive treatment of this topic is beyond the scope of the present discussion; more detailed considerations of aspects of data interpretation and application are provided by Cristofolini (1980), Cristofolini and Peri (1983), Fairbrothers and Petersen (1983), Friday (1980), Lester (1979), Lester et al. (1983), and Petersen and Fairbrothers (1983). The following discussion will consider general points and problems in the interpretation of serological data, using the foregoing references (and others) as a basis.

When the reaction is carried out in a liquid (the Boyden procedure), areas under the curves generated by plotting amount of precipitate versus antigen concentration serve as the basic source of data (Fig. 12.1). With this method, an index of serological correspondence (protein similarity) is generated by expressing the area under the curve generated by a cross reaction divided by that produced from the reference reaction (see Hillebrand and Fairbrothers, 1969; Clarkson and Fairbrothers, 1970; Pickering and Fairbrothers, 1970; Villamil and Fairbrothers, 1974, as examples). Cristofolini (1980) suggested that a simple method for computing this index is justifiable given the several unknown factors affecting the amount of turbidity produced. This formula is the mean of the values of the cross reaction A × B divided by self reaction A × A and cross reaction B × A divided by self reaction B × B, that is,

$$\frac{1}{2}\left(\frac{(A \times B)}{(A \times A)} + \frac{(B \times A)}{(B \times B)}\right)$$

The Boyden procedure has been used less frequently in recent (after 1979) serological research. The main reason for its diminished popularity is the cost in time and the relatively large amounts of antisera required to carry out the procedure. With the Boyden procedure, a single similarity value is obtained, that is, the index of serological correspondence results from the one measure, which may then be used as data. This means that the results of all reacting determinants from complex mixtures of antigens and antibodies are reduced to this one value, which is good in the sense that the value includes all reactions possible. Assume that antigens from protein extracts of two different species (A and B) are reacted with antiserum raised against antigens from species C to give similar Boyden curves. The similarity of A and B to C could be due to different shared determinants. Thus, nothing can be said about the similarity of A and B based upon their similarity to C. It should also be pointed out that while the Boyden procedure has usually been carried out with protein mixtures when employed for systematic studies, it has also been done with purified or partially purified protein, which does away with the problem of not knowing which combination of proteins is giving the reaction.

The use of double diffusion or immunoelectrophoresis has an advantage over the Boyden procedure if individual immunoprecipitating systems (i.e., IPS) are to be visualized (Figs. 12.3, 12.4, 12.5). When crude protein extracts are used, then one can determine that a minimum number of precipitin lines form when antigens of different species are reacted with the same antiserum. There are no assurances that *all* systems will be resolved by these methods. It is possible by comparing the intersection of lines from adjacent wells in the gels to ascertain whether the antigens share no determinants or some determinants, or are idential (i.e., share all determinants) relative to antideterminants present in the antiserum. When mixtures of antigens and antibodies are used in double diffusion, the number of IPS between the antiserum and antigens of a given species represents a measure of similarity. As mentioned above, in a double-diffusion gel, one can ascertain identity, partial identity, or nonidentity of the IPS of antigens (relative to the antiserum) only when they are run side-by-side in the gel. By contrast, when comparing antigens from nonadjacent wells, determination of whether the precipitin lines fuse, form spurs, or cross is not possible. A problem with side-by-side comparisons is the time and effort required if large numbers are being compared; if more than one antiserum is employed, the amount of time and effort increases a great deal.

Lacking information from side-by-side comparisons, data are gathered from double-diffusion gels by such methods as comparing the number of precipitating bands showing identity between species with the total number of bands. Also, sometimes partial identity (formation of spurs) is given an intermediate value. In all cases, the assumption of homologous proteins is based (in all but adjacent wells) on less than optimal criteria.

Presaturation (absorption) of antisera was briefly presented earlier as a method of simplifying the interpretation of gels. It may also be a very useful technique for measuring serological similarity between species when combined with other methods, primarily double diffusion. The following discussion is based largely on Lester (1979) and Lester et al. (1983). When antiserum from plants of one species (A) is reacted with antigens from another species (B), the only antibodies remaining in the antiserum are those directed to antigenic determinant sites present in A but not in B. This and other presaturated antisera from species A can then be reacted with a series of antigens from other species. One measure of the similarity of A and B is the degree to which the antigens from species B remove from antiserum of species A antibodies capable of reacting with antigens of additional test species C, D, E, and so on (Lester et al., 1983). Each so-called negative reaction, for example with antigens from species C, indicates that all antibodies capable of reacting with C have been removed when antiserum A was presaturated with antigens from B, so in reality negative reactions are a measure of the similarity of A and B. Lester et al. (1983) defined the absorption similarity coefficient based on negative reactions ($S_{\text{no.neg.}}$) as the number of negative reactions produced by a presaturated antibody system (that is, antiserum A presaturated with antigen system B) when tested with a range of antigens, divided by the number of negative reactions produced when the reference antibody system, presaturated by the reference antigen system (that is, antiserum A presaturated by antigen A) is reacted to the same systems. This means that if A and B are very similar, then the number of negative reactions will be very similar when antiserum A presaturated by antigen system B and antiserum A presaturated by antigen A are reacted with antigens from other species, or, as Lester (1979, p. 295) put it, "Thus the affinity of B to A is indicated by the degree to which antigen system B absorbs antibody system A and removes from it antibodies capable of reacting with the antigens of other taxa...."

Lester (1979) and Lester et al. (1983) indicated that one can use the number of positive as well as negative reactions as data from presaturation studies and suggested two ways to compute these scores. If, after antiserum A is presaturated by antigen system B, it is capable of reacting with antigen C, then it means that some antibodies have been left in A that can react with C despite the presaturation with B. Antiserum A could be presaturated with antigens D, E, F, and so on and then tested with antigen C to see if it will still react with C. Lester (1979) employed the absorption similarity coefficient based on the number of positive reactions (designated by ($S_{\text{no.pos.}}$), and defined this coefficient as the number of positive reactions produced by a test antigen system with an antibody system presaturated with a variety of antigen systems, divided by the number of positive reactions produced by the reference antigen system with the same presaturated antibody systems.

The absorption similarity coefficient, which is another way of measuring similarity, is computed from the total amount of precipitate ($S_{\text{ppt.pos.}}$) and is defined as the total precipitate score for all the reactions of the test antigen system with a variety of presaturated antibody systems divided by the total precipitate score of the reactions produced by the reference antigen system with the same presaturated antibody systems. For each reaction, the amount of precipitate is judged somewhat subjectively by the precipitin formed, and differences are designated in some manner. Thus the total precipitate score is derived by adding up these reactions.

Both of the similarity coefficients (i.e., $S_{no.pos.}$ and $S_{ppt.pos.}$) are based on essentially the same properties, and Lester (1979) suggested that their mean can be designated as the absorption similarity coefficient based on positive reactions ($S_{abs.pos.}$), that is, $S_{abs.pos.} = \frac{1}{2} (S_{no.pos.} + S_{ppt.pos.})$.

Lester (1979) found that with certain species of *Solanum*, the negative and positive reactions did not give uniform (or symmetrical) results, but they did give a similar ranking order for each of the antibody systems. Lester (1979) averaged the similarity coefficient for negative reactions and the coefficient for positive reactions for each of the antisera to give what was designated the mean absorption similarity coefficient, or $S_{abs.mean}$. These values produced rather symmetrical ranking orders with regard to different antigen systems with a given antiserum and various antisera with given antigen systems. A coefficient of similarity that is simpler and gives almost the same results as $S_{abs.mean}$ is what Lester et al. (1983) referred to as the mean absorption similarity coefficient based on numbers of negative and positive reactions, denoted as $S_{abs.no.mean}$. This coefficient is computed as the number of negative reactions caused by a presaturating antigen system plus the number of positive reactions it produces when it is employed as the test antigen system against the complete range of presaturated antisera, divided by the corresponding total of negative and positive reactions of the reference antigen system. This measure does not require the subjective assessment of the amount of precipitate.

It is apparent from these discussions that there have been as many ways suggested for interpreting serological data as there have been methods for measuring the serological reaction. Most serotaxonomic studies include double diffusion as one of the techniques (if not the only one) for measuring the reaction. As a result, most discussions of interpretation and scoring of serological data center on the double-diffusion technique. The methods of analysis developed by Lester and collaborators have gained wide usage in plant systematic serology.

CASE STUDIES OF THE USE OF SYSTEMATIC SEROLOGY

Discussions of systematic investigations employing serology must be selective because of space limitations; several criteria served as general guidelines for choosing particular studies. Certain investigations from more than a decade ago were chosen for discussion partly for their inherent systematic value and also because the methods may be contrasted with procedures used in more recent studies. Investigations focusing on problems at different taxonomic levels were considered, as were those demonstrating particular methods of data gathering or analysis. An attempt has been made also to cite papers not included for detailed discussion, although no concerted efforts were made to be exhaustive.

Hammond (1955), in an early serological study of the "modern" era, examined systematic relationships in the Ranunculaceae. The precipitin reaction was measured using turbidity, that is, by the generation of Boyden curves, and then comparing areas under the curves with those in the reference reactions to compute serological correspondence. Results were given for antisera from several species reacted with antigens of about 20 taxa.

Hammond (1955) examined primarily the problem of subfamilial relationships within the Ranunculaceae, that is, he was concerned with comparing serological results with relationships inferred from morphological information (primarily achenes versus follicles) and chromosome size. He found that serological correspondence did not support the recognition by early workers of two large groups within the family based on fruit type; this conclusion agrees with data gathered by later workers conducting more broadly based morphological studies. Thus the serological data were, in a broad sense, concordant with the thinking prevalent at the time of Hammond's (1955) study. Serology, in general, supported the hypothesis that the genera of Ranunculaceae having small chromosomes and large chromosomes each form two closely related groups within the family. Results for the genus *Hydrastis* indicated that it should be retained in the Ranunculaceae rather than transferred to the Berberidaceae because it showed as much serological correspondence to several other genera in the former family as they showed to each other. However, *Hydrastis* was not compared serologically with representatives of the Berberidaceae. Hammond (1955) detected no precipitate when antigens of *Paeonia* were reacted with antisera from representatives of Ranunculaceae. The author cautiously stated that, while the evidence is largely negative, the data do suggest a distant relationship between *Paeonia* and other Ranunculaceae. The genus is now generally segregated into its own family because evidence from a variety of features suggests it is distinct from Ranunculaceae (Cronquist, 1980).

The work by Hammond (1955) represents a good example of how serology can provide useful systematic data, even with problems in methods, sample sizes, and so on. Hammond (1955) presented clearly the questions being addressed in the study and stated precisely how effectively his data answered these questions. He was quick to point out deficiencies in methods, sample sizes, or other aspects of the investigation.

Over a decade later, Jensen (1968), in a classic serological investigation, carried out a more extensive study of Ranunculaceae employing methods in addition to the Boyden procedure for measuring the precipitin reaction. Jensen's results considerably extended those of Hammond (1955) and in general supported them. For example, as indicated above, Hammond (1955) suggested that *Hydrastis* should be kept in Ranunculaceae rather than placed in Berberidaceae, although it was not compared with the latter family. Jensen (1968) showed that *Hydrastis* is serologically more similar to the Ranunculaceae than to the Berberidaceae. The fact that two studies conducted over a decade apart in different laboratories on different continents produced largely similar systematic results provides confidence in the methods.

Gell et al. (1960) conducted a serological investigation of 35 species of potato (*Solanum*). Antisera were prepared against protein extracts from two species. The methods for measuring the precipitin reaction were double diffusion and immunoelectrophoresis, with presaturation also employed to simplify interpretation of patterns in gels. The data were presented in a qualitative manner and involved comparing numbers of IPS between species and whether they join in the gel. Fewer IPS or lack of them altogether were interpreted as indicating lower serological similarity between taxa.

The results obtained when antigens of other species were reacted with *S. tuberosum* antisera corresponded very well with systematic relationships inferred from morphology and crossing data. Plants in the eight taxonomic series occurring in Mexico assort into the same groups based on number of IPS. When the antiserum from *S. ehrenbergii*, first presaturated with antigens of *S. tuberosum*, was reacted with antigens of other species the results were less clear-cut relative to accepted taxonomic groupings. However, results with this antiserum were useful because in certain cases different series that were indistinguishable with antiserum from *S. tuberosum* were now separable.

The paper by Gell et al. (1960) provided both detailed discussions of methods and the rationale for using particular methods, as well as a broad and balanced perspective of the value of serological data. Their statement that serology is most useful when it can be compared with proposed relationships based on other data is particularly pleasing. If the serological information agrees, then so much the better; however, if it disagrees then the gathering of additional data is desirable. Gell et al. (1960) correctly stated that one should not always expect concordance between morphological features and serological similarity because morphological divergence could occur without detectable serological changes and vice versa.

Hillebrand and Fairbrothers (1969) employed serological data to assess relationships within the large genus (250 recognized species) *Viburnum* of the Caprifoliaceae. Twenty-three species representing eight sections of the genus were included, and antisera were raised to more than 10 species. The methods for measuring the antigen–antibody reaction were the Boyden procedure and double diffusion. The major source of data for systematic comparisons was serological correspondence from turbidity measurements, that is, the Boyden procedure. Hillebrand and Fair-brothers (1969) discussed their results from the perspective of the concepts of relationships among sections and what were thought to be the most primitive and advanced sections based on a variety of data gathered by other workers. When antiserum of *V. odoratissimum*, a member of the putatively most primitive section, was reacted with antigens of species from the same and other sections, there was an overall pattern of decreasing serological correspondence when going from the more primitive to the more advanced sections, although exceptions do exist (Fig. 12.7). Similar results were obtained when antiserum of a species from the most advanced section was reacted with antigens from other sections, with general decreasing correspondence to the presumably more primitive sections (Fig. 12.7). It will be noted that certain sections appear "out of order" with both antisera.

The study by Hillebrand and Fairbrothers (1969) attempted to assess relationships within a single large genus using serological data and then to compare results with conclusions drawn from other information. The results are significant in that they correspond with other data regarding sectional relationships in *Viburnum*. Two limitations of the study were the small number of species included, (slightly more than 10%) and the fact that systematic conclusions were largely based on one method of measuring the precipitin reaction.

Villamil and Fairbrothers (1974) used the Boyden procedure and double diffusion to ascertain whether a variable group of plants in the genus *Alnus* could be distinguished. These particular plants had been grouped

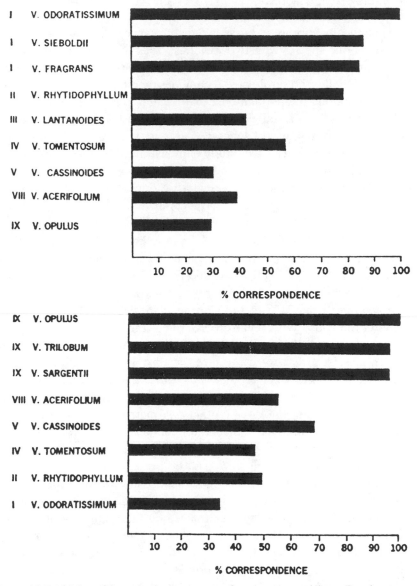

Figure 12.7 Values of serological correspondence measured from Boyden curves for different species in the genus *Viburnum*. (Top) Antiserum raised to *V. odoratissimum* reacted against antigens from other species. Numbers to left of species designate sections of the genus, with I viewed as most primitive, and increasing numbers through IX representing more advanced taxa. (Bottom) Antiserum produced against *V. opulus* reacted with antigens from other species. (With permission from G.R. Hillebrand and D.E. Fairbrothers, Bull. Torrey Bot. Club **96**:556–567. 1969, Torrey Botanical Club.)

in taxonomically various ways: as two distinct species (*A. serrulata* and *A. rugosa*), as one species with two varieties, or as only one species with no infraspecific categories recognized. Eight populations were included in the study. Briefly, the serological techniques employed did not distinguish two groups among the populations, that is, all populations were essentially the same. Villamil and Fairbrothers (1974) used their data together with other information to suggest that one species with varieties or subspecies be recognized. This study demonstrated the feasibility of using serology at lower taxonomic levels, although the data were in a sense negative because they did not distinguish groups.

Cristofolini and collaborators have conducted a wide range of serological studies in the Leguminosae (Cristofolini, 1981; Cristofolini and Chiapella, 1977, 1984; Chiapella and Cristofollini, 1980; Cristofollini and Peri, 1983, 1984 as examples). Cristofolini and Chiapella (1977) examined the tribe Genisteae and addressed questions at several taxonomic levels. Taxonomic sampling included genera from all or most of the tribes, with representatives from 31% of the European species and 22% of the total species in the tribe. With regard to sampling, the authors also cited their previous work indicating very little variation among populations of the same species. Antisera were produced against 15 species, and the serological reaction was visualized by double diffusion. Observations of spurs were used as the data indicating characters (determinants) shared by the IPS of certain species but not by others. A matrix was constructed from these data, and a matrix of correlation coefficients was computed. The data were subjected to clustering and principal component analysis. Several salient results of the study will be presented.

The serological data do not support subdivision of the tribe into subtribes because no distinct groupings occur among genera. The two genera *Genista* and *Cytisus* represent extremes, but there is essentially continuous variation between them. The genus *Chamaespartium* appears to be artificial, and serological data supported strongly the transfer of the species *C. sagittale* to *Genista*. Section *Asterospartum* of *Genista* should be treated as distinct at the generic level as the genus *Cytisanthus*; the serological homogeneity of the several species in this group is paralleled by a number of shared morphological features. The serological data of Cristofolini and Chiapella (1977) sometimes agreed with existing taxonomic concepts in the Genisteae but in other cases suggested additional study was needed to elucidate relationships.

Petersen and Fairbrothers (1979, 1983, 1985) have investigated serologically various taxa previously included in the Amentiferae. Among other methods, Petersen and Fairbrothers (1979) used turbidity analysis (Boyden procedure) and double diffusion before and after presaturation of antisera. With double diffusion, the reduction in the number of IPS after presaturation was taken as a measure of serological similarity. Results from the Boyden procedure and double diffusion were in general agreement. Salient results from Petersen and Fairbrothers (1979) were that the Myricaceae, Juglandaceae, and Fagaceae showed higher serological similarity to each other than any did to the Anacardiaceae. The conclusion was that the former three families should be retained in the subclass Hamamelidae, whereas the Anacardiaceae should not be included in this subclass.

Petersen and Fairbrothers (1983) examined the taxonomic affinities of *Amphipterygium glaucum* and *Leitneria floridana*, two taxa previously grouped in the Amentiferae, by comparing them serologically with the orders to which they have been assigned since the "breakup" of the Amentiferae. The serological methods included, among others, double diffusion with presaturated antiserum and radial immunodiffusion. The double-diffusion studies included calculating similarity coefficients for positive reactions, intensity of positive reactions, and negative reactions (Lester, 1979; see earlier discussion in this chapter). Finally, a coefficient of correlation giving the degree of correspondence of the positive and negative similarity values was computed. The results suggested that *Leitneria* is best placed in the Rutales because of its high serological cross reactivity with taxa in that order. There was no serological evidence that *Leitneria* should be placed in the Fagales, Hamamelidales, Juglandales, or Myricales. *Amphipterygium* showed the strongest serological correspondence to the Anacardiaceae, and Petersen and Fairbrothers (1983) suggested that it could be placed in that family rather than in the Julianiaceae. Young (1976), on the basis of comparative flavonoid chemistry, suggested that the Julianiaceae (containing only *Amphipterygium* and another genus) and Anacardiaceae are very closely related, and he treated the former family as a distinct subtribe within the latter family.

Petersen and Fairbrothers (1985) examined 22 species in 14 families previously placed in the Amentiferae. The method of double diffusion combined with presaturation was employed, and the data were gathered and analyzed as outlined in a previous paper (Petersen and Fairbrothers, 1983). The object of the research was to ascertain how the serological groupings of the taxa correspond to their placement in various taxonomic

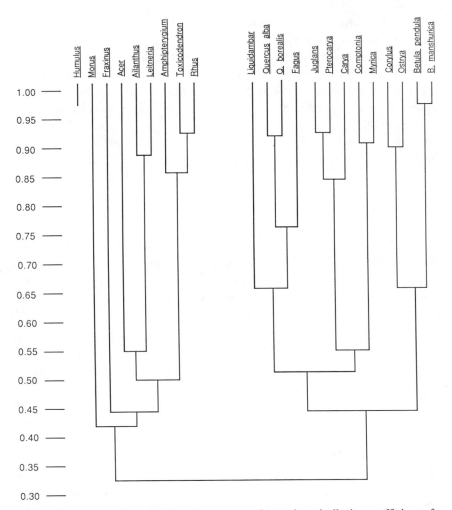

Figure 12.8 A dendrogram based on mean absorption similarity coefficients for 22 species. The serological relationships of *Humulus, Morus*, and *Fraxinus* are not clear. See text for further details and discussion. (Redrawn with permission from F.P. Petersen and D.E. Fairbrothers, Bull. Torrey Bot. Club **112**:43–52. 1985, Torrey Botanical Club.)

schemes. The results showed that most of the taxa were clustered into two groups based on serological similarity as measured by the mean absorption similarity coefficients (Fig. 12.8). The two groups in Fig. 12.8 suggest that taxa may be assigned to the two superorders Rutiflorae (cluster on left) and Rosiflorae (cluster on right). Results for the three genera *Fraxinus, Humulus,* and *Morus* were not clear from the serological data. It seems justified to conclude that the serological studies of Petersen and Fairbrothers have been most useful in placing various elements of the Amentiferae.

Jensen and collaborators (Buttner and Jensen, 1981; Jensen and Penner, 1980; Jensen and Buttner, 1981; Jensen and Greven, 1984) have carried out serological investigations utilizing partially purified storage proteins. In addition, Jensen and Grumpe (1983) and Jensen (1984) discussed various aspects of the purification and homologies of storage proteins from flowering plants. These workers suggested that the seed storage proteins of dicots consist largely of two proteins. While the homologies of the proteins have not been established among different taxa, it appears that they are very similar to (if not the same as) legumin and vicilin. In other words, the seeds of dicots contain vicilin and legumin, or proteins very similar to them. The proteins have been given different names depending on the taxa from which they were isolated, that is, nigellin for legumin-like proteins, and aquilegilin and tubliflorin for vicilin-like proteins (Jensen and Grumpe, 1983). These observations are important for serological studies because they suggest that homologous proteins may be compared across the dicots.

Jensen and Greven (1984) purified the legumin-like protein from *Magnolia tripetala* and used it to produce antisera. Antigens from the seeds of 206 species from 92 families of dicotyledonous plants were reacted with the antisera. Double immunodiffusion was used to detect the serological reactions, and several different types of reactions were recorded relative to the *Magnolia* antisera.

The results of Jensen and Greven (1984) have implications over a broad taxonomic range. Serological similarities among members of the Ranunculaceae were examined and the results compared with those obtained earlier (Jensen, 1968) using whole seed protein extracts (Figs. 12.9, 12.10). A total of six protenes (defined as number of protein reactants resolved with a method; it will correspond to a determinant if the method is capable of distinguishing all determinants) was resolved when members of the Ranunculaceae were reacted with the antisera from *Magnolia* (Fig. 12.9). Serological similarities among members of the same tribes in the family were evident. For example, Fig. 12.9 shows that pairs of genera such as *Coptis* and *Xanthorriza*, and *Thalictrum* and *Aquilegia* that have been placed in the same tribes share many of the same deter-

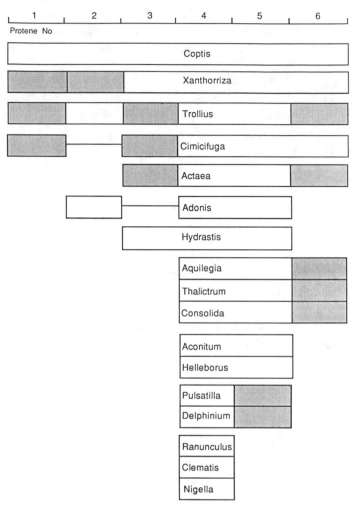

Figure 12.9 Diagram of serological similarities between legumin-like proteins from 18 taxa belonging to the Ranunculaceae when reacted with antiserum raised against the legumin-like protein of *Magnolia tripetala*. For definition of protene, see discussion in text. Six protenes were resolved with the methods employed, and the number of protenes each genus has in common with *M. tripetala* is indicated. *Coptis*, for example, shares all six, whereas *Nigella* has only one in common with *Magnolia*. Shaded areas indicate less than definitive data for a protene. (Redrawn with permission from U. Jensen and B. Greven, Taxon **33**:563 577. 1984, International Association of Plant Taxonomy.)

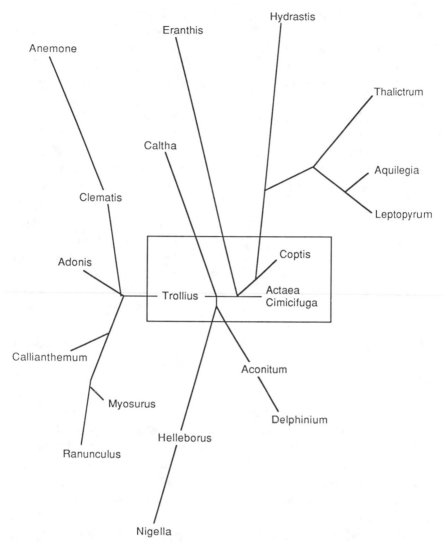

Figure 12.10 Diagram, redrawn from Jensen (1968), of serological similarities among genera of the Ranunculaceae based on total soluble seed proteins. Those genera enclosed within the rectangle are also the ones showing the most protenes in common with *Magnolia tripetala* when the legumin-like proteins are compared; compare with Fig. 12.9. (Redrawn with permission from U. Jensen and B. Greven, Taxon **33**:563–577. 1984, International Association for Plant Taxonomy.)

minants. Another interesting observation made by Jensen and Greven (1984) was that those genera sharing greatest serological similarity with the *Magnolia* antisera were also the ones Jensen (1968) viewed as occupying a "central position" serologically in the family (Fig. 12.10); these genera may represent primitive or unspecialized elements in the family.

When reacting *Magnolia* antiserum with over 200 species of dicots (and several monocots), Jensen and Greven (1984) observed only very faint reactions with 160 species. Thus, final results were based on tests with only about 55 species. While inferences of relationships at these higher taxonomic levels were based on only one seed storage protein, there are probably 20 to 30 serological determinants on this protein. These determinants provide separate characters for discriminating among taxa, meaning that single character taxonomy is not being employed. The similarities of members of other subclasses (or superorders) to the subclass Magnoliidae were ascertained. A high serological correspondence between the Magnoliidae and Hamamelidae was observed, although there was variation within the latter subclass. These data are interpreted as resulting from the plants representing primitive groups, with both having a common ancestor.

Interestingly, a rather high serological similarity was noted between the Carniflorae and Magnoliidae. As a whole, taxa in the former group exhibited fairly consistent values, but some variation was detected. While the serological results are interesting, present information does not allow for an explanation of why such high similarity was detected between what are viewed by many taxonomists as quite divergent groups of plants.

Price et al. (1987) used radioimmunoassay to study relationships among 9 of the 10 genera of the Pinaceae. Seed proteins from two genera were used to produce antisera. An immunological distance (ID) was calculated in which this value is equal to -100 log immunological similarity (IS). The IS in turn is equal to the sum of the amount of labelled antibody bound during reciprocal heterologous reactions divided by the sum of the homologous reactions. Several aspects of the results of Price et al. (1987), aside from their taxonomic implications, are of interest in the broader context of systematic serology. They found that reciprocity of the cross reactions (in which antigen and antibody are reversed) sometimes were not as good as one might hope to achieve. If, however, the reciprocal values were averaged, there was compensation for the disparate values obtained from individual reactions. Price et al. (1987) also found that individual rabbits had little effect on results because immunizing different

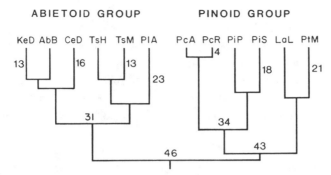

Figure 12.11   A tree of immunological distances (see text for how these were calculated) among taxa of Pinaceae using UPGMA clustering. Numbers refer to immunological distances among the taxa. Abbreviations for the taxa are: AbB, *Abies balsamea*; CeD, *Cedrus deodara*; KeD, *Keteleeria davidiana*; LaL, *Larix laricina*; PcA, *Picea abies*; PcR, *Picea rubens*; PiP, *Pinus ponderosa*; PiS, *Pinus strobus*; PiA, *Pseudolarix amabilis*; PtM, *Pseudotsuga menziesii*; TsH, *Tsuga heteophylla*; TsM, *Tsuga mertensiana*. (With permission from R.A. Price et al., Syst. Bot. **12**:91–97. 1987, American Society of Plant Taxonomists.)

animals with the same proteins produced antisera that reacted similarly.

The results of Price et al. (1987) are of phylogenetic interest because they provide a test of alternative hypotheses of relationships within the Pinaceae. One widely accepted classification divides the genera into two major groups, the so-called abietoid and pinoid groups. The former contains the genera *Abies, Cedrus, Keteleeria, Pseudolarix*, and *Tsuga*, whereas the latter consists of *Larix, Picea, Pinus*, and *Pseudotsuga*. An opposing classification divides the Pinaceae into two large groups (tribes) with *Pinus* in one group and the remaining genera in a second tribe. The latter tribe was further divided into the subtribe Laricinae (*Pseudolarix, Larix*, and *Cedrus*) with dimorphic shoots and the subtribe Abietinae (*Abies, Keteleeria, Tsuga, Picea*, and *Pseudotsuga*) with only long shoots. The immunological distance data strongly support division of the Pinaceae into the abietoid and pinoid groups rather than separating *Pinus* from the other genera and then dividing the remaining genera into sub-tribes (Fig. 12.11). The data do not support grouping *Larix, Pseudolarix*, and *Cedrus* as the subtribe Laricinae. Likewise, grouping of the five genera *Abies, Keteleeria, Tsuga, Picea*, and *Pseudotsuga* into the subtribe Abietinae is not supported by the immunological distance data (Fig. 12.11). For additional discussion of how the results from RIA support concepts of relationships among genera in the Pinaceae inferred from other data, the reader is referred to Price et al. (1987). The study of Price et al. (1987) is of interest because it represents one of the few serological investigations using radioimmunoassay. Also, the results were clear-cut in supporting one of two hypotheses of generic relationships in the Pinaceae. Lastly, it is reassuring that their data are concordant with results from double-diffusion studies by Prager et al. (1976) for the Pinaceae.

Price and Lowenstein (1989) employed radioimmunoassay to study relationships among genera belonging to the traditional families Taxodiaceae and Cupressaceae. In addition, they assessed relationships of these genera to other families of conifers. Price and Lowenstein (1989) concluded from their data that members of the traditional Taxodiaceae and Cupressaceae are best treated as members of one family. Genera from each of the traditional families are as similar to genera from other families as they are to genera assigned to the same family. The data from radioimmunoassay support Eckenwalder's (1976) proposal to merge the two families. Immunological distance data indicate that the Taxodiaceae and Cupressaceae form a lineage that is distinct from other conifers. The immunological data indicate that the genus *Sciadopitys*, which is morphologically anomalous and traditionally placed in the Taxodiaceae, is very distinct from all other conifers, including members of the Taxodiaceae and Cupressaceae.

Esen and Hilu (1989) studied relationships among subfamilies of grasses using the principal seed storage proteins of the family, the prolamins, as the antigens. The antigen–antibody reactions were measured by ELISA; in this case the enzyme was peroxidase, and absorbance at 450 nm was used to measure reactivity. In all instances, absorbance of heterologous reactions was compared with homologous reactions. The values were called the immunological similarity index (ISI), and they could range from 0 to 1 for heterologous reactions, with a value of 1 assigned to homologous reactions.

Twenty taxa from five of the seven generally recognized subfamilies were included in the study. Antisera were raised to 11 of the 20 taxa, and phenetic analyses (clustering) were carried out only for the 11 taxa for which there were antisera and thus for which ISI values for reciprocal reactions could be determined. Limited systematic inferences can be drawn from the data because of the limited taxonomic sampling, with only one taxon per tribe for all but one of the tribes examined. In general, taxa from the same subfamilies clustered together, but few new insights were provided into grass systematics. The study does demonstrate the potential of ELISA in systematic serology.

SUMMARY AND FUTURE PROSPECTS

There is no question that serological data have provided useful systematic information in many studies carried out during the past several decades. The general method has been applied most frequently at higher taxonomic levels, usually at the interfamilial level and above, but studies discussed in this chapter demonstrate that data may also be useful at lower levels. As indicated earlier, serology is a method of probing for similarities and differences in proteins, and the systematic inferences come from measures of similarities (and differences) generated from the antigen–antibody reaction.

Serological data have proved most useful for evaluating alternative hypotheses of relationships for given taxa. That is, if a particular species has been treated as a member of one family by one taxonomist and another family by a second worker, the serological reaction often has been valuable for choosing between the two alternatives. Dahlgren (1983) presented a useful discussion of many cases in taxonomically diverse groups in which serology was useful for assessing relationships. Serology, therefore, is most useful as comparative data within the context of other available systematic information rather than as independent information for constructing a classification or phylogeny. There seems little doubt that the antigen–antibody reaction will continue to be employed for these purposes.

As with any data, serology has certain limitations, which were discussed by Cronquist (1973, 1980) and Dahlgren (1983), among others; their comments serve as a basis for the following brief discussion. Serological data exist as one-to-one comparisons, and this means that if one has compared species X with species Y and Z serologically, then one can infer with confidence something about X and Y, and X and Z, but much less about the similarity of Y and Z. One problem, then, is that if many taxa are to be compared the number of comparisons becomes quite large and prohibitive. Particularly critical in this regard is the effort and expense involved in producing many antibody systems. In all fairness, it should be stated that the problem of adequate sampling is not unique to serology.

Dahlgren (1983) commented that when two taxa show a strong serological cross reaction they are viewed as closely related and that lack of a reaction or a weak reaction is taken to mean a distant relationship. Therefore, absence of or a weak reaction provides little information about relationships.

Cronquist (1980) indicated that often the taxa included in a comparative serological study are chosen based on previous, often differing, concepts of the closest relatives of the taxon under consideration. While such a choice is necessary because it is not possible to include large samples, there is always the possibility that critical taxa will not be included. Although this is a danger with any chemosystematic study, it is perhaps more of a problem with serology. With secondary chemistry, for example, if someone has studied a particular group of compounds in a taxon, then the information is in the literature, and one may compare compounds found in the taxon under study with those reported previously from the other taxon. Thus, even though there was no suspicion that the two taxa were closely related, clues from previous studies could be critical; the same situation exists for amino acid sequences of proteins and nucleic acid data. However, with serology it is necessary to cross-react the two taxa to ascertain their similarity. Therefore, while it is possible that past serological studies may offer some clues to possible similarities of the two taxa, the work still must be done again. In essence, it is not as feasible to use serological data present in the literature to infer similarity of some new taxon as it is with other kinds of chemical data.

There seems little question that serological data will continue to be used in systematic studies. The data have proved valuable as additional sources of information for choosing between alternative hypotheses of relationships of given taxa. Serology has made significant contributions to plant systematics during the past 30 years despite the small number of workers in the field. The fact that serology is not appropriate as a completely independent source of data for inferring relationships does not detract from its utility in plant systematics.

CHAPTER 13

Nucleic Acids: Introduction and Basic Methodology

INTRODUCTION

The rationale for the study of nucleic acids in plant systematics was presented in Chapter 1, and was also mentioned by Giannasi and Crawford (1986). The primary goal of the plant systematist is to produce taxonomic and/or phylogenetic schemes from any and all available data sets. The researcher assumes that the variation observed in the characters is a reflection of genetic variation. That is, for *purely* taxonomic–phylogenetic purposes variation based on other than genetic factors is of little interest. Discussions in earlier chapters on macromolecules included the nature of the genetic inferences that can be made from the data. It follows, therefore, that the optimal situation is to measure variation in the genetic material itself. Also, the genetic material in the form of DNA is passed down through generations, and thus it is the preferred material for studying the phylogeny of the organisms in which it is contained.

The purpose of this chapter is to describe some basic methodology for working with DNA for systematic purposes. The simplest approaches appropriate for the plant systematist will be emphasized. A brief discussion of sequencing will be presented; although it is not commonly employed by the systematist at present, the situation is changing, and it appears certain that emphasis on sequencing will increase. The present general discussion avoids more than one consideration of methods common to working with different kinds of DNA. The protocol for isolating DNA from plant material will depend on the kind of DNA to be studied and the types of data to be gathered. Thus, these topics will be considered briefly under the discussions of different kinds of DNA.

223

DNA–DNA HYBRIDIZATION

One method for comparative studies of nucleic acids has been DNA–DNA (or DNA–RNA) hybridizations. Simple explanations of the method were presented by Kohne (1968), Smith (1976), Harborne and Turner (1984), Giannasi and Crawford (1986), and Murphy and Thompson (1988, pp. 120–123). Britten et al. (1974) gave detailed discussions of mathematical descriptions of DNA reassociation methods for doing hybridizations. The present discussion will consider basic aspects of the method and how comparative data are generated from nucleic acid hybridizations.

Native double-stranded DNA is isolated from plants and reduced to a certain length (often 300–400 nucleotides) by mechanical shearing (Fig. 13.1). Treatment of the DNA with heat or alkali breaks the hydrogen bonds between the two strands and produces single-stranded DNA (Thompson and Murray, 1981) (Fig. 13.1). Under appropriate conditions, when the single strands of DNA (either from the same or different species) are mixed together, they will associate to form double-stranded DNA if there is sufficient complementarity among nucleotides in the two strands. The hybridization technique provides a means of assessing the similarity of DNA sequences between two species, and two measures may be employed. One involves ascertaining the fraction or propotion of the DNA of two genomes that will form hybrid duplexes under given conditions, while the other measure is the thermal stability of duplexes that form.

The basic procedure is outlined in Fig. 13.1. The single-stranded sheared DNA from one species is mixed with similarly treated but radioactively labelled DNA from another species. A considerable excess of the unlabelled DNA (the driver DNA) is added relative to the labelled (tracer) DNA, in order to ensure that the labelled DNA reassociates with unlabelled DNA. Since the probability of any two strands of DNA coming together is a function of concentration, increasing the amount of unlabelled driver DNA relative to labelled tracer DNA ensures that when labelled duplexes form they are composed of one unlabelled and one labelled strand.

The fraction of DNA in two genomes that will form duplexes has meaning only with regard to the conditions under which the reactions are carried out. The two most important variables are temperature and salt concentration, which together determine the stringency of the conditions

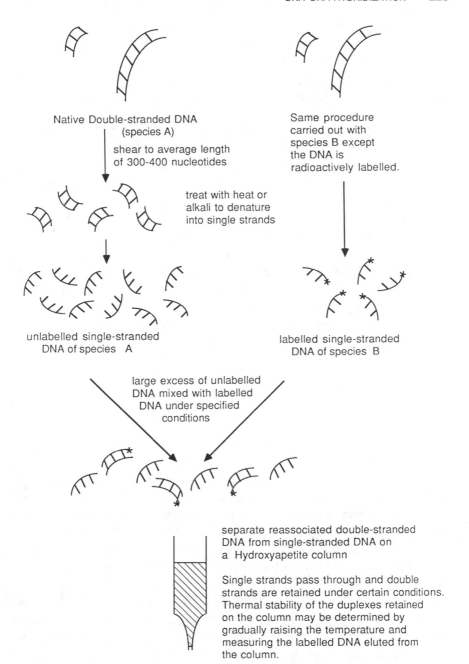

Figure 13.1 Diagram of methods for DNA–DNA hybridizations using the DNAs from different plant species.

of reassociation. This means that conditions must be uniform when the DNAs of different species are being compared in separate experiments if results are to be meaningful.

Thermal stability measurements are likewise relative in the sense that comparisons are made between the temperature at which strands of DNA from the same plant will come apart (melt) and the temperatures at which double-stranded DNA from different species will dissociate. The greater the base pair mismatching in two strands of DNA the lower the temperature at which the strands will dissociate. The thermal stability (T_m) of double-stranded DNA is the temperature at which 50% of the duplex DNA has dissociated. It appears that a lowering of thermal stability (ΔT) of 1^0C indicates a 1% mismatch of base pairs (Thorpe, 1982). The results of experiments in which DNA is renatured are usually plotted as a fraction of the DNA that has reassociated against C_0t. The latter is the concentration of DNA in moles of nucleotide per liter times the time in seconds. Since the reassociation of single-stranded DNA is a matter of chance collisions of individual strands, it follows that the concentration of DNA and the time allowed for the renaturing to occur are both critical variables.

RESTRICTION ENZYMES, GEL ELECTROPHORESIS, AND SOUTHERN BLOTS

Bacteria produce one or more kinds of restriction enzymes, which are endonucleases that cut double-stranded DNA. For our purposes, we will be concerned only with type II restriction endonucleases because each of them cuts the DNA wherever a particular sequence of bases (often six) occurs. Each enzyme will cut DNA only where the specific six-base sequence that it recognizes occur. Also, it will cut the DNA at every site where these sequences occur. The number of fragments into which a DNA molecule will be cut by a type II restriction endonuclease depends on the number of recognition sites present in the molecule (Fig. 13.2). The advantage of having these enzymes at one's disposal is that, because of the absolute specificity of the restriction enzymes, unique fragments may be produced from large complex DNA molecules and comparisons of fragments may be made (among plants, taxa, etc.).

As indicated in Chapter 1, the two strands of a double-stranded DNA molecule run antiparallel, with the 3' and 5' ends of each strand being at

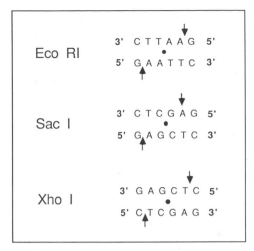

Figure 13.2   The sequence recognition sites for three restriction endonucleases. Arrows mark the sites, and the dots denote the axis of symmetry.

different ends of the molecule. Thus it may be seen that the recognition sites for restriction endonucleases are palindromic sequences (Fig. 13.2). Many different (more than 100) restriction enzymes are now commercially available for cutting DNAs at particular sequences and producing fragments from the molecules. A list of commercially available restriction endonucleases and a discussion of methods for using the enzymes are given by Brooks (1987) and Ausubel et al. (1987).

For simplest comparative purposes, the DNAs from different plants may be digested with several restriction endonucleases and the fragments separated on the basis of size by electrophoresis in agarose or polyacrylamide gels. The fragments are visualized by first staining the gels with ethidium bromide and then viewing them with transmitted UV light (see Sealey and Southern, 1982; Maniatis et al., 1982; Ogden and Adams, 1987, for discussions of gel electrophoresis of DNA). Often it is desirable to transfer the DNA from the gels to pieces of nitrocellulose filter membranes or nylon membranes to which the fragments are bound in the same pattern as in the gel. The double-stranded fragments are separated into single strands of DNA by treatment with alkali prior to transfer to the membrane. The process of transfer is referred to as blotting, and the filters containing the fragments are called Southern blots (Southern,

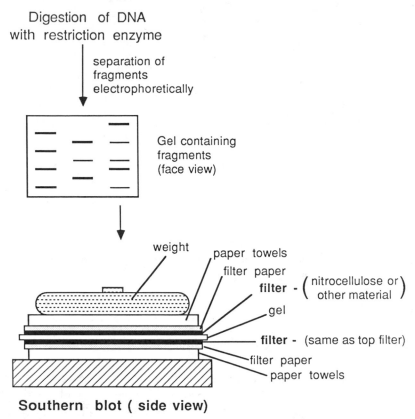

Digestion of DNA
with restriction enzyme

separation of
fragments
electrophoretically

Gel containing
fragments
(face view)

weight paper towels
filter paper
filter - $\left(\begin{array}{c}\text{nitrocellulose or} \\ \text{other material}\end{array}\right)$
gel
filter - (same as top filter)
filter paper
paper towels

Southern blot (side view)

Figure 13.3 Outline of procedure for Southern blots. The DNA fragments move from the agarose gel both up to the top filter and down to the lower filter where they are trapped. The same pattern of fragments is retained on the filter as was present in the gel.

1975). The procedure is diagrammed in Fig. 13.3, and the methods are described in Maniatis et al. (1982), Wahl et al. (1987), and Ausubel et al. (1987). Once the single-stranded DNA fragments become attached to the filters, it is then possible to carry out a number of procedures not feasible with the original gels (see next section).

FILTER HYBRIDIZATIONS

After the single-stranded DNA has been attached to nitrocellulose or nylon membranes, it is possible to carry out filter hybridizations and probe for particular fragments or look at several fragments sequentially rather than viewing them all simultaneously. Fragments of pure DNA obtained by a cloning vector (see next section) are labelled with a radioactive element such as ^{32}P by the commonly used process of nick translation or by random-oligonucleotide-primed synthesis (Maniatis et al., 1982; Ausubel et al., 1987; Meinkoth and Wahl, 1987). The filters containing the bound, single-stranded DNA are placed in a solution containing the labelled fragments. These fragments will hybridize to homologous fragments in the bound DNA on the filter. The fragments to which the labelled probes have hybridized are visualized by autoradiography, that is, the filters are exposed to x-ray film and the film developed to reveal the locations of the labelled fragments (Bonner, 1987). Thus, only certain fragments on the filter are visualized, which simplifies interpretation of gels. Also, this method allows one to compare the positions of homologous fragments from different plants on the filter.

CLONING

The availability of large amounts of pure DNA is essential for systematic (comparative) studies in which filter hybridizations using radioactively labelled DNA probes are employed. Obtaining this DNA involves cloning, that is, inserting fragments of DNA from a plant into a vector such as a plasmid. The vector replicates very rapidly after being introduced into the bacterium *Escherichia coli* and in so doing increases the amount of the inserted DNA many times over what could be obtained from the original source.

For many of the investigations carried out by systematists, it is not essential that cloning be done; in many instances suitable clone banks are available from various laboratories. That is, bacterial stocks with plasmids containing the DNA of interest are available and may be stored at low temperatures for long periods of time. Periodically, very small amounts of these stocks may be used to inoculate a nutrient medium. The bacteria multiply in the medium, as do the plasmids, and the latter produce large amounts of the DNA of interest. The plasmid DNA containing the inserted DNA may then be isolated from the bacteria and used as probes. Palmer (1986a) listed sources of clone banks for the chloroplast DNAs of flowering plants. The use of cloned DNA from species or genera other than the one on which the systematist is working is often feasible because the homologies between the DNAs of rather distantly related taxa are frequently sufficient for carrying out filter hybridizations. A detailed discussion of cloning methods may be found in a variety of laboratory manuals (see Maniatis et al., 1982 and Ausubel et al., 1987, as examples). A method called the polymerase chain reaction (PCR) will amplify DNA directly without cloning into a vector. The basic principle of the method is that in the presence of a polymerase and a primer, DNA will make copies of itself. With PCR, after a copy is made, the strands are separated by heat (i.e., the DNA is denatured). The reaction temperature is lowered, a new strand is formed, and the cycle continues. Obtaining the proper primers is important; a DNA polymerase that can withstand the high temperatures needed for denaturation is available. Engelke et al. (1988) and Saiki et al. (1985) may be consulted for examples of how the method is employed. Hartl and Clark (1989, pp. 343–344) provide a concise and lucid discussion of PCR.

DNA SEQUENCING

The use of restriction site analysis and mapping for comparing DNAs represents a method of sampling parts of the genomes of organisms for similarities and differences; the greater the number of enzymes employed the larger the sample size. The ultimate sampling strategy would be to determine the complete sequence of bases for a given piece of DNA. Sequencing has not been used frequently for systematic purposes in plants, largely because of the time, effort, and expense involved. This limits the number of taxa that can be compared, as contrasted, for example, with restriction site analysis. Despite these limitations, it seems inevitable that sequencing will be increasingly employed for generating phylogenetic hypotheses in plants. A detailed consideration of nucleic acid sequencing is beyond the scope of the present discussion; only a few brief comments will be presented. Two basic methods are employed for nucleotide sequencing. One is the so-called chemical or Maxam–Gilbert method (Maxam and Gilbert, 1980) and the other is the dideoxy or enzymatic method (Sanger et al., 1977; Sanger, 1981). A general discussion of the dideoxy method will be presented. More detailed considerations of this and the Maxam–Gilbert method are provided by Davies (1982), Ambrose and Pless (1987), and Ausubel et al. (1987).

The basic approach of the dideoxy method is to use the strand of DNA to be sequenced as a template for the in vitro synthesis of DNA. This means that the length of DNA to be sequenced must be single-stranded,

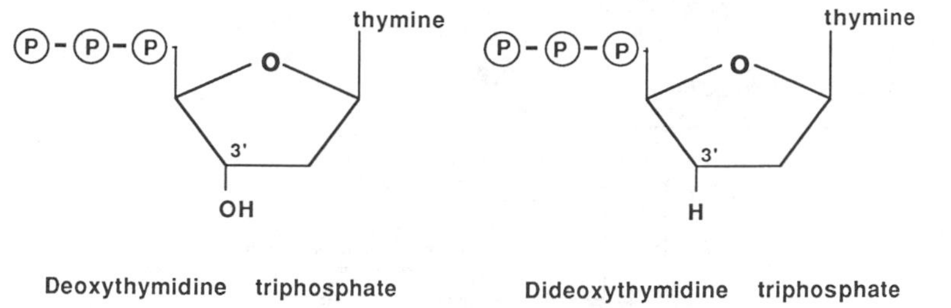

Figure 13.4   Structures of deoxythimidine triphosphate and its dideoxy analog.

which can be accomplished by cloning the piece of DNA to be sequenced into a particular phage (designated M13). After cloning, the phages released have single-stranded DNA, including the DNA to be sequenced. To this template DNA is annealed a short primer used to initiate synthesis of the complementary strand. The strands of DNA are then used in four separate reaction mixtures. Each reaction contains all of the nucleotide triphosphates (one of which is labelled with ^{32}P so that the newly synthesized strand can be separated from the "old" strand by autoradiography); in addition, *one* of the four dideoxynucleotide triphosphates is added to each of the samples. The dideoxynucleotides differ from the corresponding deoxynucleotides in the replacement of the hydroxyl group by a hydrogen atom at the 3' position (Fig. 13.4). The incorporation of a dideoxynucleotide into the newly synthesized DNA strand causes termination of the chain because the 3'-OH is necessary for the addition of more nucleotides. The incorporation of a dideoxynucleotide into the chain in any one of the four samples is a matter of chance, meaning that in each of the samples synthesis of the complementary strand will be terminated at different lengths depending on when the dideoxynucleotide is incorporated into each strand. Once the reactions have been allowed to take place, the newly synthesized strands are isolated and subjected to electrophoresis followed by autoradiography. Since the new strands contain a labelled base, they will appear in the autoradiograms (Fig. 13.5). The procedure is presented in outline form in Fig. 13.5. The critical point is that the synthesis of all new chains of DNA is initiated at the same point and the differences in lengths result only

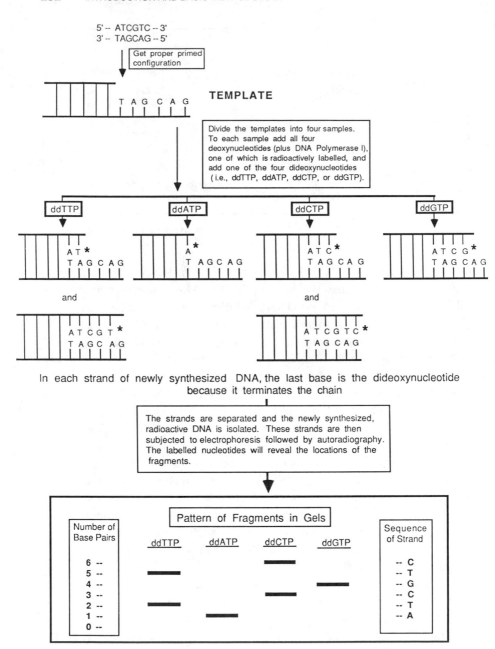

Figure 13.5 General outline for sequencing nucleic acids using the dideoxy or enzymatic method.

from the point of termination, that is, the point at which a dideoxynucleotide was incorporated.

SUMMARY

The use of restriction endonucleases that recognize specific base sequences in DNA makes it possible to carry out studies of the DNA. The fragments generated by the enzymes are separated by gel electrophoresis, and then transferred to nitrocellulose or nylon filters with the identical pattern in the gel maintained on the filter. This is called blotting or producing Southern blots. Cloned, labelled fragments may then be hybridized to fragments on the filter and the patterns visualized by audoradiography. Two methods are generally employed for sequencing DNA, the so-called chemical or Maxam–Gilbert method and the dideoxy or enzymatic method.

CHAPTER 14

The Nuclear Genome of Plants: General Features and Use in Systematic Studies

INTRODUCTION

The complex nature of the nuclear genome of plants in terms of the different kinds, arrangements, and amounts of DNA is continually being further elucidated and appreciated by molecular biologists. It is also clear that not all DNA invariably remains at a given locus in the genome; rather it may move about and insert itself at various places. A number of books and review papers have been devoted partially or totally to considerations of the complexity of the eukaryotic, or more specifically, plant genome (Walbot and Goldberg, 1979; Walbot, 1979; Flavell, 1980, 1982; Flavell et al., 1980; Thompson, et al., 1980; Thompson and Murray, 1981; Cavalier-Smith, 1985a; MacIntyre, 1985; Mitra and Bhatia, 1986; Antonov, 1986; Murphy and Thompson, 1988, pp. 105–123). The size and complexity of the plant nuclear genome can provide considerable systematic and phylogenetic information. This very nature of the genome, however, makes it difficult to extract the information. In the present chapter, a general description of the plant nuclear genome will be presented, followed by a discussion of the ways in which genome studies have contributed to systematic and phylogenetic studies of plants. The nuclear ribosomal DNA will be treated separately in the following chapter.

DNA AMOUNTS

One interesting and puzzling aspect of plant nuclear genomes (and eukaryotes in general) is the large amount of DNA present relative to that ostensibly needed for encoding proteins. Rough estimates indicate that from less than 5% to around 10% of the total DNA present in plant nuclei is needed for coding proteins (Flavell, 1980). Much has been written about this excess of DNA in eukaryotes and a number of hypotheses put forward to explain it. A detailed consideration of the topic is beyond the scope of the present discussion; the reader should consult Price (1976, 1988a,b), Cavalier-Smith (1982, 1985b,c,d), Bachmann et al. (1985), and Ohri and Khoshoo (1986) for general reviews of the topic. The excess DNA in the nuclear genome of plants represents one aspect of the so-called C-value paradox, the C value being the total amount of DNA present in the haploid nuclear genome. This value may be measured in base pairs (bp), picograms (pg), or daltons, where a picogram is equal to 0.965×10^9 bp, which in turn equals $6.1 \times 10''$ daltons. The second aspect of the C-value paradox is the large variation found in the C values among different species of plants, in some cases congeneric species (see later discussion).

Studies of relative DNA amounts are carried out using microspectrophotometry in which the nuclei are stained and their absorbance measured relative to some standard nucleus (see description in Price et al., 1980 as an excellent example of methodology). Price (1988a,b) described in detail the protocol for determining DNA content. If the genome size of the standard is known, then quantitative statements about the genome sizes of the plants under study may be made. If, on the other hand, the genome size of the standard is unknown, then statements may be made only about the relative sizes of the nuclear genomes of the experimental plants.

SINGLE-COPY AND REPETITIVE DNA SEQUENCES

The nuclear genome of plants consists of certain DNA sequences that are present once per genome. These are referred to as single-copy or unique-sequence DNA. In reality, it may be difficult to distinguish single-copy sequences from those present in two to several copies. Other sequences may occur in the genome from several up to a million times; these are designated as repetitive DNA (Flavell and Smith, 1975; Flavell, 1980; Walbot, 1979; Walbot and Goldberg, 1979; Thompson and Murray, 1981; Murphy and Thompson, 1988, pp. 106–109).

Earlier, DNA–DNA hybridization was discussed as a method for estimating the similarity of the DNAs of two species by the proportion of the DNA that will reanneal and the thermal stability of the duplexes. Also, the term C_0t was introduced and defined as the product of the concentration of DNA and the time of incubation. Studies of reassociation kinetics have also been used to ascertain the presence of single-copy and different types of repetitive DNA in nuclear genomes. The DNA from a genome is cut to a certain length and dissociated into single strands. The strands are then renatured under given conditions, much as would be done with the DNAs of two different species. Under these conditions, sequences present only once per genome have a much lower probability of "finding" or "bumping" into each other than do sequences represented several to many times. A C_0t curve can be plotted for the reassociation kinetics of the total DNA of a nuclear genome. Higher C_0t values (indicating a slower reaction) are characteristic of single- or low-copy number DNA in the genome. That is, higher concentrations of low-copy number nuclear DNA would have to react for a longer period of time in order to get reassociation, as compared with highly repetitive sequences (see excellent discussion in Lewin, 1987, Chapter 18, and Fristrom and Clegg, 1988, Chapter 3). Nuclear genomes may thus be characterized on the basis of how fast (at what C_0t values) different sequence components reassociate, with the slowest portion representing single-copy or several-copy sequences and the fastest components being the highly repetitive sequences.

Another term should be defined and discussed briefly with regard to the nuclear genome. Earlier, the term *complexity* was used to refer in a general way to the variation in DNA amounts, kinds of DNA, and so on present in the genome. Complexity is used in a more restricted way, however, to refer to the total length of sequences in the genome that are different from all other sequences. This means, for example, that all single-copy or unique sequences in a genome contribute to its complexity because they are different from all other sequences. If, by contrast, a given sequence is present 10 times, its length is counted only once in calculating genome complexity.

Other aspects of genome organization include the percentage of single-copy sequences, their lengths, and the ways in which they are arranged. The percentage of unique-sequence DNA often ranges between 20 and 70%; as absolute genome size decreases, the percentage of single-copy DNA increases (Flavell, 1980). Flavell (1980) cautioned that the proportion of DNA classed as single copy will depend on the stringency of the conditions under which the single strands are renatured. Under less stringent conditions a greater percentage of the DNA will exhibit reassociation kinetics suggesting repetitive DNA. Leutwiler et al. (1984)

showed that the very small nuclear genome of the species *Arabidopsis thaliana* has little repetitive DNA, which supports the generalization that smaller plant genomes have proportionally less repetitive DNA.

The lengths of single-copy sequences in plant genomes usually vary from 200 to several thousand bp. Murray et al. (1978), for example, inferred from renaturation data that about 80% of the unique-sequence DNA in the genome of the pea consists of lengths less than 1000 bp. Their data also suggested that a high percentage of the single-copy sequences is around 300 bp long. Flavell (1980) suggested that long sequences of single-copy DNA are likely present in most plant nuclear genomes, but in most instances they constitute a very small percentage of the total DNA. Flavell (1980) also indicated that in larger (more than 2 pg) plant genomes two-thirds or more of the single-copy DNA occurs in lengths shorter than 1400 bp. By contrast, in plants with smaller nuclear genomes a larger proportion of the single-copy sequences occur in larger stretches, (4000 bp or more). This is not surprising if it is recalled that smaller nuclear genomes contain a higher proportion of unique sequences.

Consider next the manner in which repetitive DNA is arranged in the genome. Flavell (1980) summarizes information suggesting three common patterns in plant chromosomes. One arrangement consists of repetitive DNA of 50 to 2000 bp interspersed among unique-sequence DNA some 200 to 4000 bp in length (Fig. 14.1). A second pattern has similar repetitive sequences tandemly arranged, and a third has different repeated sequences arranged in various complex permutations (Fig. 14.1). The locations of repetitive sequences on the chromosomes have been determined in certain instances (Flavell, 1980). It is known, for example, that heterochromatic regions of chromosomes at the centromeres and telomeres consist of highly repeated tandem arrays. The nucleolar organizer region of the chromosome contains a large proportion of the genes for rRNA; these are comprised of tandem arrays of sequences.

A technique that has been employed in certain plants (primarily grasses) to locate particular repetitive sequences on the chromosomes is in situ hybridization. The method involves hybridizing radioactively labelled fragments of DNA to single-stranded DNA in the chromosomes. Autoradiography then allows one to determine the location of the labelled fragments on the chromosome. Hutchinson et al. (1981) and Hutchinson (1983) described the method, while Appels et al. (1978), Gerlach and Peacock (1980), Hutchinson et al. (1980), and Jones and Flavell (1982) applied the methodology to particular plant groups. Rayburn and

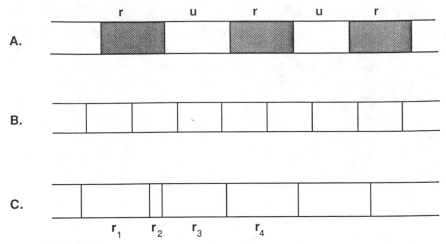

Figure 14.1   Three common distribution patterns of repetitive sequences of DNA in the chromosomes of plants. (*A*) Single copy and repetitive sequences are interspersed. u, unique or single copy; r, repetitive. (*B*) Identical or nearly identical repeating units occur in tandem arrays. Each rectangle represents a repeating unit. (*C*) Different repeating units, represented by r_1, r_2, and so on, are interspersed in various ways.

Gill (1985) used biotin-labelled rather than radioactively labelled probes for in situ hybridization with wheat chromosomes.

DNA AMOUNTS AND SYSTEMATIC RELATIONSHIPS

The discussion of DNA amounts in nuclear genomes will emphasize the interpretation of results in terms of systematic and phylogenetic relationships in particular taxa. The possible adaptive significance of the DNA amount will be considered only when appropriate within the context of the systematic question under discussion. Bennett and Smith (1976) and Bennett et al. (1982) compiled lists from the literature of nuclear DNA amounts for angiosperms. Price (1976) provided a review of DNA variation in gymnosperms and angiosperms along with discussions of the mechanisms for changing DNA amounts and the possible functional significance of the various amounts. Recent concise and useful overviews were given by Price (1988a,b).

Several examples will suffice to demonstrate the use of nuclear DNA amounts in systematic studies; no attempt is made to provide a comprehensive treatment of the topic, and the recent review by Ohri and Khoshoo (1986) should be consulted for a more detailed consideration. A series of detailed studies of nuclear DNA content in the genus *Microseris* and related genera was carried out by Price and collaborators (Price and Bachmann, 1975; Price et al., 1980, 1981a,b, 1983, 1986; Bachmann et al., 1979). While most earlier studies of DNA content did not consider possible variation within species (see Miksche, 1968, 1971; Bennett et al., 1977, as exceptions), Price et al. (1980) demonstrated variation of up to 20% within *M. douglasii*. In another study Price et al. (1981a) examined DNA content in *M. bigelovii*, which, like *M. douglasii*, consists of selfing annual plants. They found the variation of DNA content among plants of this species to be as high as 25%. Results of these two investigations demonstrated that variation may occur within species, and Price et al. (1981a) addressed the question of why reports of intraspecific variation in DNA amounts have been relatively rare. Several of the potential reasons are technological in nature. In addition, Price et al., (1981a) suggested that another reason could simply be a lack of sufficient sampling from individual plants of a species. In particular, species that are ecologically and morphologically diverse should be sampled extensively.

Price et al. (1981b) carried out a survey of DNA content in over 200 individuals of *M. douglasii* from 24 populations. Sampling was done to include plants exhibiting the geographical, ecological, and morphological range of variation in the species. They found that plants with higher DNA content occur in more mesic situations, although not all plants in these situations had higher DNA content. Within a single population, DNA amounts were correlated with precipitation, that is, high DNA content was measured in situations of greater moisture. Amounts of nuclear DNA were not correlated with morphological variation. Reasons for the correlation between moisture and DNA amounts are obscure; Price et al., (1981b) suggested that low DNA content could result in a faster mitotic cycle and thus faster development time in a moisture limiting situation. This hypothesis of DNA amounts and development times has been repeatedly suggested in the literature. However, in a more recent study, Price et al. (1986) failed to detect consistent correlations between DNA content and moisture conditions in *M. douglasii*.

The reason for the apparent prevalence of intraspecific variation in nuclear DNA content remains an open question, partly because of lack of adequate data. It may be that the level and/or pattern of variation depends on certain attributes of the species such as the amplitude of their ecological preferences. Also, variation could be associated with whether the plants are annual or perennial, and whether they are outcrossers or selfers. Price (1988a,b) is of the view that much of the intraspecific variation in DNA content probably serves an adaptive function. A great deal of correlative data suggest that this is likely the case, but direct proof is lacking.

A number of studies have related interspecific variation in DNA content with presumed evolutionary relationships. Price and Bachmann (1975) reported relative DNA amounts in the subtribe Microseridinae of the tribe Cichoreae. The genera *Phalacroseris* (one species), *Microseris* (eight species), and *Agoseris* (four species) were examined. In *Microseris* and *Agoseris*, the derived annual species have smaller DNA amounts than the perennials, with smaller chromosome size paralleling the reduction in DNA. Price and Bachmann (1975) suggested that selection for more rapid development in these annual plants may account for the smaller DNA amounts. Reduction in DNA content with evolutionary specialization was not unidirectional, however, because one presumably specialized species of *Microseris* (*M. borealis*) had the highest amount of DNA of any species examined in the genus. In addition, the putatively specialized perennial *Phalacroseris bolanderi* had the highest DNA content of any species examined. Thus, one must be cautious when interpreting DNA amount and possible evolutionary relationships within a group of plants.

The genus *Lathyrus* has been examined for nuclear DNA content in a series of studies (see Narayan and Rees, 1976, 1977; Narayan, 1982, 1983, as examples). In this genus, the fourfold difference in DNA amounts is not continuous; rather it occurs in constant increments of 3.5 to 4.0 pg. Narayan (1982, 1983) argued that the discontinuities in DNA amounts are important for the organization and the evolution of the nuclear genome in *Lathyrus*. Although the data indicated that this could be the case, it is not at all clear *why* it should be so. Narayan (1982) suggested that the evolution of *Lathyrus* has been accompanied by an increase in DNA content, although it does not appear that the genus has been the subject of a modern systematic study.

A survey of DNA amounts in 51 species and 3 subgenera of *Nicotiana* revealed a fivefold variation (Narayan, 1987). Variation among species was discontinuous, with regular increments of nuclear DNA amounts. No consistent trend of increase or reduction with speciation was detected in *Nicotiana* (Narayan, 1987). It was concluded that particular DNA amounts occur more because of fitness and genomic balance.

Poggio and Hunziker (1986) examined nuclear DNA content in the South American genus *Bulnesia*. Among diploid species, a sixfold variation was found that was positively correlated with chromosome size. The octoploid species showed a lower DNA content per genome as compared with any diploid species. The data of Poggio and Hunziker (1986) are concordant with concepts of relationships in the genus inferred from previous systematic studies. For example, prior studies suggested that the following diploid species pairs are closely related: *B. retama–B. chilensis* (highest DNA content); *B. arborea–B. carrapo* (lowest DNA values). A seventh diploid species, *B. sarmientoi*, appears as an isolated element with no close affinities, and it has intermediate DNA content. The octoploid *B. bonariensis* is likewise considered an isolated element in the genus, and it has, as indicated above, the lowest amount of nuclear DNA per genome of any species.

An attempt to correlate DNA amount with evolutionary or ecological specialization in *Bulnesia* produced two contrasting patterns. Poggio and Hunziker (1986) found the highest DNA content in the two diploid species that are most specialized morphologically and adapted to the most xerophytic environments. By contrast, the morphologically specialized *B. sarmientoi* contained the lowest DNA content per genome of any diploid species. It appears, therefore, that either increase or decrease in nuclear DNA content may be associated with specialization and evolution.

Stucky and Jackson (1975) examined relative DNA content in seven species of the tribe Astereae in an attempt to address the question of base chromosome number in the group. As discussed in Chapter 10 (when gene numbers were considered), common chromosome numbers for the tribe are $n = 4$, 5, and 9. The two hypotheses put forth to explain these numbers are that the $n = 9$ plants represent allopolyploids resulting from hybridization between plants with $n = 4$ and 5 chromosomes with subsequent chromosome doubling (Turner et al., 1961; Turner and Horne, 1964). The alternative hypothesis is that all plants are diploids with descending aneuploidy producing the lower numbers (Raven et al., 1960; Solbrig et al., 1964, 1969).

Stucky and Jackson (1975) reasoned that if the $n = 9$ plants are polyploids they should contain higher DNA amounts than the lower chromosome plants, whereas if aneuploid reduction had occurred then $n = 4$ and 5 plants should have DNA content comparable to the $n = 9$ plants. They recognized that processes might alter DNA amounts, but that these should not obscure the basic gain associated with polyploidy. Their results showed lower DNA content in one $n = 9$ species than in all other $n = 4$ and 5 taxa, and the other $n = 9$ species had lower DNA amount than all but one of the lower chromosome number taxa. Stucky and Jackson (1975) interpreted the data as support for the aneuploid reduction rather than polyploid hypothesis.

Price et al. (1984) examined relative DNA content in two species of *Coreopsis* related as progenitor and derivative. *Coreopsis nuecensoides* is the progenitor of *C. nuecensis*, with aneuploid reduction going from the former to the latter (Smith, 1974). No decrease in DNA content was found in the lower chromosome number ($n = 6$, 7) *C. nuecensis* as compared with *C. nuecensoides* ($n = 9$, 10, 11). In fact, two populations of the derivative species have about 12% more DNA than any population of the progenitor. This indicates that amplification of DNA must have occurred in *C. nuecensis* because it is otherwise difficult to understand how increase in DNA content could occur with aneuploid reduction. The data suggest that in the evolution of annual plant species DNA content does not necessarily decrease.

The basic limitation of microspectrophotometric determination of nuclear DNA content is that it measures total DNA, and the kind or kinds of DNA (single-copy, repetitive, etc.) responsible for the differences observed are not known. A number of studies have attempted to correlate DNA content with other data to provide some insight into the kinds of DNA contributing to the variation. Narayan and Rees (1976) showed that quantitative differences in DNA content in *Lathyrus* are associated with changes in the amount of repetitive DNA in the genome. In the genus *Vicia*, DNA content changes are correlated with variation in both repetitive and nonrepetitive DNA (Raina and Narayan, 1984). Other investigations have combined staining of chromosomes for heterochromatin (repetitive DNA) and measurements of DNA content (Rayburn et al., 1985; Poggio and Hunziker, 1986). In both cases, an increase in the amount of stained heterochromatin was detected in plants having higher DNA amounts.

The use of DNA content as a useful independent source of data in systematic studies has certain values and limitations. Positive correlations between DNA amount and taxonomic grouping could be taken as additional evidence for relationships as inferred from other data. The possibility cannot be dismissed, however, that increase or decrease in DNA content could take place independently even in closely related species. When only total DNA is measured, it is not possible to ascertain whether increased amounts in two species, for example, are the result of the same events in a common ancestor or have occurred independently subsequent to speciation.

One trend that does appear fairly constant is decreased DNA content in annuals of a given genus relative to perennials. Since in most genera the annual condition is viewed as derived, the data from DNA content per se are not particularly useful as independent data for inferring the direction of evolution. Among annual or perennial taxa within a genus, there appears to be no consistent clear-cut trends with regard to DNA content and presumed ecological and/or evolutionary specialization.

In summary, measurements of DNA content in plant nuclear genomes may provide useful information but appear to be of limited value as systematic data. More detailed studies of the organization of particular parts of the genome are potentially of greater utility for addressing systematic questions.

Nucleic Acid Hybridization

An early approach to the use of nucleic acids in plant systematics involved the hybridization of nuclear DNA. The method was mentioned by Alston (1967), and an early discussion of its possible taxonomic utility was given by Kohne (1968). An additional consideration of the approach for plants was presented by P.M. Smith (1976), and more recent discussions were presented by Harborne and Turner (1984) and Giannasi and Crawford (1986).

The use of nucleic acid hybridizations for ascertaining phylogenetic relationships among species is not on the increase because newer methods such as restriction analysis and sequencing are becoming more popular. It must be kept in mind that hybridization in this context refers to measurements of the proportions of the DNAs of two species that will hybridize under given conditions and the thermal stability of duplexes between strands from different species as compared with the same species (it does not refer to filter hybridizations with cloned probes as discussed in Chapter 13). There are several reasons for the decline in the use of nucleic acid hybridizations for systematic purposes. The technique is relatively time-consuming and expensive. When base pair mismatches exceed 20%, results of annealing studies become unreliable, and it is not possible to use the method for comparing more distantly related taxa. At the other extreme, the approach may not be sensitive enough to measure small differences accurately. The data from annealing studies consist of one-to-one comparisons of different taxa; thus the number of pair-wise comparisons needed for examining relationships among a number of taxa becomes quite large and impractical. In this regard, the nature of the data is similar to serology. As indicated earlier, the plant nuclear genome consists of sequences present from once up to thousands of times. This variation makes it critical that comparisons be made between DNAs of the same or very similar copy number in different species when annealing studies are carried out.

Annealing studies, while providing some measure of sequence divergence between DNAs, are not as precise as restriction analyses, or especially sequencing, for revealing the nature of the divergence. There is no indication of the locations of the divergent sequences or the number of differences at different locations. Instead, the methods provide an estimate of percentage of divergence over the entire genome or the part of the genome under consideration, such as single-copy sequences. Therefore, one could argue that if what is desired is an estimate of overall divergence between two species in, for example, single-copy DNA, then annealing studies can provide such information.

Despite limitations of DNA hybridization, it is obvious why it was the first DNA approach employed for assessing relationships: it represented the only feasible method available at that time for using DNA (other than base ratio studies, which will not be discussed; see Harborne and Turner, 1984, pp. 443–448) for comparative purposes. Recent extensive discussions of the value and limitations of DNA–DNA hybridizations for phylogenetic purposes, particularly in birds, were presented by Cracraft (1987) and Sibley et al. (1987). Sarich et al. (1989) have likewise presented a critique of data analyses and phylogenetic conclusions of DNA hybridization studies carried out by Sibley and collaborators. Despite certain controversies about data analyses, there is little question that DNA hybridizations have been useful for elucidating relationships in birds.

Several examples of the contributions of DNA annealing studies to plant systematics will be considered. Some of this discussion will be, to a certain extent, of historical interest in the sense that other methods are now more commonly employed for comparing nuclear DNAs of plants.

Among the first studies employing DNa hybridizations for examining

TABLE 14.1. Percent Homology of the DNAs of Species from the Gramineae and Leguminosae as Measured by DNA–DNA Hybridizations[a]

Labelled DNA and DNA in Agar	Heterologous Competitor	Percentage Homology
Pisum sativum (pea)	*Phaseolus vulgaris* (pole bean)	19
	Vicia villosa (hairy vetch)	48
	Nicotiana glauca (tobacco)	5–10
	Secale cereale (rye)	0–5
Vicia villosa	*Pisum sativum*	54
Secale cereale	*Hordeum bulbosum* (barley)	59, 56
	Triticum vulgare (wheat, two varieties used)	75, 74
Hordeum bulbosum	*Secale cereale*	70, 78
	Triticum vulgare	73
	Pisum sativum	10
Triticum vulgare	*Pisum sativum*	10
	Hordeum bulbosum	80
	Secale cereale	95, 90

From Bendich and Bolton, 1967.
[a] Tobacco from the Solanaceae is also included.

relationships among plants are those of Bendich and Bolton (1967) and Bendich and McCarthy (1970a,b). Bendich and Bolton (1967) carried out hybridizations among members of the grasses and legumes and included tobacco from the Solanaceae as well. Estimates of percent homology of the DNAs from these species were made from the percentage of DNA from one species that hybridized to the DNA of another species as compared with the amount hybridizing when the DNA of the same species is reannealed. The results from this early study are summarized in Table 14.1; a brief discussion of the taxonomic significance of the data is in order. The method employed in the study is competitive hybridization in which the DNA of a given species is present in agar, and labelled DNA of the same species is then reacted with increasing amounts of DNA from another species. If the DNA (heterologous) from the other species can anneal with the DNA immobilized in the agar, then it will compete with the labelled DNA (homologous) from the same species. This competition will continue and thus decrease the amount of labelled DNA that will be incorporated into duplex DNA, until the unlabelled heterologous DNA has hybridized with all the DNA it can under the conditions employed. At this point, the amount of label will not decrease with increasing amounts of unlabelled DNA from the "competing" species. Percent

TABLE 14.2. Relative Percentage Binding of DNAs and Thermal Stability Decreases of Heteroduplexes for Grasses (Cereal Grains)

DNA Bound to Filter	Relative Percentage Binding Labelled DNA				Decrease in Thermal Stability of Duplexes Labelled DNA			
	Barley	Oats	Rye	Wheat	Barley	Oats	Rye	Wheat
Barley	100	19	58	72		6.4	5.4	3.0
Oats	12	100	15	17	5.5		7.2	5.0
Rye	59	22	100	100	3.6	6.1		0
Wheat	48	16	60	100	4.0	6.1	4.1	

From Bendich and McCarthy, 1970a.

homology as measured by this method is the maximum percentage decrease in labelled DNA present in duplex form in the presence of "competitive" DNA from another species as compared with hybridizing labelled DNA to homologous unlabelled DNA in agar. All members of the legume family showed greater homology with each other than any one did with members of other families (Table 14.1). The results, however, are not very satisfying because they provided only a very crude measure of similarity. For example, pea showed a 48% homology to vetch and only a 19% homology to pole bean, whereas pea exhibited a 0–10% similarity to grasses (Table 14.1).

Bendich and McCarthy (1970s) employed both relative percent hybridizations and thermal stabilities of duplexes to examine relationships among barley, oats, wheat, and rye (Table 14.2). General agreement exists between the percent hybridization and thermal stability measurements for the different taxa, that is, the species that showed the highest percent homology also exhibited the greatest duplex stability (Table 14.2). Reciprocal hybridizations may give quite different results for percent of binding, that is, results may vary depending on which taxon is used for filter-bound DNA versus labelled DNA. For rye and wheat (Table 14.2), as much wheat DNA annealed with rye DNA on the filters as did labelled rye DNA. In the reciprocal reaction only 60% labelled rye DNA annealed with the wheat DNA on the filter as compared with the labelled wheat DNA. The reason for this finding is that the rates and percentages for duplex formation are affected by the amounts and different kinds of repetitive DNA present in the genome (Bendich and Bolton, 1967; Bendich and McCarthy, 1970a). Rye filters contained enough sequences sufficiently complementary to the labelled wheat DNA that it

TABLE 14.3. The Distribution of Seven Groups of Repeated Sequences of DNA in Grasses

Taxa	Group of Repeats						
	1	2	3	4	5	6	7
Oats	×						×
Barley	×	×				×	
Wheat	×	×	×	×			
Rye	×	×	×		×		

From Flavell et al., 1977.

could bind wheat DNA as effectively as rye DNA. When wheat DNA is bound to the filter, an equal reaction is not detected because there are sequences in rye that are not present in wheat in sufficient concentration to react with the rye DNA.

The general results of the DNA hybridization studies for these plants are in agreement with the expected results given their relationships inferred from other data, that is, wheat and rye are most closely related, with barley more distantly related and oats least closely related. In a sense, then, the results of the investigation provided some confidence in the method rather than furnishing strong confirming evidence for the relationships.

More recent studies of the genomes of cereals have been carried out using DNA annealing of repeated sequences (Flavell, 1982; Flavell et al., 1977, 1979; Rimpau et al., 1978, 1979; D.B. Smith et al., 1976; D.B. Smith and Flavell, 1974, 1977). From a systematic–evolutionary perspective, the results of Flavell et al. (1977) are of interest. Using DNA hybridizations of repetitive DNAs from barley, oats, rye, and wheat, they recognized seven basic groups of repeated sequences. Some heterogeneity was detected within groups, but this was less than the heterogeneity among the groups. The taxonomic distribution of the seven groups of repeated sequences allowed Flavell et al. (1977) to construct a phylogenetic hypothesis. All four grasses share group 1 repeats, whereas oats are unique in having group 7 (Table 14.3). Barley, rye, and wheat all share group 2 repetitive sequences while rye and wheat share group 3. Rye is unique in having group 5, and group 4 is unique to wheat (Table 14.3). The data are in agreement with the earlier studies of Bendich and Bolton (1967) and Bendich and McCarthy (1970s) with regard to relationships among the taxa. The results of Flavell et al. (1977) suggested that new families of repeated sequences arose at the time of or subsequent to the evolutionary branches to (1) oats; (2) barley, rye, and wheat; (3) rye and wheat; (4) rye; and (5) wheat (Table 14.3). Furthermore, those families or groups of repeated sequences, once introduced, are maintained through the introduction of subsequent repeated families.

Other DNA annealing studies such as those of *Cucurbita* (Goldberg et al., 1972), *Osmunda* (Stein and Thompson, 1975; Stein et al., 1979), and *Atriplex* (Belford and Thompson, 1981a,b) were discussed previously (Giannasi and Crawford, 1986) and will not be considered here. Rather, several more recent investigations will be discussed.

Nath and collaborators (Nath et al., 1983, 1984; Thompson and Nath, 1986) employed DNA hybridizations to address the age-old problem of the B-genome donor to polyploid wheats. This question was discussed in earlier chapters and will be considered in subsequent chapters as well as in summary form as a case study in Chapter 17. The three studies involved hybridization of DNAs from various diploid species of *Triticum* to either tetraploid or hexaploid taxa. Thermal stabilities of the heterologous duplexes were then used as estimates of the similarity of the presumed B-genome donors with the B-genomes present in the polyploids. The results implicated *Aegilops searsii* as the probable B-genome donor because of the greater thermal stability of DNA duplexes formed with the polyploids. Hybridizations of DNA from *T. urartu* with *T. monococcum* (A genome) and with a synthetic tetraploid containing the A and D genomes revealed that *T. urartu* has high homology with the A genome and thus could not be a B genome donor. These results will be discussed in more detail in Chapter 17, in which comparative molecular data will be presented for wheat. Suffice it to say that DNA hybridizations cast further serious doubt on *T. urartu* as the B-genome donor to polyploid wheats.

Bendich and Anderson (1983) used DNA hybridization to examine complexity (the total length of *different* sequences present) in two different components of repetitive DNA in six species, three from each of the two subgenera of *Equisetum*. They found consistent differences between the two subgenera in the complexity of the two components. The so-called component I, which makes up about 30% of the total genomes of the six species, has a complexity of 1.1 to 2.0 $\times$ 10^6 nucleotide pairs in subgenus *Equisetum*, whereas in subgenus *Hippochaete* the complexity is about 0.5 $\times$ 10^6 nucleotide pairs. Component II comprises about 20% of the genomes and has a complexity of 60 to 130 $\times$ 10^6 nucleotide pairs in subgenus *Equisetum*, while the value is 19 to 27 $\times$ 10^6 nucleotide pairs in subgenus *Hippochaete* (Bendich and Anderson, 1983). From their data, Bendich and Anderson (1983) estimated the minimum genome size of subgenus *Equisetum* to be nearly 50% larger than that of subgenus *Hippochaete*. Bendich and Anderson (1983) used inferences about genome structure and size gained from DNA annealing studies for taxonomic purposes without actually annealing the DNAs from different species.

Restriction Fragment Length Polymorphisms

The use of restriction fragment length polymorphisms (RFLPs) of plant nuclear DNA represents one of the newest methods for studying various aspects of the nuclear genome, and the method may yield useful phylogenetic data. For example, it is possible to contruct linkage maps using RFLPs and thus determine whether different species have the same order of genes in their chromosomes. Also, RFLPs may be used to estimate the number of loci controlling morphological features, that is, the RFLPs act as markers to find quantitative trait loci (QTL). Lastly, variation within and divergence among populations of the same and different taxa can be estimated, and this variation may also be employed for generating phylogenies. The use of RFLPs in plant genetics and plant breeding has been detailed in a variety of papers (Helentjaris et al., 1986; Paterson et al., 1988; Bonierbale et al., 1988; Lander and Botstein, 1989; Tanksley et al., 1988a,b, 1989).

To carry out RELP studies of the nuclear genome, total DNA is digested with different restriction enzymes, the fragments are separated by gel electrohporesis, and the fragments are then blotted onto a filter. A labelled cloned probe of some 0.5 to 2.0 kb in size is then hybridized to the filter containing the fragments, and the fragments are then made visible by autoradiography. Obtaining the cloned probes of single-copy DNA is the most difficult methodological aspect of using RFLPs of nuclear DNA. Tanksley et al. (1988a,b) provide a general methodological overview, and references cited therein can be consulted.

Although RFLPs have been used very little in plant systematics, Doebley and Wendel (1989) have provided a concise and lucid discussion of the potential value and limitations of RFLPs for systematic studies. If a probe that is employed represents single-copy DNA (i.e., is unique within the genome), then one could consider it to be a distinct locus, and variations in the banding patterns seen with the probe could be viewed as "alleles"

(Doebley and Wendel, 1989). Doebley and Wendel (1989) emphasized that one advantage this approach has over isozymes is the much larger number of loci that can be examined, which in turn will provide much greater confidence in estimates of measures of genetic diversity and genetic distance, and in constructing phylogenies. It is also possible to interpret RFLPs on the basis of the presence or absence of different fragments in different plants when the filters are hybridized to the same probe. While these fragment differences can be used to generate pheno-grams and cladograms, it is important to know that the variation is caused by gains or losses at restriction sites and not by structural changes in the genomes (Doebley and Wendel, 1989). One structural mutation could cause variation to be seen with several different restriction enzymes, and thus one would count the same mutation several times; this same problem may arise with chloroplast DNA (see discussion in Chapter 16).

Beyond providing estimates of genetic diversity and divergence and generating data for constructing phylogenies, RFLPs offer the potential of providing insights into basic questions in plant evolution. The field of cytogenetics has provided valuable insights into relationships among species and into factors that preclude gene exchange among different plants. The construction of linkage maps from RFLPs allows much more refined insights into structural differences in the nuclear genomes of two species. Studies by Bonierbale et al. (1988) and Tanksley et al. (1988a) demonstrate the utility of the method for elucidating the nature and extent of chromosomal differences among peppers, potatoes, and tomatoes. This information is valuable for determining the role of chromosomal changes in the evolution and reproductive isolation of species.

In Chapter 8, levels of divergence at isozyme loci were contrasted with morphological differences. It was emphasized, however, that, for con-generic species, quite striking morphological differenes may be based on a few major gene loci. This makes it difficult to compare genetic diverg-ence at loci encoding soluble enzymes with divergence at loci controlling morphological features. RFLPs may be used to locate gene loci that control morphological features (QTLs) (Doebley and Wendel, 1989). A discussion of the methods involved in mapping QTLs is provided by Lander and Botstein (1989), and Paterson et al. (1988) used the approach for studying several characters in tomatoes. Because many more loci can be identified with RFLPs as compared with isozymes, the RFLPs are more effective for inferring the number of loci controlling morphological features.

Song et al. (1988) employed nuclear RFLPs to study relationships among diploid and tetraploid species of *Brassica*. The present discussion will be confined to the three diploid species and their three allotetraploid species, presumably derived from hybridization among different pairs of the diploid species. The taxa were considered in Chapters 3 and 4 and will be discussed in Chapter 16; all molecular data will be compared and contrasted as a case study in Chapter 17. Song et al. (1988) digested total DNA with three restriction enzymes and then hybridized filters containing the fragments produced by the three endonucleases with 13 cloned probes (12 were nuclear and 1 was from the chloroplast). Because of the complexity of the RFLP patterns (i.e, probes hybridized to a number of different fragments), it was not possible to compare species using the preferred mutational analysis of restriction sites. Rather, it was assumed that the common occurrence of fragments in different plants reflects shared restriction sites changes. Song et al. (1988) then treated each fragment as a single character, used several wild species as the outgroups, and then carried out a cladistic analysis (using parsimony) of the data.

Although the cladograms generated from the data have rather low consistency indexes (0.31), relationships among the diploid species were well resolved. *Brassica nigra* is similar to the species used as outgroups, and *B. campestris* and *B. oleracea* share 13 derived changes in RFLPs relative to *B. nigra*. the RFLP data were quite conclusive in documenting the origins of the two tetraploid species *B. carinata* (*B. nigra* and *B. oleracea*) and *B. napus* (*B. campestris* and *B. oleracea*) because the polyploids combined fragments unique to the presumed diploid parents. The situation was not resolved for the third tetraploid, *B. juncea*, because its RFLP patterns were essentially the same as one of its diploid progenitors (*B. campestris*) and did not contain fragments of *B. nigra*, the other presumptive diploid parent. Furthermore, Song et al. (1988) detected parallel variation in RFLPs between accessions of diploid *B. campestris* and *B. juncea*. They suggested multiple origins for *B. juncea*, but the question of why no indication of the *B. nigra* nuclear genome could be found in *B. juncea* remains unexplained.

The study by Song et al. (1988) demonstrates the potential of nuclear RFLPs for studying phylogenetic relationships in plants. Also, the data for *Brassica* are concordant with most other molecular information for the taxa. With regard to the origin of the allotetraploids, RFLPs were of value for two of the three species.

While RFLPs have considerable potential for addressing a variety of important questions in plant systematics and evolution, the rate and extent of their incorporation into the repertoire of the the plant systematists remain to be determined. It seems clear the RFLPs represent a valuable source of systematic and phylogenetic data, but as Doebley and Wendel (1989) indicated, the effort and expense involved in the proper use of RFLPs represent the two largest problems with their routine use in plant systematics.

SUMMARY

The nuclear genomes of higher plants contain much more DNA than is required to encode proteins. Also, the amounts of DNA in genomes may vary several-fold, even between species in the same genus. These two factors constitute the so-called C-value paradox. Decrease in DNA amount often correlates with specialization in a genus, but this is not always the case. The nuclear genome contains some sequences present once per genome, while others occur from several to thousands of times in a genome.

Hybridization of the nuclear genomes of different species has been employed for assessing DNA similarity, and thus for inferring whether species are closely related. The thermal stabilities of DNA duplexes formed between the DNAs of different species and the proportion of two DNAs that will anneal under given conditions represent the measures commonly employed for assessing similarities. While hybridization represented the method first employed for measuring DNA similarity, it has certain limitations, and it has been superseded to a considerable degree by more recently developed approaches for comparative studies of DNA for systematic purposes.

The use of nuclear RFLPs offers considerable potential for studying both phylogenetic relationships and evolutionary processes, particularly those concerned with speciation. Chromosomal evolution can be studied at a refined level not possible with classical cytogenetics. RFLPs also offer a method for studying the genetic basis of morphological differences between plants. Hopefully, the effort and expense of generating nuclear RFLP data will not prove prohibitive to plant systematists.

CHAPTER 15

Nuclear Ribosomal DNA: Structure and Use in Systematics

INTRODUCTION

The general structure of the genes encoding nuclear ribosomal RNA (rRNA), which will be referred to as rDNA, is diagrammed in Fig. 15.1. The basic units occur as tandem repeats in the cell, with the number varying from several hundred to thousands of copies per cell in a diploid plant. These repetitive sequences are located in the nucleoli. Another array of repeated sequences, the 5S rRNA genes, occurs in plants but is not associated with the 18, 5.8, and 25S rRNA genes. In this chapter, the use of rRNA genes for systematic and phylogenetic purposes will be discussed.

METHODS OF STUDY

A variety of methods has been employed to study rDNA; they have generated different kinds of systematic data. The number of systematically oriented investigations of rDNA is relatively small, but this situation is changing as a result of recent contributions. Earlier studies of rDNAs for systematic purposes involved annealing (see Maggini, 1975; Maggini et al., 1976; Vodkin and Katterman, 1971, as examples) and will not be considered in detail. These studies suffered from some of the same limitations discussed earlier for DNA–DNA hybridizations, and it seems fair to say that they did not contribute significantly to an understanding of relationships. A potential complicating factor when rDNA is employed in the hybridizations is the different copy numbers that may be present in the species (Rogers and Bendich, 1987; Jorgensen and Cluster, 1988).

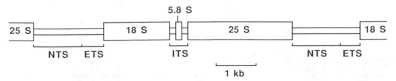

Figure 15.1   General structure of the rDNA repeating units for the 18S, 25S, and 5.8S rRNA in plants. The unit is composed of an external transcribed space region (ETS), the gene encoding for 18S rRNA, an internal transcribed spacer (ITS) on each side of the 5.8S rRNA gene, and the 25S rRNA gene. The transcribing units are separated from each other by the nontranscribed spacer (NTS).

More recent studies have involved use of restriction enzymes to cut the rDNA, separate the fragments on gels, blot the fragments onto filters, and then hybridize radioactively labelled cloned probes to the filters. It is possible to detect differences in the repeat lengths and in the restriction sites present in the rRNA genes of different plants. Variation in the length of the whole ribosomal repeat (i.e., the basic unit consisting of the nontranscribed or intergenic spacer, the 5.8S, 18S, and 25S genes with the included transcribed spacers; see Fig. 15.1), or various parts of the basic unit, may be ascertained. Lastly, sequencing of stretches of rDNA or direct sequencing of rRNA itself is the latest and most sophisticated method of study. The rDNA repeat units consist of sequences that are both highly conserved in plants (the 18S and 25S genes) and other sequences (the nontranscribed spacer) that are quite variable (see discussion in Jorgenson and Cluster, 1988). This means that depending on what part of the basic repeating unit is being studied, rDNA may be useful from the lowest taxonomic levels to comparisons at the highest levels of the taxonomic hierarchy.

VARIATION WITHIN AND AMONG POPULATIONS

Saghai-Maroof et al. (1984) studied rDNA variation in accessions (502 individual plants) of cultivated barley (*Hordeum vulgare*) and its wild progenitor *H. spontaneum*. Total DNA was digested with a restriction enzyme that cuts once on each side of the nontranscribed spacer region (referred to as the intergenic spacer or IGS in this paper). Filter hybridizations to a cloned labelled wheat rDNA repeat were used to produce autoradiograms (Fig. 15.2). The number of fragments observed for each plant varied from two to five, with the fragment at 3880 bp being invariant and presumably representing the region of the repeating unit containing the rRNA transcription units. The variable number of fragments in the size range of 4740 to 6350 bp ostensibly resulted from variation in the size of the nontranscribed spacer (NTS) or IGS. Part of this spacer region in rRNA genes is known to be made up of repeating units, called subrepeats (Jorgensen and Cluster, 1988), and in the case of barley the different fragment sizes differ by about 115 bp. Thus, the subrepeats are about 115 bp long, and the different fragment sizes reflect the number of subrepeats present (Fig. 15.2). Saghai-Maroof et al. (1984) detected more rDNA spacer length variation in *H. spontaneum* than in its cultivated derivative *H. vulgare*, whether the number of different spacer length variants or the number of different phenotypes, (that is, the number of different combinations of the variants detected in individuals) were considered. Inheritance of the spacer length variants was examined by making crosses between plants with differing phenotypes and observing segregation in the F_2 generation. Their results indicated that the spacer length variants are located at two unlinked loci and are inherited as codominants.

Flavell et al. (1986) studied variation in the nontranscribed spacer region of rDNA among 112 plants from 12 populations of the tetraploid wild wheat species *Triticum dicoccoides* occurring in Israel. They found certain populations to be variable while others were homogeneous in rDNA. Flavell et al. (1986) detected significant positive correlations between allozyme and rDNA variation within populations; they also reported a correlation between rDNA and protein variation in a population and its occurrence in a fluctuating environment. This study, like that of Saghai-Maroof et al. (1984) did not address basic systematic questions but demonstrated the practicality of the relatively simple methods for addressing problems of a systematic nature. Also, these studies documented that intrapopulational and intraspecific variation in the transcribed spacer must be considered in any systematic study.

Schaal et al. (1987) studied rDNA variation within and among populations of two subspecies of *Phlox divaricata*. No restriction site polymorphism was found, but twelve different length variants (of the rDNA units) were detected. Of interest from a systematic perspective is the fact that four length variants were found in only subspecies *divaricata* while three variants were restricted to subspecies *laphami*.

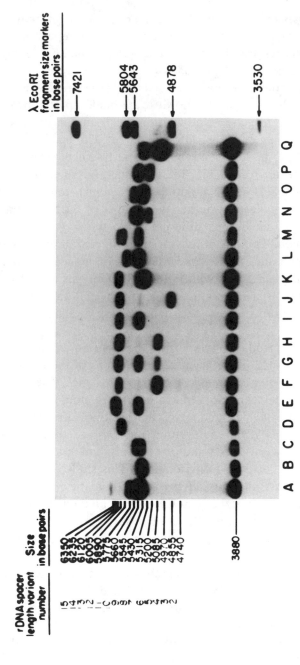

Figure 15.2 Autoradiograph of a Southern blot showing the different length variants of the nontranscribed spacer regions in *Hordeun spontaneum* and *H. vulgare*. The different size variants differ from the next variant by about 115 bp, which presumably results from the number of these subrepeating units present. The restriction enzyme employed has two sites in the repeating unit; thus the resulting fragments of 3880 bp are invariant and contain the genes for 5.8S and 25S rRNA. (Reproduced from Saghai-Maroof et al., Proc. Natl. Acad. Sci. USA 81:8014–8018. 1984. Photograph courtesy of R.W. Allard.)

Schaal and Learn (1988) presented a concise and lucid overview of the use of rDNA (essentially the nontranscribed spacer) for studying variation within and among populations. The nature and extent of rDNA variation within populations is different depending on the species. For some species (*Lupinus texensis*), as many as 11 length variants may be found within a single population, with a plant typically having 3 or 4 variants. For a number of species that have been examined, two or three length variants per plant appear to be common (Schaal and Learn, 1988). By contrast, in *Solidago altissima*, restriction site variation in the nontranscribed spacer represents the nearly exclusive source of variation at the individual, clonal, populational, and interpopulational levels. Other species, such as *Rudbeckia missouriensis*, exhibit neither length nor restriction site variation in rDNA in the six populations examined (Schaal and Learn, 1988).

In addition to the implications of the results discussed by Schaal and Learn (1988) for plant population biology, the data they present also serve as a caveat for those wishing to use variation in the nontranscribed spacer region of rDNA for systematic purposes. Namely, it is essential that variation within and among populations of a taxon be determined before making comparisons between them.

rDNA AND SYSTEMATIC STUDIES

A series of papers by various workers on rDNA structure and variation has appeared in recent years. Those discussed in some detail in an earlier review (Giannasi and Crawford, 1986) will be considered only briefly. The ribosomal genes of cereals, in particular, have been the objects of several investigations (Appels et al., 1980; Appels and Dvorak, 1982a,b; Dvorak and Appels, 1982; see also Saghai-Maroof et al., 1984, and Flavell et al., 1986, which were discussed earlier). Several studies dealt with the origins of polyploid wheats, and Fig. 15.3 will serve as a useful reference during these discussions.

Appels et al. (1980) showed that the 18–5.8–25S repeat unit in hexaploid wheat is 9 kb in size and that it has both *Eco RI* and *Bam HI* recognition sites. These workers also employed in situ hybridizations to document that most of the rDNA in hexaploid wheat is on chromosomes 1B and 6B, with much smaller amounts on 5D. The chromosomal locations of rRNA genes are not those expected, because plants that are

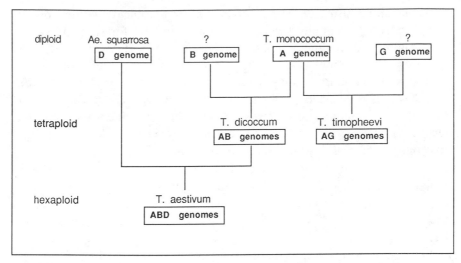

Figure 15.3 Phylogeny of polyploid wheats. The A and D genome donors are reasonably well accpeted, whereas there is less certainly about the diploids providing the B and G genomes. See text for discussion.

possible diploid donors of the A genome contain rRNA genes, and consequently one would anticipate their presence in the A genome of the hexaploid (Fig. 15.3). This finding suggests that factors other than simple hybridization have been involved in the origin of hexaploid wheat. Frankel et al. (1987) examined the distribution of rRNA genes in synthetic tetraploid hybrids between diploid species thought to be ancestral to the naturally occurring polyploids. These workers found transcriptively active rDNA in the polyploids in the same locations in the genomes as they are in the diploid progenitors. They suggested that deletion of the rDNA genes in the A genome must have occurred prior to domestication of the polyploid wheats, that is, over 10,000 years ago.

Peacock et al. (1981) studied the 18S–25S genes from hexaploid and tetraploid wheat together with the genes from diploid species representing potential donors of the A, B, and D genomes. Two fragments from the coding regions, one 0.9 kb and another 3.6 kb long, are invariant in all species. By contrast, the fragment patterns from the spacer region are variable. The hexaploid wheats (genome designation ABD, Fig. 15.3) have a major fragment of 4.4 kb with two minor variants. The tetraploid wheats, which presumably contain the genomes AB, show a very different pattern for the spacer region. All accessions except one contain the same 4.4 kb pattern typical of the hexaploids, but they have in addition 4.0 and 4.8 kb fragments. The A genome diploid species have the 4.8 kb fragment and the 4.0 kb fragment or a shorter variant of it, but they do not contain the 4.4 kb fragment characteristic of the tetraploids and hexaploids (Fig. 15.3). The latter fragment would not be expected in the A-genome diploids since it has been shown to occur in the B genome of the hexaploids. The absence of the 4.8 kb fragment in the hexaploids is puzzling, however, because it occurs in the A-genome diploids and the tetraploids. Equally strange is the fact that none of the potential B-genome donors contain the 4.4 kb fragment because this fragment *does* occur in the B genome of the hexaploid. Another perhaps even more enigmatic result is that *Aegilops squarrosa*, the generally accepted D-genome donor (Fig. 15.3), contains the 4.4 kb fragment detected in the B genome of the hexaploid wheat. Peacock et al. (1981) considered the possibility that rRNA genes from the D genome of a diploid progenitor became incorporated into the B genome of polyploid wheats. Suffice it to say that the 18S and 25S ribosomal genes offer little in the way of useful information in testing phylogenetic hypotheses for polyploid wheats.

Undaunted by the enigmatic results obtained with the 18S and 25S genes, Peacock et al. (1981) used the 5S rRNA gene to study the origin of polyploid wheats. When 5S rDNA is restricted with the enzyme *Bam HI*, two fragments are produced, one of 420 bp and the other of 500 bp. Each fragment contains one 5S RNA gene of 120 bp and spacer DNA. The 420 bp fragment is located primarily on chromosome 1B, so it could serve as a "marker" for the presence of that chromosome region. The taxonomic distribution of these fragments is as follows. All species have the 500 bp fragment. Diploid plants containing only the A genome do not have the 420 bp fragment, whereas all B-genome diploids and all hexaploid and tetraploid accessions have the 420 bp fragment. The distribution of fragments of the 5S ribosomal gene is consistent with widely held concepts on the evolution of polyploid wheats. The data do not, however, identify the species that donated the B genome.

Appels and Dvorak (1982a,b) and Dvorak and Appels (1982) examined variation in ribosomal genes within and among certain species of *Triticum*. One study (Appels and Dvorak, 1982a) examined spacer length variation within populations and among species of *Triticum*. These authors examined a 130 bp repeating unit found within the spacer region. Two of the 11 130 bp variants were sequenced, and the lowered thermal stabilities of heterologous versus homologous hybrids were employed for estimating sequence differences among the other 130 bp variants. In addition, a 750 bp nonrepeated sequence in the spacer region was studied. Different cultivars of hexaploid *T. aestivum* were assayed for sequence differences (using hybridizations) in the spacer region, and it was estimated that differences from the "standard" cultivar Chinese Spring ranged from 0.6 to 2.2% at the nucleotide sequence level (Appels and Dvorak, 1982a).

Appels and Dvorak (1982a) used the same technique for measuring the thermal stabilities of heterologous versus homologous duplexes to infer sequence differences in the 130 bp repeating unit among 19 individuals in a population of the tetraploid species *T. dicoccoides*. The thermal stabilities for most individuals were lowered about 1^0C when hybridized to the 130 bp probe from the hexaploid Chinese Spring cultivar, suggesting an average sequence divergence of 1.2% among the individual tetraploid plants from this population and the hexaploid cultivar. Employing the same methods but using the 750 bp spacer sequence gave similar ranges of sequence differences in the population of tetraploids.

Appels and Dvorak (1982b) examined relative divergence of sequences in the spacer region and the genes encoding rRNA. A total of 11 regions was studied; the methods involved hybridizing fragments from different parts of the rDNA repeating unit from six different species to labelled probes and then measuring the thermal stabilities of the heterologous duplexes compared with homologous ones. They found a 1.2 kb region preceding the 18S gene to be more highly conserved than other parts of the spacer. The transcribed spacer between the 18S and 25S genes (ETS; see Fig. 15.1) was not highly conserved, but the genes themselves were.

Doyle and Beachy (1985) examined rDNA variation in the cultivated soybean and its relatives in the genus *Glycine*. The enzyme *Xba I* cut the repeating unit only once in soybean, and the digests from this enzyme were hybridized with a cloned probe consisting of the whole 18S–25S rDNA repeat. Variation in the size of the repeating unit ranges from 7.8 to 12 kb. While no heterogeneity of repeat sizes was seen within individual plants, it was found between different accessions of a species. Variation was also found for the presence and the position of *Eco RI* sites in the nontranscribed spacer region.

Doyle and Beachy (1985) found the cultivated soybean *Glycine max* and its presumed progenitor, *G. soya* of subgenus *Soya*, indistinugishable in repeat lengths of rDNA and *Eco RI* site polymorphisms. These characters showed considerable variation in subgenus *Glycine*, as will be considered below. Doyle and Beachy (1985) emphasized that the considerable morphological differences between *G. max* and *G. soya* are associated with selection for domestication. As with many wild–domesticate pairs of taxa, the morphological differences could well have a relatively simple genetic basis, thus obscuring how similar the plants are in features not involved in the selection process. The reasons why there is so little variation in subgenus *Soja* as a whole as compared with subgenus *Glycine* are not apparent.

Within subgenus *Glycine*, certain correlations were noted between current taxonomic concepts and variation in 18S–25S rDNA, although Doyle and Beachy (1985) cautioned that small sample sizes for the taxa and the few restriction enzymes employed dictated that the taxonomic implications of their data be viewed as preliminary. *Glycine clandestina* is a morphologically variable species with several infraspecific groups recognizable on a variety of features. Three of the four accessions examined had the same *Eco RI* restriction site in the nontranscribed spacer region, but they differed slightly in repeat lengths. A fourth accession had been considered "atypical" for the species and similar to *G. tabacina* in habit and morphology. It was found that the rDNA repeat of this accession is like diploid *G. tabacina* in lacking the *Eco RI* site in the nontranscribed spacer. Also, the repeat length was shorter than in "typical" *G. clandestina*, and thus the rDNA data indicated the presence of different evolutionary lineages within what had been called *G. clandestina* (Doyle and Beachy, 1985).

The species *G. tabacina* consists of both diploid and polyploid plants, with few morphological distinctions between the two ploidy levels. Four of the five diploid accessions of *G. tabacina* lack an *Eco RI* site in the nontranscribed spacer, whereas one accession consisting of morphologically anomalous plants has the site (Doyle and Beachy, 1985). The ployploids, with one exception, are uniform in having short repeating units and three *Eco RI* restriction sites in the nontranscribed spacer. The one exception represents the only accession examined of a morphologically distinct plant from Taiwan. This plant has longer repeat units and lacks the *Eco RI* site typical of Australian polyploids. Considerable overexposure of films in producing the autoradiograms revealed a second repeat type the same as the "normal" pattern found in Australian polyploids, that is, it was the same length and had the *Eco RI* site. The presence of this repeat type in the "anomalous" Taiwanese polyplod clearly allies it to polyploids from Australia. Doyle and Beachy (1985) also emphasized that

TABLE 15.1. Repeat Lengths of 5S Ribosomal Genes Found in Populations of *Claytonia*

Species	Race	Chromosome Number	Collection Locality	Repeat Length of 5S Gene (in bp)
C. virginica	II	14	Arkansas	360
				380
				400
				440
				560
C. virginica	II	28	Texas	360
				380
				400
				440
				560
C. virginica	III	16	Indiana	335
C. virginica	III	16	Michigan	335
C. virginica	IV	30–32	Indiana	360
				370
C. virginica	IV	30–32	Indiana	360
				370
C. carolinana		16	Michigan	350
C. carolinana		16	West Virginia	350
C. carolinana		24	North Carolina	350
				400

From Doyle et al., 1984.

the presence of the *Eco RI* site in the "normal" Australian polyploids and its absence in the diploids indicates a consistent divergence between the two cytotypes.

Doyle et al. (1984) examined rDNA variation in polyploid complexes of *Claytonia* occurring in eastern North America. The complexes include the two largely allopatric species *C. virginica* and *C. caroliniana*. As reviewed by Doyle et al. (1984) and discussed in detail by Doyle (1981, 1983), infraspecific taxa are recognizable in both species. Among the factors distinguishing the races are chromosome number, morphology, flavonoid chemistry, ecology, and geographical distribution (Doyle et al., 1984). In *C. virginica*, races I, II, and III contain diploids, and the former two also have polyploid populations (Table 15.1). The diploid and polyploid members of race II are essentially indistinguishable (Doyle, 1981, 1983). Race IV consists entirely of polyploids, and the diploids from whch this race evolved are not known. *Claytonia caroliniana* consists largely of diploid plants, but polyploids are known, and they are most prevalent in the Smokey Mountains in Tennessee and North Carolina (Doyle et al., 1984).

Doyle et al. (1984) had several objectives in mind in using rDNA to study *Claytonia*, including ascertaining whether variation in rDNA could be detected in the eastern species, seeing whether variation patterns (if detected) are concordant with other available data, and determining whether molecular data could elucidate relationships not resolvable with previous methods. Variation in 5S RNA genes was examined in three races of *C. virginica* as well as diploid and polyploid populations of *C. caroliniana*. The results showed that each race of *C. virginica* has a characteristic 5S gene profile (Doyle et al., 1984; Table 15.1). Both diploid and polyploid populations of race II have the same size polymorphisms, ostensibly representing heterogeneity in spacer lengths. Each population sample of race II consisted of a number of individuals, and thus it was not possible to ascertain whether the variation resulted from different lengths within individuals or polymorphism within populations. It is interesting that the diploids and tetraploids of race II have the same pattern of variation because other data suggested that the tetraploids had originated solely from race II diploids (Doyle et al., 1984).

The repeat length observed in *C. caroliniana* is intermediate between the largest and shortest lengths found in *C. virginica* (Table 15.1). The diploid and tetraploid cytotypes of *C. caroliniana* have the same major repeat length, with a minor longer variant detected in the polyploids (Table 15.1). In summary, 5S rDNA variation in *Claytonia* is concordant with infranspecific variation found in other features.

Examination of 18S and 25S rDNAs using the same methods as employed for 5S genes revealed differences between the races in repeat lengths and/or restriction sites. For example, race III has a unique *Eco RI* restriction site in the nontranscribed spacer region. The diploid and tetraploid plants of race II have the same restriction fragments (Doyle et al., 1984), and plants of râce IV likewise exhibit unique profiles.

Diploids of *C. caroliniana* are distinct from all representatives of *C. virginica*. this is so despite the fact that sympatric populations of the two species are the same in flavonoid chemistry and cytology (Doyle et al., 1984). This means that ribosomal gene variation was useful for distinguishing the populations in cases where other approaches had failed. Disjunct populations of diploid *C. caroliniana* from the Ozarks are different from other populations of the species in rDNA despite the fact that chromosome number and flavonoid chemistry do not differentiate them. Lastly, 18S and 25S ribosomal genes indicate that a polyploid population of *C. caroliniana* from North Carolina is more similar to race IV poly-ploids of *C. virginica* than to diploids of *C. caroliniana*. Doyle et al. (1984) suggested that these data could be indicative of hybridization, a notion implied from previous work.

Doyle and Doyle (1988) followed up on the question of possible interspecific hybridization and studied *C. caroliniana* and *C. virginica* (both diploid with a chromosome number of *n* = 8) at a single locality in New York State where the two species are sympatric, with hybridization suspected on the basis of morphological considerations. Morphology and pollen stainability were examined along with repeat length and restriction site differences in rDNA. A problem with using morphology to detect hybridization between the two species is the very few features that dis-tinguish them. Doyle and Doyle (1988) determined that *C. caroliniana* and *C. virginica* differ in rDNA repeat lengths with the former having a length of 9.5 kb and the latter 10.2 kb. Also, *C. virginica* has a recogni-tion site for the restriction enzyme *Eco RI* that does not occur in *C. caroliniana*. Of the 94 plants examined from the population in which the two species occur together, 14 individuals combined the rDNA profiles unique to each of the species. Data from rDNA provided strong evidence of interspecific hybridization when other information was equivocal. Doyle and Doyle (1988) emphasized that, while hybridization between *C. caroliniana* and *C. virginica* is probably quite rare, certain naturally occurring tetraploids resemble the diploid hybrids morphologically and thus may represent allotetraploids. These polyploid plants combine the rDNA profiles of *C. caroliniana* and *C. virginica*, that is they are the same as the putative diploid hybrids. This fact strongly suggests that the tetraploids are allopolyploids.

Doyle et al. (1985) used ribosomal RNA genes to test the hypothesis of intergeneric hybridization in the Saxifragaceae. Morphological evi-dence indicated that *Tolmiea menziesii* and *Tellima grandiflora* occa-sionally produce naturally occurring hybrids. The repeat lengths of the 5S RNA genes, as determined by the pattern of fragments from enzymes with one recognition site in the genes, were about 480 bp for *Tolmiea* and 450 bp for *Tellima* (Doyle et al., 1985). The suspected hybrid was fixed for both of the repeat lengths found in the parental species (Fig. 15.4*A*). The repeat length for 18S and 25S ribosomal genes in *Tolmiea* is 11–13 kb, and it is about 9 kb in *Tellima*. As with the 5S genes, the hybrid contains both of the repeats of the parental species (Fig. 15.4*B*) (Doyle et al., 1985).

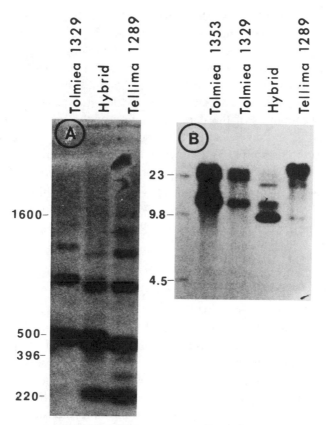

Figure 15.4 Autoradiographs of rDNA from *Tellima*, *Tolmiea*, and their hybrid. (*A*) Genomic DNAs digested with *HaeIII* and the filters hybridized to a labelled probe containing the entire 5S rDNA repeat. Left lane is *Tolmiea* with repeat a length of about 480 bp, while *Tellima* in the right lane has a repeat length of 450 bp. Note that that the hybrid in the middle lane contains both lengths. Numbers to the left denote approximate sizes of fragments. (*B*) Genomic DNAs restricted with *Eco RV* and the filter hybridized to a labelled probe containing a single 18S–25S rDNA repeat. The two lanes on the left contain DNAs from *Tolmiea*, with a repeat length of 11–13 kb, and the right lane has DNA from *Telima* with a faint fragment of 9.0 kb. The hybrid DNA in the second lane from the right has both size fragments. In the hybrid, the *Tellima* band is much stronger than the one from *Tolmiea*. (With permission from J.J. Doyle et al., Am. Jo. Bot. 72:1388–1391. 1985, Botanical Society of America.)

Zimmer et al. (1988) studied the rDNA in cultivated maize, its wild relatives the teosintes, and the genus *Tripsacum*. A total of 15 restriction enzymes was used to digest nuclear DNA; following electrophoresis and blotting, the filters were probed with labelled cloned rDNA. Despite this quite extensive study of the rRNA genes, Zimmer et al. (1988) detected so few shared variable characters that they did not attempt a quantitative phylogenetic analysis of the data. Inspection of the maps produced from the restriction endonuclease cleavage sites allowed for the rDNAs to be placed into three groups. These include (1) *Zea mays* subsp. *mays* and *Z. mays* subsp. *mexicana*; (2) *Z. luxurians, Z. diploperennis,* and *Z. perennis*; and (3) several species of *Tripsacum*. These groupings follow the generally accepted taxonomic treatment by Doebley and Iltis (1980) because *Z. mays* is assigned to one section of the genus and the other three species are placed in another section.

PHYLOGENETIC ANALYSIS OF rDNA

Sytsma and Schaal (1985b) used rDNA to examine phylogenetic relationships in the *Lisianthius skinneri* complex. Four distinct species occur in central Panama, with each known from only one or several mountain peaks in cloud forests. The fifth species, *L. skinneri*, is found at lower elevations and is widespread and patchy in distribution (Sytsma and Schaal, 1985b). Based on morphological and other considerations, there appears little question that these five species represent a monophyletic assemblage. The four endemic species are *L. aurantiacus, L. habuensis, L. jefensis,* and *L. peduncularis. Lisianthius seemannii* is closely related to the *L. skinneri* complex, yet on the basis of morphology is not a member of it, and thus it was used as the outgroup for the latter. Before discussing the rDNA results, it is worth emphasizing that rDNA represents part of a broad-based study by Sytsma and Schaal (1985a,b) of molecular variation in the *L. skinneri* complex, including investigations of isozymes and chloroplast DNA. The isozymes were discussed in Chapter 8, and the chloroplast DNA results will be considered in Chapter 16. Also, their studies should be of particular interest to plant systematists because plant material collected from natural populations in Panama was employed. Leaves were collected from plants in Panama, blotted dry, placed in bags and sealed, and shipped on wet ice to the laboratory.

Little variation was detected in the rDNA within individuals of a population and among different plants of the same population. By contrast, considerable variation was detected in the rDNA repeat length among different species. Two of the eastern populations of *L. skinneri* and one of the cloud forest species, *L. iefensis*, have a repeat length of 11.5 kb. This length probably represents the primitive (ancestral) size in the *L. skinneri* complex because it also occurs in the outgroup *L. seemannii* (Sytsma and Schaal, 1985b). A western population of *L. skinneri* and the other three cloud forest species have larger repeats (from 11.9 to 12.4 kb). All restriction site mutations in the rDNA except one were found in the nontranscribed spacer region. In all, a total of 313 restriction sites (number of sites recognized by all of the enzymes employed) accounting for 1878 bp were examined in plants from the seven populations. Eight mutations were detected, with half of them unique to single species and the other half detected in two or more species. Several insertions and deletions were also found but were not employed in the phylogenetic analysis because the lack of precision in mapping them prevented ruling out the possibility that they result from several events rather than a single one. A phylogeny, employing maximum parsimony, was constructed from the rDNA data (Fig. 15.5), and it was not necessary to postulate a single convergent restriction site mutation in the phylogeny.

Several aspects of the rDNA-based phylogeny are of interest relative to other data available for the *L. skinneri* complex. *Lisianthius jefensis* and the two eastern Panamanian populations of *L. skinneri* are indistinguishable on the basis of rDNA (Fig. 15.5), despite the fact that the two species are separable by a number of conspicuous morphological features of the leaves and flowers as well as being divergent in isozymes (Sytsma and Schaal, 1985a). Also of phylogenetic interest is the observation that the three cloud forest species *L. peduncularis, L. aurantiacus,* and *L. habuensis* share a derived restriction site change and an insertion with the western population of *L. skinneri*. These data indicate that the eastern and western populations of *L. skinneri* represent two evolutionary lines *and* that the three cloud forest species were derived from *L. skinneri* elements similar to the extant western plants. The two cloud forest species *L. aurantiacus* and *C. peduncularis* are separable only by a 100 bp length difference in rDNA (Fig. 15.5). Morphologically, these two species share several derived features but also are highly divergent in such characters as inflorescence, growth form, and flower color (Sytsma and Schaal, 1985b).

Results of the study by Sytsma and Schaal (1985b) demonstrate that

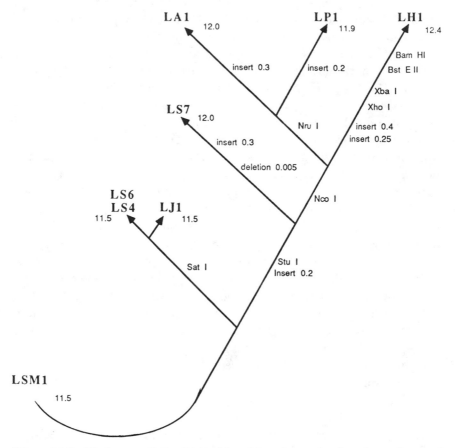

Figure 15.5 Phylogeny of the *Lisianthius skinneri* complex based on ribosomal DNA. The eight restriction site mutations are indicated by the names of the enzymes. Sizes of deletions and insertions are given along branches to populations; the number below each population gives the rDNA repeat length in kb. Species designations are LS (numbers refer to different populations), *L. skinneri*; LJ1, *L. jefensis*; LH1, *L. habuensis*; LP1, *L. peduncularis*; LA1, *L. aurantiacus*; LSM1, *L. seemannii*. (Redrawn from R.J. Sytsma and B.A. Schaal, Evolution **39**:504–508.)

divergence in rDNA is not concordant with observed morphological differences or speciation. The molecular data offer refined phylogenetic insights into the complex not attainable with morphology, namely that the cloud forest species represent two phyletic lineages, each of which was derived from different populations of *L. skinneri*.

Hamby and Zimmer (1988) used direct sequencing of rRNA from nine species of grasses (Poaceae). The sequences determined were from eight different sections in the 18S and 25S portions of the rRNA molecules, and over 1600 nucleotides were sequenced. Hamby and Zimmer (1988) detected 85 positions that were variable and informative, and from these data a phylogenetic tree was generated using parsimony. In addition, the sequence data were converted to distances among the taxa, and a phenogram was produced by clustering.

Although the number of genera (nine) was quite limited, the results of both the cladistic and phenetic analyses are of some interest when compared with each other and with generally accepted taxonomic relationships in the grasses. Of the nine genera, four (*Saccharum, Sorghum, Tripsacum*, and *Zea*) are in the subfamily Panicoideae, three (*Avena, Hordeum*, and *Triticum*) are placed in the Pooideae, and two (*Arundinaria* and *Oryza*) are traditionally classified together in the subfamily Bambusoideae. The most parsimonious tree resulting from cladistic analysis placed the four genera of the Panicoideae as a monophyletic assemblage, and the same results were obtained for the three genera in the Pooideae. Within the latter subfamily, however, cladistic analysis placed *Hordeum* and *Avena* together as more closely related to each other than either is to *Triticum*. This grouping is clearly not concordant with most views of relationships, which treat *Avena* as more distantly related to the other two genera. Lastly, cladistic analysis of the sequence data does not produce a tree in which *Arundinaria* and *Oryza* constitute a monophyletic assemblage apart from their inclusion as members of the grasses. Phenetic analysis of the data places *Oryza* within the Panicoideae, which is not in agreement with current taxonomic views.

Hamby and Zimmer (1988) indicated that additional taxonomic sampling as well as sequencing additional stretches of rRNA might elucidate certain problematical aspects of their results. Despite these limitations, their data are highly concordant with established taxonomic comcepts in the Poaceae. Clearly, considerably more data are needed before the impact of rRNA sequence data on grass systematics can be established.

SUMMARY

The genes encoding ribosomal RNA offer several advantages for systematic studies. Their presence in many copies per genome minimizes the amount of plant material needed; this may be an important factor when working with plants from natural populations. The simplicity of the methods for isolating rDNA is likewise a significant consideration for the practicing plant systematist, because many individual plants can be examined (Schaal and Learn, 1988). The use of rRNA genes for systematic purposes requires that cloned probes be available so that the fragments produced by digesting total cell genomic DNA with restriction enzymes may be visualized by autoradiography. This, however, is not a problem because clones of the ribosomal repeating units are readily available. If one wishes to carry out sequencing, the abundance of rDNA and rRNA likewise represents a distinct advantage. The highly conserved nautre of the genes encoding 18S and 25S rRNAs as contrasted with the highly variable nontranscribed spacer regions means that (depending on the methods employed) rDNA can be employed at a wide variety of taxonomic levels (Jorgensen and Cluster, 1988).

Jorgensen and Cluster (1988) considered certain limitations in using restriction enzyme analysis of rDNA for systematic and phylogenetic purposes. One factor mentioned was that the number of restriction enzymes that can be used is smaller than for chloroplast DNA because the nuclear rDNA is methylated at most CG dinucleotides and CXG (X could be any of the nucleotides) trinucleotides, and many enzymes that cleave sequences containing these di- and trinucleotides will not do so if they are methylated. This means that many variants will go undetected.

Willams et al. (1988) and Hillis and Davis (1988) discussed and debated possible problems associated with using heterologous probes to study length and restriction site variation in the nontranscribed spacer region. Because the genes for 18S and 25S rRNA are so highly conserved, it is possible to use a probe from a distantly related spaces. However, if fragments generated by a restriction enzyme contain either none or very short segments of the conserved regions, then these fragments will not hybridize to the probe, and problems in interpretation could be encountered. Use of several restriction enzymes and determining the relative positions of the sites (i.e., mapping the sites) alleviates problems of interpretation of length and restriction site variation, but caution must be exercised to ensure proper interpretation of the data.

The useful sources of data include the lengths of the repeating units in different taxa and changes in restriction sites. Digesting the rDNAs with enzymes with only one site in the repeating unit allows for ascertaining the length of the units. If sequencing is done, then the nucleotide sequences themselves represent the data.

Because rDNA is inherited biparentally, it is useful for studying hybridization and reticulate evolution in plants. In this regard, it is sometimes of greater utility than chloroplast DNA (see Chpater 16). Although relatively few studies are presently available employing ribosomal genes for systematic and phylogenetic purposes, those that have been done attest to their value and potential. These factors, combined with the relative simplicity of the methods, suggest that an ever-increasing number of studies will incorporate length and restriction site data from rDNA.

There seems little question that the sequencing of the highly conserved regions encoding rDNA will be an area of increasing activity in the future. The results of such studies should be phylogenetically useful at higher taxonomic levels.

CHAPTER 16

The Chloroplast Genome and Plant Systematics

INTRODUCTION

Chloroplast DNA (cpDNA) is being used more and more frequently in systematic and phylogenetic studies of plants. A number of recent reviews have discussed the value of the chloroplast genome for inferring relationships in plants (Curtis and Clegg, 1984; Giannasi and Crawford, 1986; Palmer, 1985a, 1986b, 1987; Zurawski and Clegg, 1987; Palmer et al., 1988). In addition, reviews dealing with the organization and expression of the chloroplast genome, and the evolution of the genome per se, have appeared (Whitfield and Bottomley, 1983; Palmer et al., 1984; Gillham et al., 1985; Palmer 1985b,c; Clegg et al., 1986; Zurawski and Clegg, 1987). In this chapter, the structure and organization of the chloroplast genome in land plants will be reviewed very briefly. The advantages of cpDNA for studies of plant systematics and evolution will be considered, and then examples of the systematic applications of cpDNA data will be presented.

STRUCTURE AND ORGANIZATION OF THE CHLOROPLAST GENOME

Detailed considerations of the organization fo cpDNA are presented elsewhere (Whitfeld and Bottomley, 1983; Palmer et al., 1984; Gillham et al., 1985; Palmer, 1985b,c; Zurawski and Clegg, 1987); thus, a brief discussion will suffice. Palmer (1985c) presented a table of the sizes of cpDNA molecules in land plants (and algae, but these will not concern us here); the size range is from 127 to 217 kb. Palmer (1985c) emphasized,

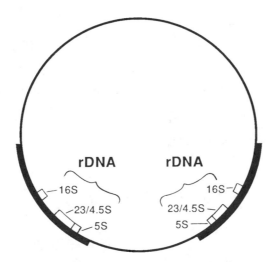

Figure 16.1 Generalized diagram of the chloroplast DNA molecule in higher plants. Heavy lines denote the inverted repeat containing the genes for rRNA. See references in text for more detailed descriptions of the chloroplast genome.

however, that this range distorts the uniformity found in most land plants because the low end of the scale is occupied by particular legumes, and ony two species account for an additional 55 kb in size. The majority of angiosperms examined, therefore, have a narrow range in chloroplast genome size, between 135 and 160 kb.

Chloroplast chromosomes occur as closed circles (Fig. 16.1). In contrast to nuclear genomes, there are very few repeated sequences in the cpDNA of land plants. One repeat is present in all land plants investigated (except one group of papilionoid legumes and certain genera of the Pinaceae); it is a large inverted repeat containing the genes encoding rRNA (Fig. 16.1). The repeat ranges in size from 10 to 76 kb in land plants (see Table II in Palmer, 1985c), and it divides the chloroplast molecule into large and small single-copy regions (Fig. 16.1). Additional genes have been identified in the chloroplast genome; Palmer (1985c) and Zurawski and Clegg (1987) provided reviews while Ohyama et al. (1986) and Shinozaki et al. (1986) determined the complete sequences for the genomes of the liverwort *Marchantia polymorpha* and tobacco (*Nicotiana tabacum*), respectively.

Other than the large inverted repeats, the structural organization of the chloroplast chromosome of land plants, like its size, is rather highly conserved. The cpDNA molecule has been relatively free of large deletions, insertions, transpositions, and inversions during evolution (Palmer, 1985b, 1987). When sizable inversions occur, they often characterize certain higher taxa (see later discussion). Deletions and insertions are often quite small (several to several hundred base pairs) and usually cause few problems of detection and interpretation. This is true unless they are in "hotspot" regions of the molecule where many length mutations have occurred, making assessment of homologies difficult (Plamer, 1985c; see also comments later in this chapter).

ADVANTAGES OF CHLOROPLAST DNA FOR PLANT SYSTEMATICS

The chloroplast molecule is present in many copies per cell, making it easy to isolate in sufficient quantities from very small amounts of plant material. As few as 10 g fresh weight is adequate for obtaining partially purified cpDNA. Methods employing total DNA require less than 1 g fresh weight (Doyle and Doyle, 1987; Doyle and Dickson, 1987). All the molecules within individuals appear to represent a homogenous assemblage, that is, there is no evidence of heterogeneity in size or structure within a plant.

The small size of the cpDNA molecule makes it possible to visualize on a single agarose gel all the fragments produced by digestion with many of the common six-base restriction enzymes (see Fig. 16.2 as an example) (Palmer, 1987). This is ideal because it often allows the systematist to make at least preliminary comparisons of differences between species and to gain valuable data with minimal time, effort, and material. Visualization of fragments in gels in UV light after staining with ethidium bromide is feasible only if the cpDNA has been partially purified. Palmer (1986a) presented a concise and complete discussion of methodology. Even for studies of DNAs of cogeneric species, it is increasingly common for blots and filter hybridizations to be used for critical comparisons (Sytsma and Gottlieb, 1986a,b; Doebley et al., 1987; Jansen and Palmer, 1988, as examples).

The homogeneity of DNA molecules within individuals is an important advantage to the systematist because it means there is no problem with ascertaining the different size classes (or other differences) present in individual plants. The structural homogeneity within individuals also extends, in most instances, to different individuals of the same population and different populations of the same species (Palmer, 1987; Banks and

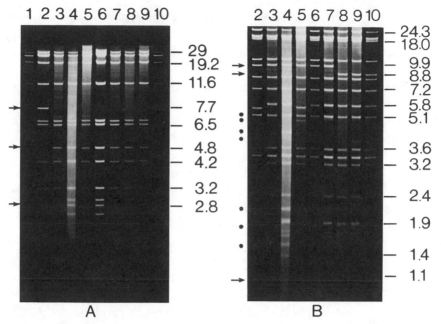

Figure 16.2   Photographs of cpDNA fragments from *Lycopersicon* and *Solanum* separated by electrophoresis in 0.8% agarose gels. (*A*) Digests of the restriction enzyme *Kpn I*. Note that samples 1 and 2 have a fragment of 7.7 kb that is lacking in samples 3–10. The latter DNAs have fragments of 4.8 and 2.8 kb that are lacking in DNAs 1 and 2. These two patterns are best interpreted as DNAs 3–10 having a restriction site for *Kpn I* that is lacking in samples 1 and 2. (*B*) Fragments produced by the endonuclease *Sac I*. Numbers on the right refer to sizes in kb of the different fragments, which were determined by reference to digests of phage DNA of known sizes. Arrows to the left of each gel denote differences in fragment sizes resulting from restriction site mutations. Sample numbers represent species as follows: 1, *Solanum lycopersicoides*; 2, *S. juglandifolium*; 3, *Lycopersicon peruvianum*; 4, *L. chilense*; 5, *S. pennellii*; 6, *L. hirsutum*; 7, *L. chmielewskii*; 8, *L. esculentum*; 9, *L. pimpinellifolium*; 10, *L. cheesmanii*. Samples 2–7 have a 9.9 kb fragment that is lacking in DNAs 8–10. Samples 8–10 have fragments of 8.8 and 1.1 kb instead. The different patterns result from a restriction site for *Sac I* that is present in samples 8–10 but not in DNAs 2–7. (Reproduced from J.D. Palmer and D. Zamir, Proc. Natl. Acad. Sci. USA **79**:5006–5010. 1982. Photograph courtesy of J.D. Palmer.)

Birky, 1985). The significant factor for the plant taxonomist interested in working at the species level or higher is that cpDNA is useful because of the minimal variation encountered within and among conspecific populations. This topic wil be considered in more detail later in the chapter.

The lack of frequent structural changes (inversions, transpositions, deletions, and insertions) in the chloroplast genome makes it relatively easy to work with in comparative studies. This is true because restriction pattern differences between species usually result from mutations at restriction sites rather than from structural changes. If the latter were frequent it would be difficult, tedious, and time-consuming to do comparative studies involving a number of taxa. Also, because changes such as large inversions are rare, they may prove to be valuable phylogenetically for identifying monophyletic assemblages (see later discussion).

The rate of sequence change in cpDNA appears to be rather conservative compared with single-copy nuclear DNA in plants, and may be about the same as the plant mitochondrial genome, although the data on which such statements are made are very scanty (see discussions in Palmer, 1985b, 1987, and citations therein). Analyses of data by Wolfe et al. (1987) suggest that the silent substitution rate for plant mitochondrial DNA is only one-third of the rate for cpDNA, but that the rate for cpDNA is about half the rate for nuclear DNA. The relatively slow rate of nucleotide substitutions in cpDNA minimizes the problem of parallel and convergent evolution when comparing genomes of congeneric species. This slow rate of evolution may also be a problem in some cases because variation is inadequate to resolve relationships among all species. It is possible in some cases to do valid comparisons among distantly related genera within a large and diverse family such as the Asteraceae (Michaels et al., 1987; Jansen and Palmer, 1988; Palmer et al., 1988 as examples). Also, different parts of the cpDNA molecule evolve at different rates, making it possible to compare taxa at various hierarchical levels by using different parts of the molecule.

As indicated earlier, the chloroplast genome is composed almost entirely of single-copy or unique-sequence DNA. This makes it easier to employ in systematic studies because the problem of determining the evolutionary history of the DNA sequences relative to the phylogeny of the organisms is not encountered.

STUDIES OF CHLOROPLAST DNA AT THE INTRA- AND INTERPOPULATIONAL LEVEL

Preliminary results suggest that variation of cpDNA is minimal within and among populations of the same species. Studies of variation at the intra- and interpopulational levels have been few in number, with even fewer concentrating on the subspecific level. Additional studies focused at these levels are needed before more definitive statements can be made about variation.

Scowcroft (1979) examined cpDNA from nine populations of *Nicotiana debneyi* by restricting the DNAs with the enzyme *Eco RI*. Chloroplast DNA was extracted from bulked young plants from these populations; thus there was no attempt to determine intrapopulational variability (Scowcroft, 1979). Plants from seven populations displayed one pattern when fragments were separated on gels whereas those from the other two populations had a different pattern. One pattern had two small fragments, the combined size of which equalled a large fragment unique to the other pattern. The different patterns result from the presence of an *Eco RI* restriction site in one group of plants that is lacking in the other, that is, a restriction site was either gained or lost by mutation. While individual plants were not studied, Scowcroft (1979) commented that no heterogeneity was detected in gels containing fragments of DNA isolated from up to 15 plants. This finding suggests very strongly that there is no intrapopulational variation for the restriction site difference.

Palmer and Zamir (1982) used 25 restriction enzymes and detected low levels of cpDNA polymorphism in the species *Lycopersicon peruvianum*. They used single plants for each of the six populations of the species examined; thus no attempt was made to assess intrapopulational variation.

Clegg et al. (1984) studied cpDNAs from twelve collections of *Pennisetum americanum* (pearl millet) representing the range of morphological variability and geographical distribution of the species. No variation was detected among any of the collections when the fragments were blotted onto filters and the filters hybridized to several labelled cloned fragments from the pearl millet genome.

Palmer et al. (1985) found restriction site variation among individuals of a population of *Pisum humile*. Certain individuals had a fragment of 12 kb when their DNAs were digested with the enzyme *Hpa I* whereas other plants lacked this fragment but contained instead fragments of 8.5 and 3.4 kb. The two chloroplast genomes differ, therefore, in that the one with the two small fragments has a restriction site for *Hpa I* that is lacking in the other molecules. Also, different lengths in cpDNAs were found among individuals within the same population of *P. humile* (see discussion later in this chapter).

Banks and Birky (1985) examined 100 plants from 21 populations of *Lupinus texensis*. Eighty-eight of the plants produced the same pattern of fragments for each of the seven restriction enzymes. In one population, the cpDNA of one of the two plants examined had fragments of 9.9 and 3.7 kb when digested with the enzyme *Bcl I*, whereas the common (referred to as the wild type by Banks and Birky, 1985) pattern had a single fragment of 13.7 kb and lacked the two smaller fragments. This finding suggests that the DNA with the two smaller fragments had a restriction site for *Bcl I* not present in the wild-type cpDNAs. One of six plants examined from another population lacked a restriction site for the enzyme *Bgl II* that was present in the common forms of cpDNA.

In one population, all 11 plants examined had cpDNAs that differed from the wild-type DNAs in several ways. First, the same loss of a site for *Bgl II*, as noted above, was seen in the DNAs of all plants. Additional variations in the cpDNAs of plants from this population were due to a deletion rather than restriction site change, and will not be discussed further here.

Wagner et al. (1987) examined cpDNA variation in 153 individuals from 63 populations of lodgepole pine (*Pinus contorta*) and 210 plants from 68 populations of jack pine (*P. banksiana*). Although only two restriction enzymes were employed, these workers detected two different fragment patterns for *Sal I* and 13 distinct patterns for *Sst I* in the two closely related species.

Neale et al. (1988) studied cpDNA variation within and among 30 populations (245 accessions) of wild barley (*Hardeum vulgare* subsp. *spontaneum*) in Israel and Iran. They surveyed about 2000 bp in the chloroplast molecules and detected polymorphisms at three restriction sites and one insertion–deletion. More than one plant was examined in 26 of the 30 populations; 15 populations were monomorphic, and 11 were polymorphic for at least one of the three restriction sites. Some geographic pattern was noted for certain cpDNA types in Israel, although the different cpDNAs are strictly allopatric. Neale et al. (1988) emphasized that, while cpDNA diversity is minimal in wild barley, erroneous phylogenetic conclusions could be drawn if it were assumed that no polymorphism exists within the species and thus that one sample is representative of the whole species.

Rieseberg et al. (1988) detected some cpDNA variation among populations of *Helianthus bolanderi* and *H. annuus*. The greatest divergence occurs between the serpentine and weedy races of *H. bolanderi*, where, depending on which populations are compared, they differ by eight or nine restriction site mutations. Within each of the races of *H.*

bolanderi, however, no variation was observed among populations of the weedy race (Rieseberg et al., 1988). Since it is possible that these two races are best treated as distinct species (Rieseberg, personal communication), it may be that infraspecific variation in cpDNA is essentially nonexistent in *H. bolanderi*. Three restriction site mutations were found within *H. annuus*.

Soltis et al. (1989) included 28 populations of the species *Heuchera micrantha* in a survey of cpDNA variation. This species is variable morphologically, and several varieties accommodate the variation; also, both diploid and tetraploid plants occur. Fourteen restriction site mutations and three deletions were detected. Four of the mutations were found in single populations while the other 10 occurred in 2 to 6 populations. This study represents the highest level of diversity found in the cpDNA of a single species.

Studies of variation within and among populations of plants of the same species suggest that variation in cpDNA restriction sites is quite small. This no doubt is a reflection of the low rate of mutation in the chloroplast genome and illustrates that it is a rather conservative molecule (Palmer, 1987). The studies discussed in this section indicate quite clearly that one cannot assume a priori that the cpDNA from one plant or one population is representative of the entire species, that is, in most instances in which interpopulational sampling has been done, some diversity has been detected. It seems particularly desirable to obtain cpDNAs from plants representing the range of morphological variation and geographical distribution of a species. If subspecific taxa have been recognized, then it is important to sample each of them. Just how much lack of infraspecific sampling affects interspecific comparisons will no doubt vary for different groups of plants.

CHLOROPLAST DNA AND THE STUDY OF DIVERGENCE AMONG CONGENERIC SPECIES

A number of studies, particularly earlier ones, employed a single restriction enzyme for digesting cpDNAs with the resulting fragments compared after separation in agarose gels (see Vedel et al., 1976, 1978; Atchison et al., 1976; Lebacq and Vedel, 1981, as examples). While these studies offered some useful data, it is highly desirable to use several restriction enzymes, with the more the better. The reason for this is that the more enzymes employed the more restriction sites surveyed, and thus the larger the sample of the chloroplast genome. Sequencing the molecules would be most desirable, but given the time, effort, and expense, this procedure is not feasible for comparative systematic studies. As Palmer (1987) emphasized, fragment patterns in gels produced by restriction enzymes may result in ambiguities because of insertions and deletions, and additional (i.e., mapping) studies are needed to clarify the situation. This will be considered in more detail later when specific studies are discussed.

Restriction site analysis of cpDNA may be used to investigate divergent evolution among species as well as to study reticulate evolution, with the latter case often involving the origin of allopolyploid species. Selected examples of each situation will be discussed.

Palmer and Zamir (1982) examined the cpDNAs from eight species of *Lycopersicon* and two species of *Solanum* thought to be closely related to *Lycopersicon*. Twenty-five restriction enzymes were used to digest the DNAs from a total of 15 populations, and from these a total of 39 different phenotypes (fragment patterns) was produced (Fig. 16.2). Fourteen of these patterns may be used for phylogenetic analysis because they are shared by the DNAs of two or more populations; the remaining 25 are unique to the DNA of a single population and therefore are not informative phylogenetically (Palmer and Zamir, 1982). The phylogeny based on shared mutations is shown in Fig. 16.3. The tree was constructed on the basis of parsimony and requires the assumption of only one convergent mutation. The tree was rooted by assuming that change has been fairly constant in each lineage, so that the number of mutations from the base of the tree to the terminal points are nearly equal in each case (Palmer and Zamir, 1982). A tree with the same topology is obtained if two species of *Solanum* are used as the outgroup (J.D. Palmer, personal communication).

A noteworthy aspect of the results of this study is the low level of variation detected among the cpDNAs of the different species. For certain enzymes, all species were identical, with the largest total number of site differences for any two DNAs being 20 (Palmer and Zamir, 1982). From the restriction site data, the maximum percent sequence divergence was estimated for any pair of DNAs as only 0.7. The highly conservative nature of the cpDNAs of the species of *Lycopersicon* and *Solanum* minimizes the problem of parallel gain or loss of restriction sites.

Several interesting phylogenetic results were obtained from the study by Palmer and Zamir (1982). First, the three species with red-orange

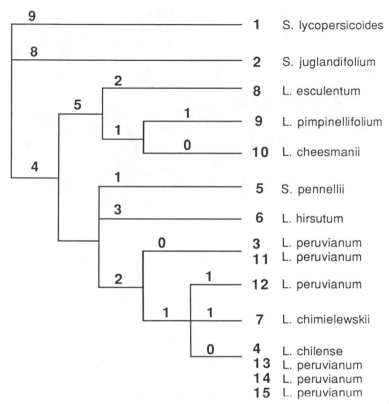

Figure 16.3 Phylogeny based on restriction site analyses of chloroplast DNAs from *Lycopersicon* and three species of *Solanum*. Numbers before species names indicate different accessions, and those on each branch of the diagram give the number of mutations unique to the branch. (Redrawn from J.D. Palmer and D. Zamir, Proc. Natl. Acad. Sci. USA **79**:5006–5010. 1982.)

fruits emerged as a monophyletic group with five restriction site muta-tions specific to *L. esculentum* (cultivated tomato), *L. pimpinellifolium*, and *L. cheesmanii* (Fig. 16.3). Two of the three species of *Solanum* included in the study clearly are quite distinct from members of *Lyco-persicon*, with *S. lycopersicoides* and *S. juglandifolium* each having a large number of specific restriction site differences (Fig. 16.3). By contrast, the cpDNA of *S. pennellii* suggests that it belongs in *Lycopersicon*, which is in line with other data gathered and summarized by Rick (1979). In fact, D'Arcy (1982) formally transferred this species to *Lycopersicon*. The three species *L. chilense, L. chmielewski,* and *L. peruvianum* are all

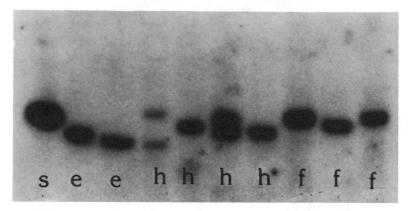

Figure 16.4 Autoradiograph showing length polymorphism in the chloroplast genomes of *Pisum*. Total cell DNA of all accessions (except the one in the left lane, which consisted of purified cpDNA) was digested with *XhoI*. The fragment on the left is 3.2 kb. Size varies from 3.0 to 3.25 kb. Note that two accessions of *P. humile* contain two different size forms. Abbreviations for species: *s*, *P. sativum*; e, *P. elatius*; h, *P. humile*; f, *P. fuluvum*. (Photograph courtesy of J.D. Palmer.)

grouped on the basis of two unique restriction site mutations (Fig. 16.3). Within this complex, relationships are not clear because the variation encountered in *L. peruvianum* includes all the variation detected in the other two species. Palmer and Zamir (1982) cautioned that because cpDNA is usually inherited maternally in angiosperms it is possible that a species could originate by hybridization (possibly followed by back-crossing to one of the parents) and appear morphologically quite unlike the species whose chloroplast genome it contains.

Palmer et al. (1985) examined 30 cpDNAs in the genus *Pisum*. These included the wild species *P. elatius*, *P. fulvum*, and *P. humile* as well as the cultivated pea, *P. sativum*. *Pisum elatius* and *P. humile* will cross readily with cultivated pea, and current taxonomic thought recognizes them as members of the same variable species. By contrast, *P. fulvum* is more distantly related to and reproductively isolated from the other three taxa.

The cpDNAs of *Pisum* contrast with those in *Lycopersicon* (Palmer and Zamir, 1982) because of the much more frequent occurrence of length mutations, that is, insertions and deletions (Fig. 16.4). Also, an inversion was detected, and these structural changes will be discussed as

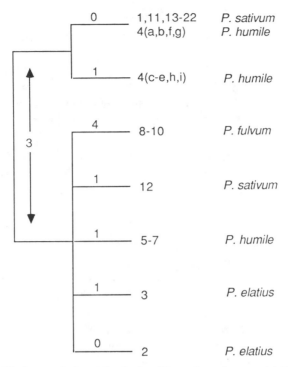

Figure 16.5 Phylogenetic hypothesis for *Pisum* based on restriction site mutations of chloroplast DNAs. Numbers along branches and at arrow indicate numbers of restriction site mutations. Numbers at ends of branches refer to accessions
of a species while letters denote individuals of accession four of *P. humile*.
(Redrawn with permission from J.D. Palmer et al., Genetics 109:195–213. 1985,
Genetics Society of America.)

they relate to the elucidation of phylogenetic relationships in *Pisum*. A
total of 11 restriction site mutations was detected.

Restriction site mutations were used to construct phylogenies for
Pisum, with 9 of the 11 mutations informative while the remaining 2 are
unique to one DNA each (Palmer et al., 1985). One of the most parsimonious trees is shown in Fig. 16.5, and it is significant that there is no need
to postulate a single parallel or convergent mutation. *Pisum fulvum* is
clearly distinct from the other species on the basis of restriction site
differences. This is in agreement with relationships inferred from crossing
studies.

Length mutations were used to group several different DNAs. For example, all samples of *Pisum fulvum* had a 200 bp insertion not found in other DNAs, which further supports the distinctive nature of this species (Palmer et al., 1985). In addition, a 1200 bp deletion characterized the DNAs of this species. Different accessions of *P. sativum* exhibited distinctive length mutations; of particular interest was the occurrence of two length classes in the DNAs of different individuals from the same population of *P. humile*. The DNAs of five individuals differed from the other four by at least three independent deletion–insertion changes. The combined cpDNA data suggest that the cultivated *P. sativum* originated from *P. humile* from northern Israel because the DNA of one accession of *P. humile* from this area shares three unique restriction site mutations with populations of *P. sativum* (Fig. 16.5).

The study of cpDNAs from *Pisum* illustrates that structural differences may be useful phylogenetically if changes are infrequent enough that it seems reasonable to assume that they are monophyletic, as appeared to be the case with certain length mutations in *Pisum*. By contrast, if there are "hotspots" in particular areas of the molecule, similar deletions and/or insertions could occur independently in different DNAs. Detailed studies beyond the scope of most systematic investigations would be necessary to ascertain whether the length mutations in different DNAs are identical and thus probably the result of a single event. In the absence of data, it is best not to incorporate the changes into the construction of a phylogeny (Palmer et al., 1985).

Sytsma and Schaal (1985b) used cpDNA to study phylogeny in the so-called *Lisianthus skinneri* complex occurring in Panama. Details of the plants were presented earlier in chapters when the isozyme rDNA data were discussed. A total of 12 mutations was detected in the cpDNAs from the populations representing six species in the *L. skinneri* complex. Shared mutations were superimposed on a phylogeny generated from rDNA, and this tree is consistent topologically with the one generated from the rDNA data (Fig. 16.6). No shared mutations were found for the eastern populations consisting of *L. skinneri* (LS4, LS6), the other *L. skinneri* population (LS7), and *L. jefensis* (LJ1), making it impossible to ascertain branching patterns for these populations. By contrast, the other species share several mutations, thus making it possible to resolve branching patterns (Fig. 16.6). The latter situation, with derived mutations shared by the three cloud forest species, (*L. auantiacus*, *L. habuensis*, and *L. peduncularis*) is in agreement with data from rDNA (see Chapter 15). Sytsma and Schaal (1985b) noted that the branch giving rise to *L.*

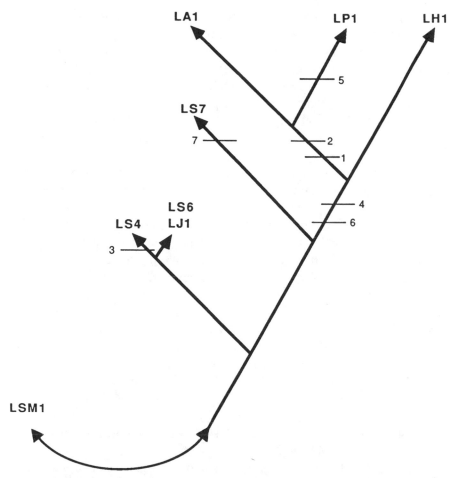

Figure 16.6 Restriction site mutations for chloroplast DNAs of the *Lisianthius skinneri* complex superimposed upon the phylogeny generated from ribosomal DNA (see Fig. 15.5). Note that populations LS4, LS6, LJ1, and LS7 share no unique mutations. The three remaining populations do share mutations, and their relationships could be resolved using cpDNA. See text for additional discussion. Species designations are: LS, *L. skinneri* (numbers denote populations); LA1, *L. aurantiacus*; LH1, *L. habuensis*; LJ1, *L. jefensis*; LP1, *L. peduncularis*; LSM1, *L. seemannii*. (Redrawn from K.J. Sytsma and B.A. Schaal, Evolution **39**:594–608. 1985.)

habuensis contains a number of rDNA restriction site mutations but lacks any cpDNA mutations. Data from cpDNA resolved phylogenetic relationships among some species in the *L. skinneri* complex, but lack of variation precluded resolution of relationships among all species. Because more mutations were detected in the rDNAs, it afforded greater resolving power than cpDNA in this group of plants.

Doebley et al. (1987) employed 21 restriction enzymes to digest chloroplast DNAs from four species and three subspecies of the genus *Zea* and from one accession of the presumably closely related genus *Tripsacum*. The species included *Z. diploperennis, Z. luxurians,* and *Z. perennis* from section *Luxuriantes*, and four subspecies of *Z. mays* from section *Zea*. These plants include maize (*Z. mays* subsp. *mays*) and its wild relatives, the teosintes. Of the 580 restriction sites examined, 24 were variable in one or more cpDNAs. In addition, five length (insertion–deletion) mutations were found (Doebley et al., 1987). This total of 29 mutations was used to construct a phylogeny (using parsimony) for maize and its relatives. The species of *Tripsacum* was employed as the outgroup. A total of 30 steps was needed to account for the 29 mutations used to construct the tree shown in Fig. 16.7. The cpDNA data place species of *Zea* into two groups corresponding exactly to the sections recognized by Doebley and Iltis (1980). *Zea luxurians* appears closely related to *Z. perennis* and *Z. diploperennis* on the basis of cpDNA, which is concordant with other lines of evidence (Doebley et al., 1987). Only one restriction site mutation separates *Z. perennis* and *Z. diploperennis*, which is not surprising because the former species is viewed as an autopolyploid derivative of the latter (Doebley et al., 1987).

The cpDNA data do not distinguish among subspecies of *Z. mays* (Fig. 16.7); in addition, the variation detected in *Z. mays* does not correspond with the accepted taxonomic designations of the subspecies (Fig. 16.7). Doebley et al. (1987) suggest that this nonconcordance could be due either to convergent evolution of certain restriction sites or reticulate evolution (hybridization) between the subspecies of *Z. mays*. Given the generally conservative nature of the chloroplast genome, the reticulate evolution hypothesis appears to be the more attractive one. The cpDNA data do strongly implicate either subsp. *mexicana* or subsp. *parviglunis* as the ancestor of maize (Fig. 16.7).

Coates and Cullis (1987) examined cpDNAs in 10 species of *Linum*, which included 8 species from section *Linum* and 2 species from two other sections. Four of the 8 species in section *Linum* were treated as part

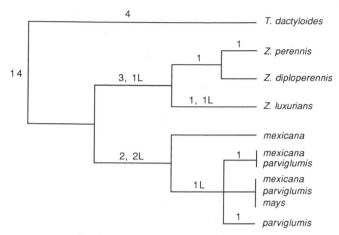

Figure 16.7 Phylogenetic tree (Wagner parsimony) for the genus *Zea* based on restriction site changes (denoted by numbers) and length mutations (indicated by L) in the chloroplast DNAs. The four species represented are *Tripsacum dactyloides* (as the outgroup), *Z. perennis, Z. diploperennis, Z. luxurians,* and *Z. mays.* All other names (*mays, mexicana,* and *parviglumis*) are subspecies of *Z. mays.* (Redrawn with permission from J. Doebley et al., Genetics **117**:139–147. 1987, Genetics Society of America.)

of the so-called "*perenne*" group, whereas the other four were designated as the "non-*perenne*" group. Labelled cloned probes from the chloroplast genome of *L. usitatissmum* were hybridized to filters containing fragments resulting from the digestion of the cpDNAs of the other species with several restriction enzymes.

Within section *Linum*, gains or losses were detected at 20 restriction sites, and four insertions (relative to *L. usitatissimum*) were found (Coates and Cullis, 1987). A phylogeny was constructed based on the presence of shared mutations, including length changes (Fig. 16.8); only two of the 24 changes result from convergence. The cpDNA data clearly distinguish two groups within section *Linum*; they correspond exactly to the "*perenne*" and "non-*perenne*" assemblages. Perhaps the most noteworthy difference between the two groups is a 13 kb insertion–deletion, with the larger genome present in the "*perenne*" group. When Coates and Cullis (1987) attempted to compare the cpDNAs from section *Linum* with two species from two other sections, the genomes were so distinct that mutational analyses were not possible. Also, the two species from the other sections had chloroplast genomes very distinct from each other.

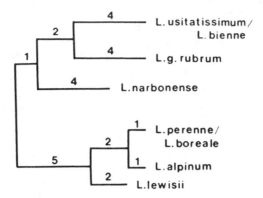

Figure 16.8   Dendrogram of relationships among species of *Linum* based on restriction site changes in their chloroplast genomes. *L. g. rubrum* designates *L. grandiflorum* var. *rubrum*. Numbers indicate numbers of mutations common to species at ends of the branches. (Redrawn with permission from D. Coates and C.A. Cullis, Am. Jo. Bot. **74**:260–268. 1987, Botanical Society of America.)

Thus, restriction site analyses were not suitable for comparing the cpDNAs of species from different sections of *Linum*.

Several more recent studies of chloroplast DNA have dealt primarily with noncultivated plants. Sytsma and Gottlieb (1986a,b) assessed evolutionary relationships in a section of the genus *Clarkia* to ascertain the phylogenetic position of the monotypic genus *Heterogaura* within *Clarkia*. Their results are particularly interesting because, as indicated in several earlier chapters, *Clarkia* has been the subject of detailed biosystematic (Harlan Lewis and collaborators) and electrophoretic (Leslie Gottlieb and coworkers) studies.

Heterogaura heterandra exhibits a variety of unusual features of the flowers and fruits, and appears so distinct from plants assigned to *Clarkia* that its status as a separate genus has not been seriously questioned (see Sytsma and Gottlieb, 1986a for a brief description of these characters). The distinctive morphology of the species has precluded determination of its relationship to *Clarkia*, that is, it is different in so many respects that it has not been possible to establish with reasonable certainty the species of *Clarkia* to which it is most closely related.

A total of 119 mutations was detected for *Heterogaura*, members of *Clarkia* section *Peripetasma*, and two additional species from two other sections of *Clarkia* (Sytsma and Gottlieb, 1986a). Fifty-five of these mutations are shared by two or more species (i.e., are phylogenetically

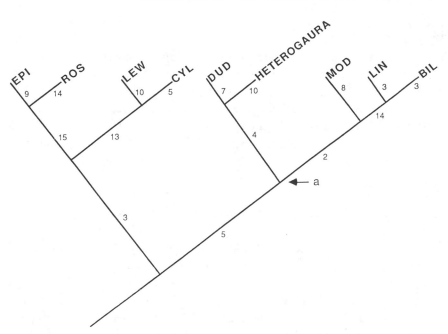

Figure 16.9 Phylogeny (most parsimonious tree) for *Clarkia* section *Peripetasma* and *Heterogaura heterandra* based on chloroplast DNA restriction site mutations. Numbers along branches refer to shared mutations for all taxa distal to the point on the tree where the number appears. For example, *C. lewisii* and *C. cylindrica* share 13 mutations, while each species has 10 and 5 unique mutations, respectively. Less parsimonious trees placing the branch to *Heterogaura* above or below the node designated *a* cannot be rejected statistically. However, placing the branch to *Heterogaura* below all species of sect. *Peripetasma* can be rejected statistically. See text for discussion. BIL, *biloba*, CYL, *cylindrica*, DUD, *dudleyana*, EPI, *epilobioides*, LEW, *lewisii*, LIN, *lingulata*, MOD, *modesta*, ROS, *rostrata*. (Redrawn from K.J. Sytsma and L.D. Gottlieb, Proc. Natl. Acad. Sci. USA **83**:5554–5557. 1986.)

informative). The most parsimonious tree produced from the restriction analyses is shown in Fig. 16.9. Numbers refer to the mutations occurring along a particular node of the tree. *Heterogaura* shares four restriction sites with *Clarkia dudleyana* and appears most closely related to it. A word of caution is indicated, however; Sytsma and Gottlieb (1986a) showed that the placement of *Heterogaura* above or below the node *a* cannot be rejected statistically (Fig. 16.9). However, placing *Heterogaura* at the base of section *Peripetasma* can be rejected statistically, and thus at worst the genus can be placed reliably within a group of four species of *Clarkia*.

Results of the study of *Heterogaura* and *Clarkia* are interesting and in several ways provocative for the plant systematist (Sytsma and Gottlieb, 1986a). First, the data demonstrate that the genus is a derivative of *Clarkia*, as had been suspected, but its morphological divergence had precluded demonstrating this fact conclusively or indicating to what species it is most closely related. Sytsma and Gottlieb (1986a) suggested that the genera consisting of one, two, or several species do not necessarily have to occur at a basal position with regard to larger genera. Rather, the less speciose genera may arise from within the more speciose groups. Lastly, while biochemical and molecular data are useful for producing more reliable phylogenetic hypotheses than is normally possible with morphology, at the same time these phylogenies may place the traditional taxonomist in a dilemma with regard to classification.

Sytsma and Gottlieb (1986b) examined the phylogenetic relationships in *Clarkia* section *Peripetasma* using cpDNA (Fig. 16.9) and compared hypotheses generated from these data with inferences made from prior electrophoretic studies. The eight diploid species of this section had been assigned to three subsections by Lewis and Lewis (1955). Subsection *Lautiflorae* contains *C. biloba*, *C. dudleyana*, *C. lingulata*, and *C. modesta*, subsection *Micranthae* is composed only of *C. epilobioides*, and subsection *Peripetasma* includes *C. cylindrica*, *C. lewisii*, and *C. rostrata*.

The close relationship of *C. epilobioides* and *C. rostrata* as revealed by cpDNA would not be expected from morphological considerations; the two species had been placed in separate subsections. Odrzykoski and Gottlieb (1984) found that most species of *Clarkia* contain duplicated genes for both the cytosolic and plastid isozymes of 6-phosphogluconate dehydrogenase (6PGD; see Chapter 10 for discussion); thus the two duplications represent the ancestral condition in the genus. The three species of subsection *Peripetasma* and *C. epilobioides* express only one gene for cytosolic 6PGD, the hypothesis being that gene silencing occurred in the ancestor common to these four species. In addition, one of the genes for the plastid form of 6PGD had been silenced in *C. epilobioides* and *C. rostrata*; however, because these two species were placed in separate subsections there was some question as to whether loss of duplicate gene expression occurred once in an ancestor common to the two species or was the result of two independent events (Odrzykoski and Gottlieb, 1984). The cpDNA data provide strong, independent support for a close relationship between *C. rostrata* and *C. epilobioides* and add considerable strength to the view that loss of expression of the plastid 6PGD isozyme occurred only once in a common ancestor.

The restriction site analyses revealed that *C. biloba* and *C. lingulata* differ by only six mutations, and these data, suggestive of a close relationship between the species, agree with results from morphology, cytogenetics and enzyme electrophoresis (Lewis and Roberts, 1956; Gottlieb, 1974a).

Evidence from cpDNA also suggests that in certain instances morphological differentiation between taxa is not a reliable indicator of phylogenetic relationships. One example is *C. rostrata*, which is much more similar morphologically to *C. cylindrica* and *C. lewisii* than it is to *C. epilobioides*. Because *C. rostrata* and *C. epilobioides* share a more recent common ancestor than either has with the other two species (Fig. 16.9), the latter species must therefore have diverged considerably in morphology from the common ancestor, whereas *C. rostrata* has undergone little change by comparison (Sytsma and Gottlieb, 1986b). An even more dramatic example of this unevenness in morphological divergence is the case already discussed of *Heterogaura heterandra*.

A question discussed in Chapter 10 concerns the distribution of the duplication of the gene for cytosolic phosphoglucoisomerase (PGI) in *Clarkia*. The only species lacking the duplication in section *Peripetasma* is *C. rostrata*. Lack of the duplication could be explained on the basis of the species never having had it, and thus probably being misplaced in the section, or the duplication may have been silenced. The totality of evidence suggested silencing, and the cpDNA supply strong data in support of this hypothesis.

In summary, cpDNA has provided the basis for constructing a phylogeny totally independent of the extensive data already available for *Clarkia* section *Peripetasma*. In certain instances, the data substantiate prior concepts, in other cases they allow for choosing between alternative hypotheses, and in still other situations they suggest relationships contrary to those that had been generally accepted. A discussion of the uses and contributions of various molecular data to the understanding of evolution in *Clarkia*, particularly section *Peripetasma*, will be presented in Chapter 17.

Yatskievych et al. (1988) employed restriction site analyses of cpDNA to examine relationships within and among the related fern genera *Phanerophlebia, Cyrtomium*, and *Polystichum*. A basic question about these ferns has been whether the genera *Cyrtomium* and *Phanerophlebia* are monophyletic, that is, do they have a common ancestor within *Polystichum* or did they originate from different elements within *Polystichum*? The results indicated clearly that *Cyrtomium* and *Phanerophlebia* originated from different elements in *Polystichum*.

CHLOROPLAST DNA AND STUDIES OF HYBRIDIZATION AND POLYPLOIDY

The genus *Brassica* has been the subject of various types of molecular systematic studies; some of these investigations were discussed in earlier chapters, and they will be summarized in Chapter 17. No doubt one reason for the attention given to these plants is their economic significance; they include, among others, cauliflower, cabbage, broccoli, turnip, and black mustard. The generally accepted phylogenetic relationships are presented in Fig. 16.10; three diploid species are postulated to be ancestral to three allotetraploid species. In the same year and the same journal, two groups of workers published results of restriction site analyses of cpDNAs in *Brassica* (Erickson et al., 1983; Palmer et al., 1983). Both laboratories reached similar conclusions for the most part, but the Palmer et al. (1983) study will be considered in more detail because it included more restriction enzymes, and the results were presented in terms of mutational changes at sites for specific restriction enzymes. This type of analysis and presentation allows for the generation of phylogenetic hypotheses based on precise restriction site mutations and for estimating genetic divergence between cpDNAs in a quantitative manner.

Palmer et al. (1983) examined 22 DNAs with 28 restriction enzymes; all DNAs except radish (*Raphanus sativa*) and white mustard (*B. hirta*) were from the three diploid and three tetraploid species of *Brassica* (Fig. 16.10). Several small (50 to 400 bp) deletions and insertions were detected but were not included in the phylogenetic analyses because of their small size and the attendant problems of establishing the identities of the fragments. Also, particular places or regions in the genomes may be "hotspots," and the frequency of occurrence of length mutations increases the probability of parallel or convergent origins for insertions or deletions (Palmer et al., 1983). A total of 50 mutations was analyzed in the 22 DNAs, with 31 of them resulting from 15 less frequently cutting enzymes while 19 mutations were seen with the 13 enzymes that had more

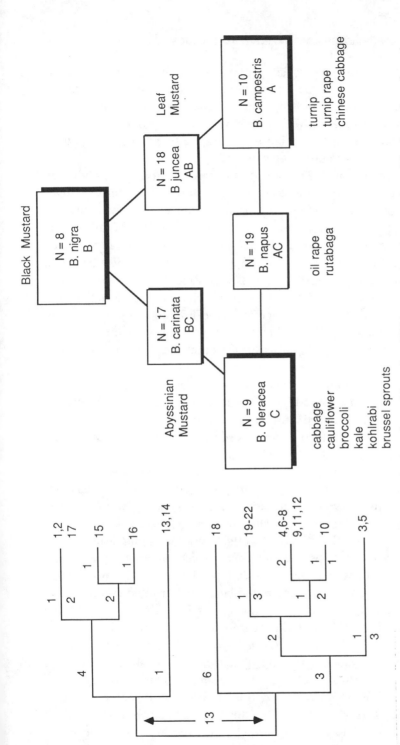

Figure 16.10 (Left) Phylogeny of *Brassica* species based on chloroplast DNA restriction site mutations. Numbers above branches refer to mutations of rarer cutting enzymes unique to the branches, while numbers below branches designate unique mutations generated by the more frequently cutting restriction enzymes. Numbers at ends of branches refer to population designations for species and are as follows: 1, 2, *B. carinata*; 3–5, *B. napus*; 6–8, *B. juncea*; 9–12, *B. campestris*; 13, 14, *B. hirta*; 15–17, *B. nigra*; 18, *Raphanus sativa*; 19–22, *B. oleracea*. (Right) Hypothesis of phylogenetic relationship among diploid and tetraploid species of *Brassica* based primarily on cytogenetic data. Letters below names denote genomes of each species. (Redrawn with permission from J.D. Palmer et al., Theor. Appl. Genet. **65**:181–189. 1983, Springer-Verlag, Inc.)

recognition sites in the chloroplast genome. Fragments generated by the less frequently cutting enzymes were used to group four major lineages within *Brassica* cpDNAs, and the more frequent cutters produced fragments for resolving variation within each of the major groups. No attempt was made to analyze all the variation among the four lineages produced by the more frequently cutting enzymes (Palmer et al., 1983).

In certain instances it was possible to interpret the differences between DNAs solely by inspection and comparison of fragment patterns in agarose gels. In other cases, additional procedures, including filter hybridizations, were needed to assess evolutionary directionality of mutational changes (Palmer et al., 1983). These methods included labelling the largest fragment for the mutation in question, and then hybridizing it to a filter containing other DNAs with a similar-sized fragment *and* DNAs with two smaller fragments whose total size is about the same as the large fragment. If the large labelled fragment hybridizes to similar-sized fragments in other outgroup DNAs but to both of the smaller fragments in one of the ingroup DNAs, then this suggests that the derived condition is the gain of a restriction site (Palmer et al., 1983).

The total of 50 mutations included 40 phylogenetically informative mutations, that is, ones shared by two or more DNAs. The less frequently cutting enzymes produced four lineages consisting of (1) *B. carinata* and *B. nigra*; (2) *B. hirta*; (3) *Raphanus sativa*; and (4) *B. campestris, B. juncea, B. napus*, and *B. oleracea* (Fig. 16.10). Only 41 independent restriction site mutations are required to account for the distribution of the 40 mutations in the tree (Fig. 16.10). The tree was rooted by assuming a constant rate of evolution in the chloroplast genome, which dictates that the number of restriction site changes be approximately the same in the different lineages. Palmer et al. (1983) indicated that using radish (*Raphanus sativa*) DNA as the outgroup, which would be the typical procedure in a cladistic analysis, would produce a phylogeny that is "lopsided" in the sense that there would be 14 to 18 rooted mutations for DNAs 1, 2, and 13–17 (i.e., top lineage in Fig. 16.10) whereas the remaining *Brassica* DNAs would have only four or five rooted mutations. Use of cpDNAs of two genera (*Arabidopsis* and *Capsella*) from other tribes in the Cruciferae as outgroups indicates that radish is best treated as a member of *Brassica* (J.D. Palmer, personal communication).

At the diploid level, *Brassica nigra* is more highly divergent from *B. campestris* and *B. oleracea* in cpDNA than the latter two are from each other (Fig. 16.10). The cpDNA data clearly implicated *B. nigra* as the maternal parent of *B. carinata*. As discussed by Palmer et al. (1983) and considered in earlier chapters, this finding is in agreement with most other molecular data. Likewise, cpDNA implicates *B. campestris* as the maternal parent of *B. juncea*. The situation with regard to the maternal parent of the third allopolyploid, *B. napus*, is not so clear. There appears little doubt from other data that the diploid parents of this species are *B. campestris* and *B. oleracea*. The chloroplast genome of *B. napus* (or some accessions of it) is distinct from its two putative parents. Most interestingly, Palmer et al. (1983) showed that the two diploid species *B. campestris* and *B. oleracea* share two derived restriction site mutations not detected in *any other species*, including *B. napus*. This indicates that the cpDNA of *B. napus* must have diverged before the two shared derived mutations evolved in each of its parents. The reason for the different cpDNA found in *B. napus* has been suggested as introgressive hybridization with some as yet unidentified species (Palmer et al., 1983). It also should be mentioned that analysis of ribosomal DNA in the plants of *B. napus* with the "weird" cytoplasm shows very clearly the presence of specific sequences from each of the parents, which documents conclusively the hybrid origin of the tetraploid species. This example from *Brassica* shows a limitation of the use of the maternally inherited chloroplast DNA for documenting the hybrid origin of plants. The contributions of different molecular data toward understanding relationships in *Brassica* will be compared and contrasted in Chapter 17.

In earlier chapters various molecular approaches applied to questions of the phylogeny of polyploid wheats were discussed; these will be summarized and evaluated in Chapter 17. Particular interest has centered on the cultivated hexaploid *Triticum aestivum* and several tetraploid species including *T. araraticum, T. dicoccoides, T. dicoccum,* and *T. timopheevii*. Two groups of workers used restriction site analyses of chloroplast DNAs to investigate the origins of polyploid wheats (Ogihara and Tsunewaki, 1982; Tsunewaki and Ogihara, 1983; Bowman et al., 1983). The proposed phylogeny for certain species of polyploid wheats is given in Fig. 16.11 and will be the central focus for the discussion.

One question involves the identity of the B-genome donor to the tetraploid *T. dicoccum*, and thus also to the hexaploid *T. aestivum* (Fig. 16.11). Considerable evidence from a variety of sources has suggested that the B-genome donor of the tetraploid *T. dicoccum* and hexaploid *T. aestivum* is also the cytoplasmic donor. This means that a diploid species with a chloroplast genome similar or identical to that found in the poly-

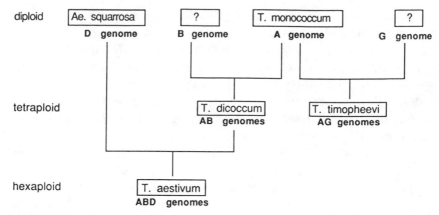

Figure 16.11   Phylogeny of polyploid wheats. Donors of the A and D genomes are generally accepted while the donors of the B and G genomes are much less certain. See text for discussion.

ploids would be a strong candidate as the B-genome donor. Tsunewaki and Ogihara (1983) found that the chloroplast genome of an accession of *Aegilops longissima* (a diploid) is identical to the two polyploid wheats; this species is therefore considered to be the source of the B genome. The population of *Ae. longissima* studied by Bowman et al. (1983) has cpDNA distinct from the polyploid wheats and from the accession examined by Tsunewaki and Ogihara (1983). These results show that intraspecific variation exists in *Ae. longissima* and also make it possible to ascertain which population (or populations) of the species may have donated the B genome. Significantly, Bowman et al. (1983) found variation among populations in the cpDNAs of several species of *Aegilops*. These results demonstrate the need for sampling more than one population of a species for cpDNA variation.

Another question regarding the phylogeny of polyploid wheats concerns the origin of the G genome in the tetraploid *Triticum timopheevii* (Fig. 16.11). One group of workers (Ogihara and Tsunewaki, 1982; Tsunewaki and Ogihara, 1983) suggested from their data that the donor of the chloroplast genome (and thus the G genome) to *T. timopheevii* (including the two additional species *T. araraticum* and *T. zhukovskyi*) is *Aegilops aucheri*. This latter species was not included by Bowman et al. (1983), but they concluded that *Ae. speltoides* is the most likely donor of the G genome. There appears to be some question as to whether *Ae. aucheri* and *Ae. speltoides* should be recognized as distinct species, so it is likely that the results of the two studies are concordant with regard to the G-genome donor. Surveys of additional populations are needed to ascertain the extent of variation in cpDNA in these diploid species and to attempt to determine more precisely the G-genome donor. The two studies are in agreement in showing that the two groups of tetraploid wheats (the *T. dicoccoides* versus *T. timopheevii* groups) are not monophyletic. Rather, one contains the B genome of *Ae. longissima* whereas the other has the G genome from a complex involving *Ae. aucheri* and *Ae. speltoides.*

Although the significant conclusions derived from the results of Tsunewaki and Ogihara (1983) and Bowman et al. (1983) remain unaffected, certain aspects of their studies could be improved to increase further the impact of the data (J.D. Palmer personal communication). First, many fragment differences among cpDNAs are caused by small insertions and deletions. This means that the same length mutation will be seen with different restriction enzymes, and counting the fragment alteration for each enzyme as a separate difference between DNAs exaggerates their divergences. Bowman et al. (1983) were aware that many of the observed differences represented length mutations while this apparently was not the case for Tsunewaki and Ogihara (1983). As indicated above, this does not call into serious doubt the conclusions reached with regard to which diploid species donated the B and G genomes to the polyploids. It should be emphasized, however, that if one were making quantitative comparisons of divergence among cpDNAs of various species, the procedure (of counting the same length mutations several times) could seriously distort the results.

Tsunewaki and Ogihara (1983) digested the DNAs with seven restriction enzymes and Bowman et al. (1983) cut them with four endonucleases. It would be desirable to increase considerably (perhaps to 20 or more) the number of restriction enzymes employed because it would allow a higher percentage of the genomes to be probed for differences. If more enzymes had been employed, more mutations undoubtedly would have been detected among DNAs. This in turn may have allowed a cladistic analysis of the DNAs in terms of shared, derived mutations. While the chloroplast genomes of *T. aestivum* and *Ae. longissima* are identical for the number of enzymes employed, it is not impossible (although highly unlikely) that they are primitively identical (J.D. Palmer personal communication), that is, their cpDNAs could represent the ancestral condition for all of the *Aegilops* and *Triticum* species under study. If, by contrast, the chloroplast genomes of the two species could be shown to share several derived restriction site mutations, then any uncertainty would be eliminated. In Chapter 17, a discussion integrating data from different molecular methods for polyploid wheats will be given, and the cpDNA results will be compared and contrasted with other molecular data.

Wendel (1989) employed variation in cpDNA restriction sites to examine the origin of the five species of tetraploid cultivated cottons in the New World. Cytogenetic data demonstrated that these species are allotetraploids with one genome (A) donated by an Old World diploid and the other (D) coming from a New World diploid. Wendel (1989) examined cpDNAs from 26 populations, including the two A genome diploid species, 10 of the 13 D genome diploids, and all known New World tetraploid species.

The chloroplast genomes of the tetraploids are quite similar to Old World diploid species, with five restriction site mutations distinguishing them. By contrast, these same two groups have cpDNAs that differ from the New World diploids at the same 28 restriction sites. This finding indicates that the maternal parent of the New World tetraploid cotton was an Old World plant, that is, the A genome source was also the chloroplast genome donor. Because the chloroplast DNAs of the A genome species *G. arboreum* and *G. herbaceum* were indistinguishable, it was not possible to determine which species is the parent of the tetraploids. Wendel (1989) found 12 restriction site mutations among A genome diploids and tetraploid New World cottons, but five of these are unique to single species and are therefore phylogenetically uninformative. Of the seven informative mutations, three serve to group the diploids while two occur only in the tetraploids. This very low level of divergence argues in favor of hypotheses of a recent origin for tetraploid cottons (perhaps within the last 2 million years), as opposed to a much older, possibly Cretaceous, origin for them.

In a study mentioned earlier in the chapter, Soltis et al. (1989) surveyed populations of diploid and tetraploid (autotetraploids) plants of *Heuchera micrantha* for variation at cpDNA restriction sites. The results show quite clearly that autotetraploids have arisen several times independently in this species because diploid and tetraploid populations share different restriction site mutations. Soltis et al. (1989) presented convincing arguments that distribution of cpDNAs were not caused by hybridization between diploid and tetraploid plants.

Reiseberg et al. (1988) tested a much-cited hypothesis of introgressive hybridization using cpDNA. Heiser (1949) suggested that introgression of genes from *Helianthus annuus* into a race of *H. bolanderi* restricted to serpentine soil produced a weedy race of *H. bolanderi*. Data from cpDNA reject the hypothesis of the hybrid origin of weedy *H. bolanderi* because its chloroplast genome is distinct from those of both the serpentine race of *H. bolanderi* and *H. annuus*. This fact suggests that the weedy *H. bolanderi*, rather than being of recent hybrid origin, may be a more ancient taxon.

CHLOROPLAST DNA AND STUDIES AT HIGHER TAXONOMIC LEVELS

There are three methods of using cpDNA at higher taxonomic levels. One, that has already been discussed, is analysis and mapping of restriction enzyme cleavage sites. A second approach is survey for structural mutations in the chloroplast molecules of different taxa. The third method involves sequencing different parts of the chloroplast genome and then comparing sequences from different taxa.

Mutational analyses of restriction enzyme cleavage sites may be employed effectively for generating phylogenies among genera within a family. J.D. Palmer and collaborators have carried out extensive mapping studies of restriction site mutations in the cpDNAs of genera representing all tribes of the Asteraceae (Palmer et al., 1988). Representatives of 57 genera of Asteraceae were studied with 11 restriction enzymes. From these enzymes a total of 926 restriction site mutations was mapped, with 328 of them being phylogenetically informative. That is, these 328 mutations were shared by two or more but not all taxa. These data were then subjected to analysis by parsimony to produce a number (6 or 12 depending on the program employed) of equally parsimonious phylogenetic trees with about 1315 steps. A majority rule consensus tree was generated via bootstrap analysis (see Chapter 1) in which 1318 steps were required and there is 30% homoplasy (Fig. 16.12). It is worthy of mention that the homoplasy present in these cladograms based on cpDNA is considerably lower than the percent homoplasy in cladograms using morphological features, even when employed at lower taxonomic levels.

Several aspects of the results of Palmer et al. (1988) are of considerable phylogenetic interest with regard to the tribes of composites. The

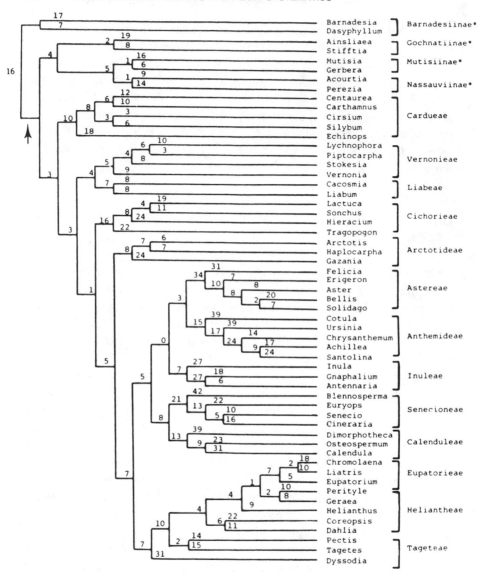

Figure 16.12   A phylogeny of the Asteraceae based on restriction site mutations in cpDNA. This is a majority rule consensus tree produced by the bootstrap method. Numbers at nodes and along each lineage designate the number of endonuclease cleavage mutations. Brackets delimit tribes, and the four subtribes of the Mutisieae are indicated by asterisks. The arrow designates the taxonomic distribution of the large inversion in the cpDNA molecule. (With permission from J.D. Palmer et al., Ann. Missouri Bot. Gard. **75**:1180–1206. 1988, Missouri Botanical Garden.)

subfamily Asteroideae (which includes the bottom eight tribes in Fig. 16.12) is a monophyletic group based on the phylogeny generated from cpDNA. By contrast, the remaining tribes of the Cichorioideae represent a paraphyletic and not a monophyletic group (Fig. 16.12). If one looks at the tribal level, three (Heliantheae, Mutisieae, and Tageteae) of the 14 tribes are paraphyletic, based on cpDNA phylogeny. These general results are of particular interest when compared with Bremer's (1987) cladistic analysis of the Asteraceae based on morphological features. Morphology, like cpDNA, indicates that the Asteroideae is monophyletic. Bremer (1987) concluded that the Liabeae and Vernonieae are monophyletic tribes, which is concordant with molecular results (Fig. 16.12). Both data sets indicate that the Mutisieae is paraphyletic. In some instances, the same phylogenetic relationships are not indicated by morphology and cpDNA. For example, Bremer's (1987) results placed the Eupatorieae near the Astereae, while the analysis of Palmer et al. (1988) suggested that the Eupatorieae is closest to the Heliantheae (Fig. 16.12). What is most intriguing about the results from morphology and cpDNA data is how similar the resulting cladograms are, not how different they are. Both Bremer (1987) and Palmer et al. (1988) commented that additional analyses might prove illuminating in resolving relationships in this large and diverse family of flowering plants.

Whether restriction analyses will prove to be as useful in other families of flowering plants as it is in the Asteraceae remains to be determined. It may well be that length mutations and other structural changes as well convergent restriction site changes will preclude effective use.

As indicated earlier in this chapter, major structural changes are rare in the chloroplast genome; this is fortunate because it enhances the feasibility of utilizing restriction site analyses for systematic studies. Since these structural changes occur so infrequently, they are of potential phylogenetic significance when they are detected. This aspect of cpDNA has not been employed extensively in plant systematics, but it is a most promising area. Several of the better studied situations will be discussed briefly (see Palmer, 1986b, 1987; Palmer et al., 1988 for more extensive discussions).

The identity of the ancestral group within the Asteraceae is among the most fundamental and intriguing evolutionary questions in the family. While various tribes have been proposed as the primitive element on the basis of morphology, there has been no consensus of opinion because of problems of convergence associated with adaptive evolution in this large and diverse group. Jansen and Palmer (1987a,b) surveyed all the recognized tribes of Asteraceae and several supposedly closely related families for variation in cpDNA. Their preliminary studies had detected a large inversion in certain representatives of the family, and the distribution and nature of this inversion have now been examined in some detail.

Jansen and Palmer (1987a,b) found a 22 kb inversion in all members of the Asteraceae surveyed except three genera of the subtribe Barnadesiinae of the Mutisieae. The cpDNAs of the nine other families, all of which have been suggested as closely related to the Asteraceae at various times by various workers, likewise lack the inversion. Thus, composites may be divided into one lineage with the inversion and another lacking it. Because the inversion has not been detected in any other family, its absence is viewed as the primitive or ancestral condition. If this is true, it follows that the Barnadesiinae of the Mutisieae can be considered as the basal or ancestral group of the Asteraceae. The alternative hypothesis that the inversion occurred in the ancestor of the Asteraceae and was reversed cannot be dismissed totally although it is less parsimonious than the other explanation.

Jansen and Palmer (1988) carried out an extensive study of restriction site variation in the Mutisieae as a test for the two alternative explanations for the different genome arrangements found in the Asteraceae. Members from other tribes of the family and from the families Dipsacaceae and Rubiaceae served as outgroups. Phylogenetic analyses of restriction site data clearly implicated the Barnadesiinae as the ancestral group within the composites. This notion in turn supports the hypothesis that the Barnadesiinae lacks the inversion because it never had it, that is, lack of the inversion is the ancestral condition within the Asteraceae. The paper by Jansen and Palmer (1988) demonstrates how restriction site changes and structural differences in the chloroplast genome may provide complementary data for inferring phylogenetic relationships.

Rare structural changes in the cpDNA are of potential phylogenetic value in the Leguminosae. One of these is the loss or deletion of one of the inverted repeats in certain papilionoid legumes (Palmer et al., 1987). Loss of the inverted repeat has occurred in two genera of the Pinaceae (Strauss et al., 1988) in addition to the legumes. Thus, the loss has probably occurred only twice in land plants, with its deletion in the legumes representing one event (Palmer et al., 1987). Lavin et al. (1988) reported that of the 101 species of legumes (representing all three subfamilies) examined for the inverted repeat, it was absent only from several tribes of papilionoid legumes. These include the Cicereae, Galegeae, Trifolieae, and Vicieae, all of which share a number of morphological and chromosomal features. Very few disagreements exist between the molecular and other data, one example being *Wisteria*, which represents the only genus lacking the inverted repeat in the tribe (Milletieae) in which it is traditionally placed. There is some question, however, whether *Wisteria* is properly placed in the tribe.

A 50 kb inversion and the transfer of a gene from the chloroplast to the nucleus have been detected in all three subfamilies of the Leguminosae (Palmer et al., 1987; unpublished data of J. Palmer and J. Doyle, discussed in Palmer et al., 1988). These features, while appearing to be characteristic of the whole Leguminosae, have not been detected in certain closely related families (Palmer et al., 1989). Additional families remain to be surveyed for the inversion and gene loss, so it may be possible to identify the progenitor of the legumes.

Other major structural changes in chloroplast genomes and their potential phylogenetic utility were considered by Palmer et al. (1988), and the reader is referred to this discussion.

Sequencing particular genes in the chloroplast genome appears to be the method of choice for inferring phylogenetic relationships at higher taxonomic levels. Advantages of sequence data over restriction site mapping were given by Palmer et al. (1988) and include the following. When comparing restriction site data at higher taxonomic levels, there is a greater danger of cleavage sites being lost by base changes at different positions within the recognition sequence, that is, if a restriction enzyme recognizes a six-base sequence, then changes at any of the six positions will result in site loss. Genes are often more conservative than the chloroplast genome as a whole and are thus of greater utility at higher taxonomic levels. Also, length mutations and other structural changes are not a problem with sequencing, as they may be with restriction site analyses. As always, however, there is a trade-off; the greater time and effort involved with sequencing precludes sampling as many taxa as is possible with endonuclease cleavage analyses.

The gene encoding the large subunit of rubisco (rbcL) is the only chloroplast gene for which sequences are known from enough species so that phylogenetic inferences may be made from the data. Even so, there are still so few species sequenced that the ultimate value of the data remains to be determined. It may be seen in Fig. 16.13 that the phenogram based on rbcL sequences is in general agreement with relationships

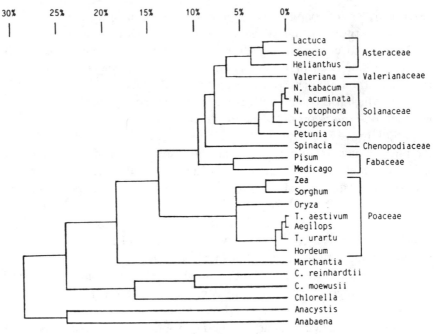

Figure 16.13   Phenogram of relationships based on rbcL sequence divergence. Percentages indicate sequence divergences. (With permission from J.D. Palmer et al., Ann. Missouri Bot. Gard. **75**:1180–1206. 1988, Missouri Botanical Garden.)

inferred from other data. It will be interesting to see the accumulation of additional sequences for rbcL.

SUMMARY

Restriction site analyses of cpDNA have provided valuable phylogenetic information in different groups of flowering plants. The relatively conservative evolution of the chloroplast genome often allows a straightforward analysis of changes at restriction sites when comparing congeneric species. This is so because parallel losses and gains of restriction sites are rare among congeners, and the construction of phylogenetic trees based on maximum parsimony is straightforward. Most taxonomists work at the level of interspecific variation, which makes cpDNA restriction analyses especially attractive for inclusion in taxonomic studies. A major shortcoming of this approach in certain genera is that lack of variation precludes the resolution of relationships for all taxa. By contrast, the results of Coates and Cullis (1987) for *Linum* indicate that within this genus divergence among chloroplast genomes in different sections has been so great that restriction site changes cannot be analyzed in the same simple manner employed for many genera. With regard to studies at higher taxonomic levels (intergeneric, interfamilial, and higher) restriction site analyses will likewise probably prove to be of limited value because of accumulated length changes and/or convergent site gains or losses. Such analyses, however, will have to be determined for each group of plants; Palmer et al. (1988) demonstrated the value of restriction site mutations for studying phylogenetic relationships among the tribes of the Asteraceae. The method of choice for studying these higher order relationships will almost certainly be sequencing of particular parts of the chloroplast genome. Sequencing the gene for the large subunit of rubisco appears to be a feasible approach for generating phylogenies at higher taxonomic levels (Ritland and Clegg, 1987; Zurawski and Clegg, 1987; Palmer et al., 1988).

Structural changes (inversions) in the chloroplast genome have been detected in several large families of flowering plants. The most completely studied of these is an inversion in the Asteraceae, and its distribution within the family has provided valuable insight into the ancestral group in the composites. Deletion of the large repeat in certain legumes represents another good example of the value of structural changes. The eventual contributions of these inversions to phylogeny at higher levels of the taxonomic hierarchy remain to be determined, but the possibilities look exciting and promising.

CHAPTER 17

Case Studies in Molecular Systematics and General Comments on the Contributions and Future Applications of Molecular Data to Plant Systematics

INTRODUCTION

This final chapter serves as a summary for the book. First, case studies of particular plant groups that have been examined using various methods to generate a variety of molecular data will be considered. The different data sets presented in earlier chapters for the plants will be compared, contrasted, and integrated. Thus it may be shown that, depending on the plant group, different methods may provide equally useful and concordant data, may differ in utility for resolving a particular question, or may produce nonconcordant data. The plants chosen for these case studies are the polyploid wheats, the *Brassica* polyploid complex, and the genus *Clarkia*, with emphasis on section *Peripetasma*.

The last part of this chapter will attempt to provide an overview of the contributions of molecular data to plant systematics as well as some thoughts on active areas of research in the future.

CASE STUDY 1—POLYPLOID WHEATS

Because a wide array of molecular approaches have been applied to studies of the origin and evolution of polyploid wheats, they will serve as one of the case studies to illustrate the impact of different kinds of molecular data on phylogenetic interpretations.

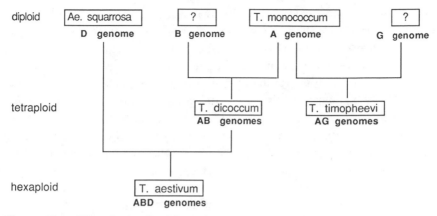

Figure 17.1 Hypothesized origins of polyploid wheats. See text for additional details.

In Chapter 3, the use of seed protein electrophoresis for elucidating relationships among the polyploid wheats was discussed. The outline of hypothesized relationships among the diploid, tetraploid, and hexaploid species is shown in Fig. 17.1. As mentioned earlier, different taxonomic treatments have been accorded these plants, including the placement of all of them in *Triticum* or some species in *Aegilops*. Also, various combinations of wild species and their derivative cultivated species are included in different studies. These factors, however, do not alter the basic evolutionary situations outlined in Fig. 17.1.

Johnson (1972a), using banding patterns of seed proteins in polyacrylamide gels, suggested that the hexaploid *T. aestivum* contains the genomes of *T. dicocuum*, a tetraploid cultivar, and a diploid wild species *A. squarrosa*. This hypothesis was based on the interpretation that the protein profiles of the hexaploid wheat represent a summation of protein bands present in *T. dicoccum* and *Ae. squarrosa*. These electrophoretic banding patterns agree with the generally held view on the donors of the A and D genomes to hexaploid wheat (Fig. 17.1). At the time that these studies were carried out in the early 1970s, the origin of the B genome in the tetraploid (and also hexaploid) wheats represented a more controversial problem than the identities of the A- and D-genome donors (Johnson, 1972b, 1975; Caldwell and Kasarda, 1978). A variety of species in *Triticum* and *Aegilops* has been postulated as the B-genome donor to the polyploids, with most of the evidence coming from morphology and cytogenetics. Johnson (1972a,b, 1975) considered that the seed protein banding patterns implicated a species of *Triticum* (likely *T. urartu*) as the source of the B genome to *T. dicoccum* and *T. aestivum*. Also, Johnson (1972a,b, 1975) said that his data argued against a species of *Aegilops* (including *Ae. longissima*) being the B-genome parent. Caldwell and Kasarda (1978) suggested that *T. urartu* could have donated one genome to *T. dicoccum* and *T. aestivum*, but, on the basis of their data, did not dismiss the possibility that a species of *Aegilops* could have contributed one genome as well. These workers, like Johnson, concluded that *Ae. longissima* was not the source of the B genome.

Triticum timopheevii is a tetraploid species, and the source of the so-called G genome present in this species has been a matter of discussion. One question is whether *T. timopheevii* and another tetraploid species (*T. dicoccum*) represent a monophyletic group, that is, containing the same two genomes (Fig. 17.1). Both Johnson (1975) and Caldwell and Kasarda (1978) reported different electrophoretic banding patterns for these tetraploid species. Johnson (1975) interpreted his data to suggest that two different biotypes of *Triticum urartu* contributed their genomes to the two tetraploids. By contrast, Caldwell and Kasarda (1978) viewed their seed protein data as less than definitive with regard to this question.

Comparisons of protein bands (visualized with a general protein stain) in gels, while suggesting possible genome donors to polyploid wheats, appeared to be less than definitive, with two laboratories expressing different levels of confidence in their data.

At about the same time as seed protein electrophoresis was being conducted on diploid and polyploid wheats, Chen et al. (1975) and Hirai and Tsunewaki (1981) carried out electrofocusing studies of rubisco in the same taxa. The banding patterns are shown in Fig. 17.2, where it may be seen that all species display the same small subunit band while there are two patterns for the large subunits. *Aegilops speltoides* is the only diploid species with the same large subunit pattern as the polyploids. A variety of data indicates that *T. monococcum* is the A-genome donor to both *T. dicoccum* and *T. timopheevii*. Since *T. monococcum* does not have the same large subunit pattern as the polyploids *and* the large subunit is maternally inherited, then the B-genome donor to the polyploids is also the large subunit donor. It follows, therefore, that *Ae. speltoides* can be viewed as the most likely source of the B genome to the two tetraploid species *T. dicoccum* and *T. timopheevii*.

The rubisco data are not concordant with Johnson's phylogenetic interpretations of seed protein profiles because the former data set clearly

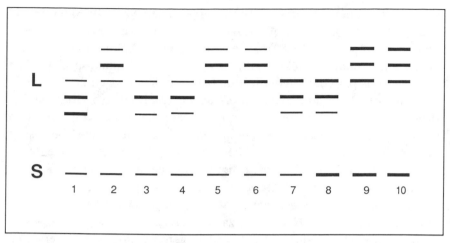

Figure 17.2 Drawing of large and small subunits of rubisco in species of *Triticum* and *Aegilops*, and reciprocal interspecific hybrids as resolved by electrofocusing in polyacrylamide gels. L designates the large and S the small subunit. 1, *Ae. squarrosa*; 2, *Ae. speltoides*; 3, *T. urartu*; 4, *T. monococcum*; 5, *T. dicoccum*; 6, *T. aestivum*; 7, *T. boeoticum*; 8, Hybrid with *T. boeoticum* the egg parent and *T. dicoccoides* the pollen plant; 9, Hybrid with *T. dicoccoides* the egg parent and *T. boeoticum* the pollen parent; 10, *T. dicoccoides*.

implicates *Ae. speltoides* as the B-genome donor while the latter suggests *T. urartu* as the source. Caldwell and Kasarda (1978) viewed their electrophoretic data as less than definitive and suggested that either *T. urartu* or *Ae. speltoides* could have contributed the B genome. The rubisco banding patterns did not distinguish between the B-genome tetraploid *T. dicoccum* and the G genome *T. timopheevii*, suggesting that they do in fact contain the same genomes and are monophyletic. Recall that differences between the two species were seen with seed proteins.

Clearly, the two types of electrophoretic data give conflicting data on the origin of polyploid wheats. The rubisco patterns appear simple to interpret because there are so few bands and the conservative nature of the molecules allows one to place the patterns into one of two groups. This placement, on the one hand, simplifies data interpretation but it does so at the cost of resolving differences among the taxa. By contrast, the complexity and variation in the seed protein profiles, while allowing for greater resolution of taxa, also cause problems in interpreting identities of bands among taxa. Suffice it to say that both approaches, while offering some insights, provided less than conclusive data.

In Chapter 14, the DNA–DNA hybridization studies of Nath and coworkers (Nath et al., 1983, 1984; Thompson and Nath, 1986) on diploid and polyploid wheats were discussed. These workers hybridized both unique and repetitive DNA from several species of diploids that represent putative B-genome donors to both tetraploid and hexaploid species. Most emphasis was placed on hybridization between diploids and the tetraploid *T. dicoccum*. Nath and collaborators also carried out hybridizations between a synthetic tetraploid known to contain the A and D genomes on the one hand and possible B-genome diploids on the other. The results suggest that *T. urartu*, which had been implicated as the B-genome donor by seed protein electrophoresis, is not the source of this genome. Rather, the results of Nath et al. (1983, 1984) and Thompson and Nath (1986) indicate high homology between *T. urartu* and the A genome, which was inferred from DNA–DNA hybridization between *T. urartu* and the synthetic tetraploid containing the A and D genomes. These workers concluded from the thermal stabilities of heteroduplexes that *A. searsii* (or *T. searsii*) is the source of the B genome in tetraploid *T. dicoccum* and hexaploid *T. aestivum*. The results of the DNA–DNA hybridizations by Nath and collaborators cast further doubt on *T. urartu* as the source of the B genome in polyploid wheats. It is of interest to note that *Ae. longissima*, which was among the diploids included in the hybridization studies, showed the least homology with tetraploid *T. dicoccum*.

Studies of ribosomal DNA (rDNA) in diploid and polyploid wheats (Appels et al., 1980; Peacock et al., 1981) were discussed in Chapter 15. The main point emerging from these investigations is that repeat lengths for the 18S–25S rRNA genes (combined with in situ hybridizations to locate the genomes containing the rRNA genes) failed to provide evidence on the source of the B genome to polyploid wheats. One problem with using the ribosomal genes is that they have apparently moved to different genomes and/or been deleted during the evolution of the polyploids. Thus simple comparisons of rDNA repeat lengths in putative diploid progenitors and polyploid derivatives are not possible. For example, A-genome diploids contain rRNA genes, and in synthetically produced tetraploids the A genomes likewise have rDNA (Frankel et al., 1987). Polyploid wheats, however, lack rDNA in their A genomes, suggesting deletion of these genes during evolution of the polyploids. Equally enigmatic is the presence of a 4.4 kb fragment (generated with a restriction enzyme) in the B genome of polyploids (tetraploids and hexaploids); the fragment is lacking in *all* diploids that are potential sources of this genome.

Although the 18S–25S rRNA genes are not useful for ascertaining the parentage of polyploid wheats, an important point is illustrated by these results. The value of molecular data (and indeed all data) for inferring evolutionary relationships, particularly when some type of reticulate process such as hybridization has occurred, will depend on how little evolutionary change has occurred subsequent to a critical event such as hybridization. Clearly, the rDNA of polyploid wheats has undergone rather extensive changes subsequent to the origin of the species.

In the previous chapter, the results of restriction endonuclease analyses of cpDNAs in diploid and polyploid wheats were discussed. Work in two different laboratories (Bowman et al., 1983; Tsunewaki and Ogihara, 1983) resulted in similar interpretations of the diploid progenitors of the polyploids. The cpDNA of *T. urartu* is similar if not identical to the A-genome species *T. monococcum*, which is concordant with other molecular data (including electrofocusing patterns of rubisco and thermal stabilities of heteroduplexes in DNA hybridizations) in dismissing *T. urartu* as the B-genome donor to *T. dicoccum* and *T. aestivum*. The cpDNA data are in agreement with the DNA hybridization results of Nath and coworkers in arguing that *T. urartu* is indeed an A-genome species. Both laboratories concluded from their cpDNA data that the tetraploids *T. dicoccum* and *T. timopheevii* do not constitute a monophyletic group because the latter contains the G (and not B) genome, the source of which is *Ae. speltoides* or a species closely related to it. The results of Tsunewaki and Ogihara (1983) identified *A. longissima* as the B-genome donor, while Bowman et al. (1983) were not able to find a diploid cytoplasm that matched the cytoplasms of *T. dicoccum* and *T. aestivum* because of variation in the cpDNAs of *Ae. longissima*. Only cpDNA data, of the various molecular approaches, distinguished the genomes of the tetraploids *T. dicoccum* and *T. timopheevii* and in addition identified the probable sources of each of the genomes.

Of the various molecular approaches used to study wheats, only seed protein profiles provided data that proved to be phylogenetically erroneous at worst and highly inconclusive at best. All additional molecular data provided insights into various aspects of relationships among the diploid and polyploid wheats. None of the data, exclusive of the seed proteins, was conflicting. The weight of additional molecular information over a period of about a decade provided progressively more refined answers to long-standing and difficult phylogenetic problems in the wheats.

CASE STUDY 2—*BRASSICA*

A variety of molecular methods has been applied to the study of relationships among three diploid species of *Brassica* and their allopolyploid derivatives. Thus, these plants can serve as the second of the case studies illustrating the contributions of different molecular data to a phylogenetic problem. The hypothesis of relationships with regard to the origin of the polyploid brassicas is shown in Fig. 17.3.

Seed protein electrophoresis was considered in Chapter 3, where several papers were discussed (Vaughan et al., 1966, 1970; Vaughan and Denford, 1968; Yadava et al., 1979). Regarding relationships among the three diploid species, different interpretations of the seed protein profiles were given depending on the study. For example, Vaughan et al. (1966) suggested that *B. campestris* and *B. nigra* are the two most closely related diploid species. By contrast, Vaughan and Denford (1968), on the basis of seed protein banding patterns, concluded that *B. campestris* and *B. oleracea* are more closely related to each other than either is to *B. nigra*. It is difficult to tell from the study by Yadava et al. (1979) just which two diploid species they considered most closely related. However, if one attempts to interpret their drawings of seed protein profiles, then *B. nigra* and *B. oleracea* appear most similar. Vaughan and Denford (1968) stated that each species has few unique bands; rather , it is the combination of bands that distinguish the taxa. Clearly, seed protein profiles were of limited value in distinguishing the three diploid species.

The allopolyploid species, in most instances, display diagnostic bands (few as they are) from their presumed diploid progenitors. However, given the very few bands unique to diploid species, the evidence on the allopolyploids is not strong. Vaughan and Denford (1968) interpreted two bands in *B. carinata* as identical to ones occurring in *B. campestris*, which is unexpected because the latter species is not viewed as a parent of the former (Fig. 17.3). Interestingly, Yadava et al. (1979) went further in suggesting that *B. campestris* is one of the parents of *B. carinata*. Protein profiles also led Vaughan and Denford (1968) to suggest that *B. juncea* contains a band identical to one in *B. oleracea*, which would not be predicted from the suggested parentage of *B. juncea* (Fig. 17.3).

Seed protein electrophoresis does not discriminate the diploid species of *Brassica* very effectively, and consequently the method was of limited value in documenting the diploid parentage of the allopolyploids. The problem of interpreting and comparing protein profiles is illustrated by

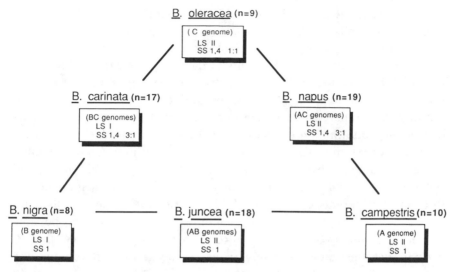

Figure 17.3 Proposed phylogenetic relationships among species of *Brassica*. Chromosome numbers are given in parentheses after each species name. Each box contains the genome designation, and the large (LS) and small subunit (SS) polypeptides of rubisco for each species. See text for additional details and discussion.

the fact that different studies of the same plants using the same methods came to different conclusions of relationships among the diploid species. Likewise, the putative presence of protein bands from the "wrong" diploids in two of the polyploids remains unexplained but may result from problems in interpreting homologous protein bands.

Electrofocusing patterns of the large and small subunits of rubisco in *Brassica* were discussed in Chapter 4 (Gatenby and Cocking, 1978; Uchimiya and Wildman, 1978; Robbins and Vaughan, 1983). Two large subunit patterns were found in the diploids (Fig. 17.3), with *B. campestris* and *B. oleracea* having one pattern and *B. nigra* the other. This finding (concordant with one but nonconcordant with two seed protein studies) suggests that *B. campestris* and *B. oleracea* are the most closely related diploid species. The small subunits of rubisco offer little insight into relationships among the diploids because they only differ in *B. oleracea*, which has one band in addition to the one it shares with the other two species (Fig. 17.3). Gatenby and Cocking (1978b) suggested that the additional band in *B. oleracea* is the result of its polyploid nature (see discussion in Chapter 4), but if this is indeed the case, it still does not help to clarify relationships with *B. campestris* and *B. nigra.*

Because the large subunit of rubisco is maternally inherited, it is useful for ascertaining the maternal parents of the polyploids. It may be seen in Fig. 17.3 that the directions of the crosses can be determined for the two polyploids *B. carinata* and *B. juncea,* but not for *B. napus* because its two putative diploid parents have the same subunits. While the large subunits can provide information on the maternal parent, they cannot document additivity of parental genomes, and thus they are in a sense of limited value for studies of hybridization. Lack of divergence among the diploid species in their small subunits limits to some extent their value for inferring the parents of the polyploids. If, however, one combines information from the large and small subunits for *B. carinata,* it is possible to document its diploid parentage. Reference to Fig. 17.3 shows that *B. carinata* has the same large subunit as *B. nigra,* while it contains the unique small subunit band from *B. oleracea.* This fact indicates that *B. nigra* is the maternal and *B. oleracea* the paternal parent of *B. carinata.* Presence of the small subunit 4 in *B. napus* indicates that *B. oleracea* is the paternal parent, and thus *B. campestris* is the maternal parent.

Vaughan et al. (1966) carried out serological studies (double diffusion and immunoelectrophoresis) of the three diploid species of *Brassica* and concluded that *B. campestris* and *B. oleracea* are more similar than either is to *B. napus.* This finding agrees with the majority of seed protein and rubisco data concerning relationships among the three diploid species.

The nuclear RFLP data of Song et al. (1988) for *Brassica* were discussed in Chapter 14, in which it was shown that *B. campestris* and *B. oleracea* are the two most closely related diploid species while the third diploid, *B. nigra,* is a distantly related, more primitive element. RFLPs conclusively documented the parentage of the two allotetraploid species *B. carinata* and *B. napus* because they contain fragments unique to each of their presumed diploid parents. These data are concordant with generally accepted views on the parentage of these two polyploids (Fig. 17.3). The third polyploid, *B. juncea,* contains only fragments from one diploid parent (*B. campestris*) and not the other presumed parent (*B. nigra*).

Palmer et al. (1983) and Erickson et al. (1983) carried out restriction site analyses of cpDNAs in the diploid and polyploid species of *Brassica.* Both investigations came to very similar conclusions, but the study of Palmer et al. (1983) will be emphasized because the results were presented in terms of mutational changes at restriction sites, and the data were analyzed quantitatively. The cpDNAs of *B. campestris* and *B.*

oleracea are much more similar to each other than either is to *B. nigra*, that is, 0.3% estimated sequence divergence versus 2.4% divergence. It may be seen in Fig. 16.10 that *B. nigra* is in one lineage and the other two diploid species are part of another distinct lineage within *Brassica*. These data are concordant with certain information from seed protein profiles and with serology regarding relationships among the diploid brassicas. Also, it will be recalled that *B. campestris* and *B. oleracea* have the same banding pattern for the maternally inherited subunit of rubisco.

The cpDNAs of the polyploids are likewise fully concordant with the rubisco data regarding the directions of the crosses (i.e., the maternal parent) that produced the species. The chloroplast genome of *B. carinata* and *B. nigra* are essentially indistinguishable, indicating that the latter species and not *B. oleracea* is the maternal donor (Fig. 16.10). Because the cpDNA of *B. juncea* is identical or nearly identical to that of *B. campestris* and not *B. nigra,* it indicates that *B. campestris* is the maternal parent (again in agreement with the large subunit data; see Fig. 17.3). As with the rubisco data, it is not possible to use cpDNA mutational analysis to indicate the direction of the initial cross. Thus, while the maternal inheritance of cpDNA can indicate the direction of a cross in some instances, it is not possible to detect additivity, as can be done with biparentally inherited characters encoded in the nuclear genome.

Nearly all of the molecular data that have accumulated for *Brassica* are concordant in suggesting that two of the diploid species are more similar to each other than either is to a third species. Of the methods employed, only cladistic analyses of nuclear RFLPs and cpDNA mutations produced explicit phylogenies showing that *B. campestris* and *B. oleracea* are monophyletic while *B. nigra* represents a distinct lineage.

CASE STUDY 3—*CLARKIA* WITH EMPHASIS ON SECTION *PERIPETASMA*

As indicated in earlier chapters, the elegant biosystematic data gathered by Harlan Lewis and collaborators on the genus *Clarkia* have provided the evolutionary framework within which Leslie Gottlieb and coworkers have carried out a variety of molecular studies. The central, but not sole, focus of this discussion will be section *Peripetasma* (apparently now called section *Sympherica*; see Sytsma and Smith, 1988) because more data are available for it than for other sections.

Enzyme electrophoresis was carried out on two species of section

Peripetasma (*C. biloba* and *C. lingulata*) by Gottlieb (1974a) to test the hypothesis of Lewis and Roberts (1956) that the latter species is a derivative of the former. The allozyme data are concordant with the Lewis and Roberts (1956) hypothesis because populations of the two species exhibited nearly the same genetic identities (ca. 0.90) as populations of the same species (Gottlieb, 1974a). Also, *C. lingulata,* the presumed derivative species, has about half the level of observed heterozygosity as *C. biloba.* Such a level would be expected if the derivative species contains a limited extraction of the variation present in its progenitor.

Extensive surveys for isozyme number in *Clarkia* have been conducted by Gottlieb and coworkers (see references and discussions in Gottlieb, 1986, 1988). The rationale for these studies was presented in Chapter 10, in which it was indicated that a minimal conserved isozyme number occurs for many enzymes in diploid vascular plants. Any increase in number, therefore, results from gene duplication at the diploid level or from polyploidy. The distribution of gene duplications for isozymes in *Clarkia*, taken from Gottlieb (1988), is presented in Table 17.1. The phylogenetic distribution of the duplications in *Clarkia* may be appreciated by comparing Table 17.1 with Fig. 17.4; distribution of the cytosolic phosphoglucoisomerase (PGI) duplication is shown in Fig. 17.4, where it may be seen that this phylogeny requires ony one duplication event, whereas an earlier hypothesis would have dictated several independent events. The duplications for cytosolic and plastid 6-phosphogluconate dehydrogenase (6-PGD) characterize the entire genus except for four species in section *Peripetasma*, thus indicating that the duplications originated in an ancestor of *Clarkia* and that subsequent silencing has occurred; this situation will be discussed later. Both triose phosphate-isomerase (TPI) duplications occur throughout *Clarkia*, but the cytosolic TPI duplication is widespread throughout the Onagraceae, whereas the extra plastid isozyme is limited to *Clarkia*, indicating a much more recent origin for the plastid as compared with the cytosolic form (Gottlieb, 1988).

The phylogenetic distribution of PGM duplications in *Clarkia* represents a more complex situation than the other enzymes (Fig. 17.4, Table 17.1). The loss of the "extra" plastid PGM in two species, *C. concinna* (section *Eucharidium*) and *C. lassenensis* (section *Rhodanthos*), in distantly related sections is best viewed as independent losses. The situation with regard to the cytosolic PGM duplication is not as clear with regard to relationships among sections. Its presence in all species of sections *Gode-*

TABLE 17.1. Distribution of Duplicate Isozymes in Diploid Species of *Clarkia* and in *Heterogaura*[a]

Section and Species	PGI	PGM		6-PGD		TPI	
	Cy	Pl	Cy	Pl	Cy	Pl	Cy
Eucharidium							
C. breweri	2	2	1	2	2	2	?
C. concinna	2	1	1	2	2	2	?
Fibula							
C. bottae	2	2	1	2	2	2	2
Peripetasma							
C. cylindrica	2	2	1	2	1	2	2
C. lewisii	2	2	1	2	1	2	2
C. epilobioides	2	2	1	1	1	2	2
C. rostrata	1	2	1	1	1	2	2
C. biloba	2	2	1	2	2	2	2
C. dudleyana	2	2	1	2	2	2	2
C. lingulata	2	2	1	2	2	2	2
C. modesta	2	2	1	2	2	2	2
Heterogaura							
H. heterandra	2	2	1	2	2	?	2
Phaeostoma							
C. xantiana	2	2	1	2	2	2	2
C. unguiculata	2	2	1	2	2	2	2
Godetia							
C. imbricata	1	2	2	2	2	2	2
C. nitens	1	2	2	2	2	2	2
C. speciosa	1	2	2	2	2	2	2
C. williamsonii	1	2	2	2	2	2	2
Myxocarpa							
C. mildrediae	1	2	2	2	2	2	2
C. virgata	1	2	2	2	2	2	2
Rhodanthos							
C. arcuata	1	2	2	2	2	2	2
C. lassenensis	1	1	1	2	2	2	2
C. amoena	1	2	1	2	2	2	2
C. franciscana	1	2	1	2	2	2	2
C. rubicunda	1	2	1	2	2	2	2

[a] The number 1 designates a single isozyme for a species while 2 indicates duplicate isozymes. Cy designates a cytosolic and Pl a plastid isozyme. Original sources of the data may be found in Chapter 10 and in Gottlieb (1988). PGI, phosphoglucoisomerase; PGM, phosphoglucomutase; 6-PGD, 6-phosphogluconate dehydrogenase; TPI, triose phosphate isomerase.

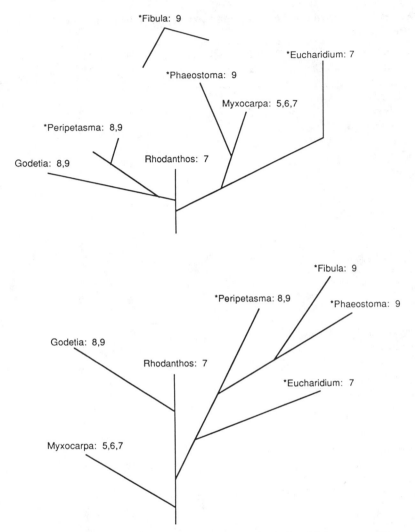

Figure 17.4 Two hypotheses of phylogenetic relationships among the seven sections of the genus *Clarkia*. (*Top*) Modified from Lewis and Lewis (1955). (*Bottom*) Modified from H. Lewis (1980). Asterisks denote sections with the duplicated gene for cytosolic PGI; numbers following sections indicate gametic chromosome numbers. (Redrawn with permission from U. Jensen and D.E. Fairbrothers, "Proteins and Nucleic Acids in Plant Systematics." 1983, Springer-Verlag, Berlin, Heidelberg.)

tia and *Myxocarpa* is generally congruent with the present hypothesis of a close phylogenetic relationship between the two sections (Fig. 17.4, bottom). It must be noted, however, that with the branching patterns shown in Fig. 17.4, one could not account for the extra isozyme by a single duplication event. The presence of the cytoplasmic PGM duplication in *Clarkia arcuata* of section *Rhodanthus* is more difficult to explain phylogenetically. It is possible that this species represents a lineage from which sections *Godetia* and Myxocarpa evolved (Fig. 17.4), or alternatively, the duplications may represent two independent events.

Sytsma and Smith (1988) analyzed restriction site mutations in the cpDNAs of representative species from all sections of *Clarkia*. A total of seven restriction enzymes was employed, and 51 site mutations were detected, 23 were cladistically informative. A species of *Epilobium* was used as the outgroup, and the resulting cladograms were either rooted using this outgroup, or midpoint rooting was employed, which requires the assumption of equal mutational rates in the lineages. Consensus trees were then constructed from the 15 or 100 (depending on the analysis) most parsimonious trees (Fig. 17.5).

The results of this study indicate that section *Godetia* is either monophyletic and the sister group to the rest of *Clarkia* (Fig. 17.5*A*), or it forms a lineage with section *Eucharidium* (Fig. 17.5*B*). It will be recalled (Table 17.1) that section *Godetia* lacks the PGI duplication, while section *Eucharidium* has the additional isozyme. It may be seen that both cladograms fail to resolve relationships among most sections of *Clarkia*, including particularly those with the PGI duplication. Thus these preliminary data [Sytsma and Smith (1988) emphasized that additional sampling of restriction enzymes and species are needed] from cpDNA restriction site analyses have neither confirmed nor rejected the hypothesis that sections with the PGI duplication represent a monophyletic assemblage in *Clarkia*. Sytsma and Smith (1988) pointed out that the primary problem with cpDNA data for determining relationships among sections of *Clarkia* is the lack of derived mutations shared by two or more but not all sections. They suggested that an early and rapid divergence of the sections in the genus could account for the pattern now seen in which cpDNA restriction site mutations are more abundant within rather than among sections. Suffice it to say that, at the generic level, cpDNA data have not yet provided a rigorous test of hypotheses of phylogenetic relationships among sections in *Clarkia*.

Consider next phylogeny within *Clarkia* section *Peripetasma*, which is

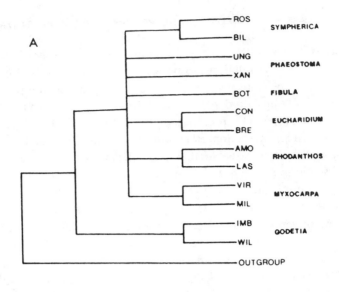

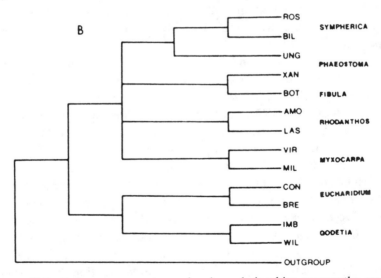

Figure 17.5 Strict consensus trees showing relationships among the seven sections of *Clarkia* based on restriction site mapping (section *Symperica* is the same as section *Peripetasma*). Species designations are: ROS, *C. rostrata*; BIL, *C. biloba*; UNG, *C. unguiculata*; XAN, *C. xantiana*; BOT, *C. bottae*; CON, *C. concinna*; BRE, *C. breweri*; AMO, *C. amoena*; LAS, *C. lassenensis*; VIR, *C. virgata*; MIL, *C. mildrediae*; IMB, *C. imbricata*; WIL, *C. williamsonii*. (A) Tree rooted with the genus *Epilobium* as the outgroup. This represents a consensus of the 100 most parsimonious trees. (B) Tree generated by omitting the outgroup and using midpoint rooting. This is the consensus tree of the 15 most parsimonious trees. (With permission from K.J. Sytsma and J.F. Smith, Ann. Missouri Bot. Gard. **75**:1217–1237. 1988, Missouri Botanical Garden.)

viewed as an advanced element within *Clarkia* (Fig. 17.4); all but one species in this section have the cytoplasmic PGI duplication. It may be seen in Table 17.1 that *C. rostrata* lacks the duplication (as was discussed in Chapter 10) and that four species express only one isozyme for cytosolic 6-PGD while two of these four species likewise lack the duplication for plastid 6-PGD. The basic question about *C. rostrata*, after it was discovered that it had only one cytoplasmic PGI isozyme, was whether it never has had the duplication and had been misplaced in section *Peripetasma* or whether silencing of one of the genes had occurred (Gottlieb and Weeden, 1979). This question was answered decisively by cpDNA restriction site analyses carried out by Sytsma and Gottlieb (1986a,b), and it may be seen in Fig. 17.6 that *C. rostrata* clearly is a "good" member of section *Peripetasma*, meaning that loss of duplicated gene expression has occurred in *C. rostrata*.

Lack of duplicate gene expression for cytoplasmic 6PGD in four species of section *Peripetasma* and its absence of the duplication for plastid 6PGD (Table 17.1) raised several interesting phylogenetic questions (Odrzykoski and Gottlieb, 1984). Because the three species *C. cylindrica*, *C. lewisii*, and *C. rostrata* had been placed in their own subsection *Peripetasma* and were thought to be closely related, it seemed quite possible that gene silencing occurred in an ancestor common to the three species. The situation with regard to *C. epilobioides* was not as clear because it is a morphologically distinct species that had been placed in its own section, *Micranthae*. It seems seasonable to envision that the lack of duplicate expression of cytoplasmic 6PGD in *C. epilobioides* occurs because it is in the same lineage as the three species of section *Peripetasma*. However, the common lack of the plastid 6PGD duplication in *C. epilobioides* and *C. rostrata* is more difficult to explain as a loss of expression in a common ancestor, since it would mean that previous concepts of phylogenetic relationships for the species were in error when *C. rostrata* and *C. epilobioides* were placed in different subsections. Because common absence of a feature is not strong phylogenetic evidence relative to the joint presence of a character, the isozyme data were less than definitive. CpDNA restriction site data showed without doubt that loss of each of the duplications represents single events (Sytsma and Gottlieb, 1986a,b) (Fig. 17.6). It is particularly impressive that 15 mutations support the hypothesis that *C. epilobioides* and *C. rostrata* comprise a monophyletic group (Fig. 17.6). It now is obvious that if subsection *Peripetasma* includes *C. rostrata*, the subsection is paraphyletic. Another interesting aspect of the cpDNA phylogeny is that *C. biloba* and *C.*

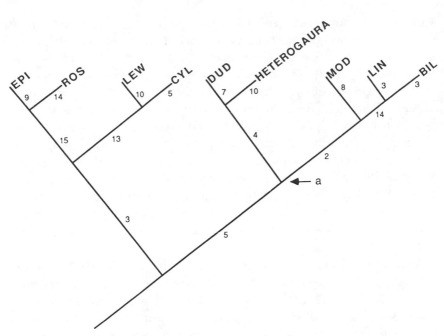

Figure 17.6 Phylogeny (most parsimonious tree) for *Clarkia* section *Peripetasma* and for *Heterogaura heterandra* based on chloroplast DNA restriction site mutations. Numbers along branches refer to shared mutations for all taxa distal to the point on the tree where the number appears. For example, *C. lewisii* and *C. cylindrica* share 13 mutations, while each species has 10 and 5 unique mutations, respectively. Less parsimonious trees placing the branch to *Heterogaura* above or below the node designated by *a* cannot be rejected statistically. However, placing the branch to *Heterogaura* below all species of section *Peripetasma* can be rejected statistically. BIL, *biloba*; CYL, *cylindrica*; DUD, *dudleyana*; EPI, *epilobiodes*; LEW, *lewisii*; LIN, *lingulata*; MOD, *modesta*. (Redrawn from K.J. Sytsma and L.D. Gottlieb, Proc. Natl. Acad. Sci. USA **83**:5554–5557. 1986.)

lingulata form a monophyletic group and that fewer species-specific mutations (autapomorphies) occur in these species than in any other species of section *Peripetasma*. These data are concordant with the very low level of isozyme divergence detected between the taxa (Gottlieb, 1974a).

In *Clarkia*, cpDNA data have not yet confirmed or rejected hypotheses of relationships among the sections inferred from morphology and gene duplications for isozymes. Within section *Peripetasma*, however, restriction site analyses of cpDNAs have provided unambiguous choices among alternative phylogenetic hypotheses generated by isozyme numbers.

CONTRIBUTIONS AND FUTURE APPLICATIONS OF MOLECULAR DATA TO PLANT SYSTEMATICS

This section contains my own thoughts on and evaluations of the impact that molecular data have had on plant systematics and the future contributions of molecular studies to the field. More specific comments on the various methods discussed in this book will be presented, and then more general, wide-ranging thoughts will be given.

Seed Protein Electrophoresis

It is evident from prior discussions earlier in this chapter and in Chapter 3 that seed protein profiles, while having provided useful data, do have certain limitations in plant systematics. One of the basic problems is establishing homologies of bands in the sometimes complex patterns generated in gels. The patterns can be used to demonstrate differences among plants (populations, species, etc.), but the data generated from the profiles must be represented by some sort of similarity measure rather than by a measure of genetic similarity or difference. Seed protein electrophoresis has been replaced in large measure by enzyme electrophoresis, and it appears that this trend will continue.

Electrophoretic Studies of Rubisco

Banding patterns of rubisco, usually achieved by electrofocusing, have proved useful in plant systematics. Given that the large subunit is maternally inherited whereas the small subunit is inherited biparentally, the electrophoretic method may be a powerful method for assessing the origin of allopolyploid taxa. Several examples illustrating this point were given in Chapter 4. The primary factor limiting use of rubisco electrofocusing is the rather conservative nature of the patterns, with several congeneric species often being the same. Not only does this compromise the method for studying divergence, but it in turn can prevent complete resolution of the parents of polyploid taxa. Nonetheless, the contributions of the method for resolving the species serving as the maternal and paternal genome donors to polyploid species attests to its value. Future studies using electrofocusing of rubisco will likely center on similar kinds of phylogenetic questions.

Enzyme Electrophoresis

Considerable space has been devoted to enzyme electrophoresis in this book, and while this emphasis reflects my own interests, it also attests to the popularity of this method among plant systematists. They early studies by Leslie Gottlieb demonstrated that the utility of allozyme data was not restricted to population biology but that the information could be very powerful for making inferences on relationships among congeneric species. The ability to quantify genetic variation and divergence represents the primary value of allelic frequency data. Additional advantages of enzyme electrophoresis are the ability to examine large numbers of individual plants at relatively low cost. One of the disadvantages of allozymes, as pointed out by Doebley and Wendel (1989), is that the number of loci that can be sampled is relatively limited. With this limitation, the question arises as to the future use of enzyme electrophoresis compared with restriction fragment length polymorphisms (RFLPs) of genomic DNA. Given the time, labor, and expense currently necessary to use RFLPs, it seems safe to say that allelic frequency data generated from enzyme electrophoresis will be used in plant systematics for the forseeable future. The method is most effectively applied to the study of congeneric species or species in closely related genera. At higher taxonomic levels, species may have few if any electromorphs (presumably alleles) in common, and assessments of homology based on electrophoretic mobility becomes particularly tenuous.

Use of isozyme number for assessing phylogenetic relationships has been most effective at the sectional and generic levels. In certain instances, duplications ostensibly representing single events have been observed within families. The phylogenetic value of gene duplications rests upon the assumption that additional isozymes detected in different diploid taxa are the result of one event occurring in a common ancestor of the taxa. The assessment of whether the extra isozyme is the manifestation of a single duplication in turn rests largely on the taxonomic distribution of isozyme numbers. However, despite this limitation, the distribution of a duplication in a genus or among genera can stimulate a critical reappraisal of phylogenetic relationships.

It is apparent that duplicated genes can also be silenced, which means that the derived condition for any group of plants may be the lower conserved isozyme number and not the higher number. This requires that the isozyme number in the outgroup of the taxa under consideration be determined. Isozyme number, therefore, while providing useful phylogenetic information, does not serve as a powerful independent source of data. Another limitation of any one duplication is that it divides taxa into two groups, those that have it and that do not. Thus the resolution that can be achieved in any phylogeny based on gene number is limited. The number of gene duplications in any group of diploid plants does not appear adequate (and duplications may not be differentially distributed among taxa) to produce a very highly resolved phylogeny. Despite these limitations, studies of isozyme number may prove useful for suggesting certain phylogenetic relationships in plants and perhaps for prompting a reevaluation of relationships.

In summary, the future looks bright for studies utilizing enzyme electrophoresis because the method is relatively simple and inexpensive, and a wide variety of questions in plant systematics can be addressed with the data. The method is not as "glamorous" as it once was because DNA studies have assumed that role. Rather, enzyme electrophoresis has taken its place as one of the valuable data sources in the arsenal of the plant systematist.

Amino Acid Sequencing

As indicated in Chapter 11, amino acid sequencing really ushered in the application of "hard core" molccular data to plant systematics. The potential impact of the sequence data on concepts of phylogenetic relationships in the flowering plants was the primary reason that sequence studies were more widely discussed and debated than other kinds of molecular data such as enzyme electrophoresis. While there is little doubt that amino acid sequence data will continue to be generated and employed for phylogenetic purposes, the method is being replaced by nucleic acid sequencing because the latter is faster and easier. While this trend will continue, it does not detract from the contributions of amino acid sequence data to the study of plant phylogeny. The major impact of amino acid sequencing on plant systematics has not been its phylogenetic insights into the flowering plants per se; rather it has caused plant systematists to think more critically about the manner in which angiosperm phylogeny should be studied. Issues such as assesment of homologies among structures, quantification of differences among taxa, and alternative methods of data analysis were brought into sharper focus as a result of the efforts of Donald Boulter.

Systematic Serology

The rather long discussion (in Chapter 12) on systematic seroloy of plants provides ample evidence for the number of serological studies that have been conducted, albeit by a relatively small group of dedicated workers. Some investigations centered on the relationships of taxa whose affinities had been debatable, and in most instances serological data provided compelling reasons for choosing among alternative hypotheses of relationships. As pointed out in Chapter 12, one of the problems with serological data is that they exist as one-to-one comparisons. This means that one really cannot apply the results of earlier studies to present investigations in the same way one compares known nucleotide sequences with newly generated sequences. Rather, one can only say that species A and B are more similar to each other than they are to species C. If one wants to investigate the relationships of species D and E to the other three species, then one must carry out the experiments and make direct comparisons because nothing can be said a priori about D and E relative to A, B, and C from previous results for the latter three taxa.

Serology will continue to make useful contributions to systematics, but problems with the methodology (necessity of using animals), the nature of the data, and the analysis of the data mitigate against its increased popularity.

DNA

The use of DNA, since it represents the new and expanding area of molecular systematics, is the most difficult to assess with regard to future contributions. It seems fair to say that at the present time the most widely used approach by plant systematists per se is restriction site analyses of chloroplast DNA. This is probably the case because the methodology is fairly straightforward, and the work of Jeffrey Palmer and collaborators has demonstrated rather dramatically the utility of the data for inferring phylogenetic relationships among congeneric species or among related genera. The cladograms constructed from analyses of restriction site mutations typically have very low homoplasy and high consistency indices relative to cladograms generated from morphological characters. These kinds of studies will become rather routine as systematists construct phylogenetic hypotheses for the species in their particular genera, and so on. Onc potential limitation in using cpDNA at the generic level is the lack of sufficient variation. Thus, while the conservative evolution of cpDNA removes convergent site gains and losses as a problem, this same conservatism can prevent complete resolution of relationships among all the species being studied.

Restriction site changes and differences in repeat lengths in rDNA [both of which result from differences in the nontranscribed spacer (NTS) region] will continue to prove useful for studies at lower taxonomic levels. These kinds of rDNA data will be most valuable at the populational and species level, but the high rate of base substitutions and changes in repeat lengths in the NTS makes the data of less value at these higher levels. Sequence data from the highly conserved coding regions of rDNA will be applied increasingly in phylogenetic studies at higher taxonomic levels.

The phylogenetic value of rare major structural changes (primarily inversions) in cpDNA has been demonstrated for several large families of flowering plants, the most thoroughly documented case being the elegant study of the Asteraceae by Robert Jansen and Jeffrey Palmer. This one inversion provided insights into the one age-old question of the ancestral tribe of the composites. Loss of the inverted repeat in certain papilionoid legumes represents another example of the phylogenetic utility of a major structural change in the cpDNA genome. Chloroplast rearrangements are somewhat analogous to gene duplications for isozymes with regard to phylogenetic utility. They both apparently represent rare events, and they both divide the taxa into two groups. This is both a strength and a weakness of cpDNA structural differences as phylogenetic information. On the one hand, the ancestral and derived elements can be identified, but on the other hand the data divide the taxa under consideration into only two groups. Once the structural mutation has been used phylogenetically, other data may then be applied for finer resolution.

The study of relationships at higher taxonomic levels will rely on sequencing particular kinds of DNA that evolve at the appropriate rate for the taxonomic group being studied. The gene encoding the large subunit of rubisco, for example, appears to be useful at higher taxonomic levels (among distantly related genera within a family and higher). As mentioned earlier, the coding regions of rDNA likewise should provide useful sequences at higher taxonomic levels.

THE FUTURE OF MOLECULAR SYSTEMATICS

Given that molecular systematics, primarily DNA, presently represents the "hot" area in systematics, there is the temptation to treat it as the most important and significant area of the discipline. This temptation is perhaps greater for macromolecules (particularly DNA) than it was in the past for other "hot" new data sources such as cytogenetics and micromolecules and for new types of data analysis such as phenetics and cladistics. Molecular biology, the field that furnishes methods for generating data that can be applied to systematic questions, has had and will continue to have a much larger impact on science and society in general than cytogenetics, secondary chemistry, cladistics, or phenetics ever had. Biotechnology is a household word, whereas these other terms never came close to achieving that status. As molecular biology continues to flourish, newer methods that produce more sophisticated data ever more rapidly and inexpensively will be continually available to the plant systematist. There is always something beguiling about new kinds of data (particularly if one is not thoroughly familiar with them!) because of the possibility that they will solve problems that have proved refractory to all other approaches. Another reason for suspecting that DNA will continue to be a most active and fashionable area of systematic research rests on the fact that it is the genetic material. All other kinds of data employed in plant systematics are used with the underlying assumption that the variation measure used in making taxonomic or phylogenetic decisions is a reflection of genetic variation. If one accepts this basic assumption, it is difficult to argue against using DNA in plant systematics, nor should one argue against its use.

The look into the crystal ball at future macromolecular studies will focus almost entirely on DNA, since the potential of other molecular data discussed in this book seems established in large measure. Enzyme electrophoresis, for example, has become integrated into systematics, and its advantages and limitations are now evident as a result of numerous electrophoretic studies. Such is not yet the case with DNA.

Except for morphology, the plant systematist has always been largely dependent on other areas for the development of methods that generate new data *or* for new methods of data analysis. This is certainly true with DNA; the methodology has developed for molecular biology. A critical step in the incorporation of basic molecular methods into plant systematics is the availability of persons who can simplify and adapt the methods to the needs and resources available to plant systematists. These persons may be trained molecular biologists whose interests in and appreciation of systematics provide them with the proper perspective of what is needed if the method is to be of systematic utility, or, alternatively, they may be organismal botanists who take postdoctoral positions with molecular biologists and thereby learn to adapt molecular methods to problems in plant systematics.

If molecular systematics is to progress and mature, it will require the cooperation and collaboration of open-minded molecular biologists *and* open-minded plant systematists. In addition, it is apparent that the day of the well-rounded young plant systematist who can apply all techniques to his/her taxonomic group is over. There are simply not enough hours in the day to apply all the methodology available. What is critical, however, is that the young systematist have some appreciation of the potential value and limitations of data generated by various molecular methods. This requires that doctoral students in plant systematics be introduced to molecular methodology and data during their graduate training even if they never use it in their research.

Up to the present date, data have often been used to generate phylogenies. Indeed, the immense popularity of cpDNA restriction site analysis among plant systematics rests largely on the unambiguous phylogenies that can be generated from the data. In the future, rather than having the generation of phylogenies serve as the primary goal of molecular studies, it should serve as the *starting point* for evolutionary studies of taxa. This is particularly true for congeneric species. One can look at character changes between sister species, and attempt to understand the genetic bases of species differences. The use of nuclear RFLPs as markers for quantitative trait loci (QTL) may provide refined insights into morphological and other features distinguishing species. Although one of the important functions of plant systematists is to produce phylogenetic hypotheses, it is critical that, with better means of generating phylogenies, they not just stop with the phylogenies. It must be remembered that RFLP analysis of cpDNA, while very useful phylogenetically, tells the systematist very little, if anything, about adaptive evolution.

An issue that has not yet been addressed in any detail is whether DNA divergence can be helpful for inferring relative ages of taxa. This is, of course, the old molecular clock controversy. By carefully selecting the plants for study it may be possible to gain some insight into this problem. Could it be that molecular evolution, like morphological evolution, can be somewhat episodic?

Because this book is about molecular systematics, I quite naturally am enthusiastic about the contributions the data have made to systematics and feel that more exciting insights are yet to come. Molecular information cannot, however, be studied within a vacuum if one really desires not only to classify plants, but also to understand the processes and forces that create new species. The future does not belong only to the macromolecules, but it belongs to anyone sufficiently enlightened to apply the data to long-standing problems in all aspects of plant systematics and evolution. The only limitations to progress are lack of vision and fear of new methods and ideas. This does not mean that I am advocating uncritical acceptance of new concepts; rather I am saying that one must be willing to examine them critically. When a new method or new kind of data is introduced into plant systematics, there are often two groups that retard its inital acceptance. One is represented by the overzealous advocates of the method and the other by systematists who pretty much dismiss the novel approach out of hand. Fortunately, the vast majority of plant systematists are open-minded scientists and realize that everyone ultimately benefits from the incorporation of new data into the continuing effort to understand the evolution of plants.

■■■■ REFERENCES

Adams, E.N. 1972. Consensus techniques and the comparison of taxonomic trees. Syst. Zool. 21:390–397.

Alston, R.E. 1967. Biochemical systematics. Evol. Biol. **1**:197–385.

Altosaar, I., B.A. Bohm, and R. Ornduff. 1974. Disc-electrophoresis of albumin and globulin fractions from dormant achenes of *Lasthenia*. Biochem. Syst. Ecol. **2**:67–72.

Ambrose, B.J.B., and R.C. Pless. 1987. DNA sequencing: chemical methods. Meth. Enzymol. **152**:522–538.

Antonov, A.S. 1986. Molecular analysis of plant DNA genomes: conserved and diverged DNA sequences. In: S.K. Dutta, ed. "DNA Systematics," Vol. II: "Plants." Boca Raton, FL: CRC Press, Inc., pp. 21–43.

Appels, R., and J. Dvorak. 1982a. The wheat ribosomal DNA spacer region: its structure and variation in populations and among species. Theor. Appl. Genet. **63**:337–348.

Appels, R., and J. Dvorak. 1982b. Relative rates of divergence of spacer and gene sequences within the rDNA region of species in the *Triticeae*: implications for the maintenance of homogeneity of a repeated gene family. Theor. Appl. Genet. **63**:361–365.

Appels, R., C. Driscoll, and W.J. Peacock. 1978. Heterochromatin and highly repeated DNA sequences in rye (*Secale cereale*). Chromosoma **70**:67–89.

Appels, R., W.L. Gerlach, E.S. Dennis, H. Swift, and W.J. Peacock. 1980. Molecular and chromosomal organization of DNA sequences coding for the ribosomal rRNAs in cereals. Chromosoma **78**:293–311.

Atchison, B.A., P.R. Whitfeld, and W. Bottomley. 1976. Comparison of chloroplast DNAs by specific fragmentation with EcoRI endonuclease. Mol. Gen. Genet. **148**:263–269.

Ausubel, F.M., R. Brent, R.E. Kinston, D.D. Moore, J.G. Seidman, J.A. Smith, and K. Struhl, eds. 1987. "Current Protocols in Molecular Biology." New York: Wiley.

Ayala, F.J. 1982. The genetic structure of species. In: R. Milkman, ed. "Perspectives in Evolution." Sunderland, MA: Sinauer, pp. 60–82.

Ayala, F.J., and J.A. Kiger. 1984. "Modern Genetics," 2nd ed. Menlo Park: Benjamin–Cummings.

Bachmann, K., K.L. Chambers, and H.J. Price. 1979. Genome size and phenotypic evolution in *Microseris* (Asteraceae, Cichorieae). Plant Syst. Evol. **129**: 119–134.

Bachmann, K., K.L. Chambers, and H.J. Price. 1985. Genome size and natural selection: observations and experiments in plants. In: T. Cavallier-Smith, ed. "The Evolution of Genome Size." Chichester: Wiley, pp. 267–276.

Banks, J.A., and C.W. Birky, 1985. Chloroplast DNA diversity is low in a wild plant *Lupinus texensis*. Proc. Natl. Acad. Sci. USA **82**:6950–6954.

Bayer, R.J. 1985a. Investigations in the evolutionary history of the polyploid complexes in *Antennaria* (Asteraceae: Inuleae). I. The *A. neodioica* complex. Plant Syst. Evol. **150**:143–163.

Bayer, R.J. 1985b. Investigations into the evolutionary history of the polyploid complexes in *Antennaria* (Asteraceae: Inuleae). II. The *A. parlinii* complex. Rhodora **87**:321–339.

Bayer, R.J. 1988. Patterns of isozyme variation in western North American *Antennaria* (Asteraceae: Inuleae) I. Sexual species of sect. *Dioicae*. Syst. Bot. **13**:525–537.

Bayer, R.J., and D.J. Crawford. 1986. Allozyme divergence among five diploid species of *Antennaria* (Asteraceae: Inuleae) and their allopolyploid derivatives. Am. J. Bot. **73**:287–296.

Belford, H.S., and W.F. Thompson. 1981a. Single copy DNA homologies in *Atriplex*. I. Cross reactivity estimates and the role of deletions in evolution. Heredity **46**:91–108.

Belford, H.S., and W.F. Thompson. 1981b. Single copy DNA homologies in *Atriplex*. II. Hybrid thermal stabilities and molecular phylogeny. Heredity **46**: 109–122.

Bendich, A.J., and R.S. Anderson. 1983. Repeated DNA sequences and species relatedness in the genus *Equisetum*. Plant Syst. Evol. **143**:47–52.

Bendich, A.J., and E.T. Bolton. 1967. Relatedness among plants as measured by the DNA–agar technique. Plant Physiol. **42**:959–967.

Bendich, A.J., and B.J. McCarthy. 1970a. DNA comparisons among barley, oats, rye and wheat. Genetics **65**:545–565.

Bendich, A.J., and B.J. McCarthy. 1970b. DNA comparisons among some biotypes of wheat. Genetics **65**:657–573.

Bennett, M.D., and J.B. Smith. 1976. Nuclear DNA amounts in angiosperms. Phil. Trans. Roy. Soc. Lond. B **274**:227–274.

Bennett, M.D., J.P. Gustafson, and J.B. Smith. 1977. Variation in nuclear DNA in the genus *Secale*. Chromosoma **61**:149–176.

Bennett, M.D., J.B. Smith, and J.S. Heslop-Harrison. 1982. Nuclear DNA amounts in angiosperms. Proc. Roy. Soc. Lond. B **216**:179–199.

Bonierbale, M.W., R.L. Plaisted, and S.D. Tanksley. 1988. RFLP maps based on a common set of clones reveal modes of chromosomal evolution in potato and tomato. Genetics **120**:1095–1103.

Bonner, W.M. 1987. Autoradiograms: ^{35}S and ^{32}P. Meth. Enzymol. **152**:55–61.

Booth, T.A., and A.J. Richards. 1978. Studies in the *Hordeum murinum* L. aggregate: disc electrophoresis of seed proteins. Bot. J. Linn. Soc. **78**:115–125.

Bosbach, K. 1983. Rubisco as a taxonomic tool in the genus *Erysimum* (Brassicaceae). In: U. Jensen and D. Fairbrothers, eds. "Proteins and Nucleic Acids in Plant Systematics." Heidelberg: Springer-Verlag, pp. 205–208.

Boulter, D. 1972. Amino acid sequences of cytochrome c and plastocyanins in phylogenetic studies of higher plants. Syst. Zool. **22**:549–553.

Boulter, D. 1973a. The use of amino acid sequence data in the classification of higher plants. In: G. Bendz and J. Santesson, eds. "Chemistry in Botanical Classification." New York: Academic Press, pp. 211–216.

Boulter, D. 1973b. The molecular evolution of higher plant cytochrome c. Pure Appl. Chem. **34**:539–552.

Boulter, D. 1974. The evolution of plant proteins with special reference to higher plant cytochromes c. Curr. Adv. Plant Sci. **8**:1–16.

Boulter, D. 1980. The use of amino acid sequence data in phylogenetic studies with special reference to plant proteins. In: F.A. Bisby, J.G. Vaughan and C.A. Wright, eds. "Chemosystematics, Principles and Practice." London: Academic Press, pp. 235–240.

Boulter, D., D.A. Thurman, and B.L. Turner. 1966. The use of disc electrophoresis of plant proteins in systematics. Taxon **15**:135–142.

Boulter, D., J.A. U. Ramshaw, E.W. Thompson, M. Richardson, and R.H. Brown. 1972. A phylogeny of higher plants based on the amino acid sequences of cytochrome c and its biological implications. Proc. Roy. Soc. Lond. B **181**:441–455.

Boulter, D., J.T. Gleaves, G.B. Haslett, D. Peacock, and U. Jensen. 1978. The relationships of 8 tribes of the Compositae as suggested by plastocyanin amino acid sequence data. Phytochemistry **17**:1585–1589.

Boulter, D., D. Peacock, A. Guise, T.T. Gleaves, and G. Estabrook. 1979. Relationships between the partial amino acid sequence of plastocyanin from members of ten families of flowering plants. Phytochemistry **18**:603–608.

Bowman, C.M., G. Bonnard, and T.A. Dyer. 1983. Chloroplast DNA variation between species of *Triticum* and *Aegilops*. Location of the variation on the chloroplast genome and its relevance to the inheritance and classification of the cytoplasm. Theor. Appl. Genet. **65**:247–262.

Branlard, G. 1983. Study of genetic determination of 20 gliadin bands. Theor. Appl. Genet. **64**:155–162.

Brehm, B., and M. Ownbey. 1965. Variation in chromatographic patterns in the *Tragopogon dubius–pratensis–porrifolius* complex (Compositae). Am. J. Bot. **52**:811–818.

Bremer, K. 1987. Tribal interrelationships of the Asteraceae. Cladistics **3**:210–253.

Bremer, K. 1988. The limits of amino acid sequence data in angiosperm phylogenetic reconstruction. Evolution **42**:795–803.

Britten, R.J., D.E. Graham, and R. Neufeld. 1974. Analysis of repeating DNA sequences by reassociation. Meth. Enzymol. **29**:363–418.

Brooks, J.E. 1987. Properties and uses of restriction endonucleases. Meth. Enzymol. **152**:113–129.

Brown, A.H.D., and B.S. Weir. 1983. Measuring genetic variability in plant populations. In: S.D. Tanksley and T.J. Orton, eds. "Isozymes in Plant Genetics and Breeding," Part A. Amsterdam: Elsevier, pp. 219–239.

Brown, A.H.D., A.C. Matheson, and K.G. Eldridge. 1975. Estimation of the mating system of *Eucalyptus obliqua* L'Herit. by using allozyme polymorphisms. Aust. J. Bot. **23**:931–949.

Brown, C.R., and S.K. Jain. 1979. Reproductive system and pattern of genetic variation in two *Limnanthes* species. Theor. Appl. Genet. **54**:181–190.

Brown, J.W.S., T.C. Osborn, F.A. Bliss, and T.C. Hall. 1981. Genetic variation in the subunits of globulin-2 and albumin seed proteins of french bean. Theor. Appl. Genet. **60**:245–250.

Bryan F.A., and Soltis, D.E. 1987. Electrophoretic evidence for allopolyploidy in the fern *Polypodium virginianum*. Syst. Bot. **12**:553–561.

Bulinska-Radomska, Z., and R.N. Lester. 1985. Relationships between five species of *Lolium* (Poaceae). Plant Syst. Evol. **148**:169–175.

Burnouf, T., R. Bouriquet, and P. Poullard. 1983. Inheritance of glutenin subunits in F_1 seeds of reciprocal crosses between European hexaploid wheat cultures. Theor. Appl. Genet. **64**:103–107.

Buttner, C., and U. Jensen. 1981. Homologization of storage proteins from *Aguilegia vulgaris* and *Digitalis purpurea*. Biochem. Syst. Ecol. **9**:251–256.

Caldwell, K.A., and D.D. Kasarda. 1978. Assessment of genomic and species relationships in *Triticum* and *Aegilops* by PAGE and by differential staining of seed albumins and globulins. Theor. Appl. Genet. **52**:273–280.

Carey, K., and F.R. Ganders. 1987. Patterns of isozyme variation in *Plectritis* (Valerianaceae). Syst. Bot. **12**:125–132.

Carr, G.D., and D.W. Kyhos. 1981. Adaptive radiation in the Hawaiian silversword alliance (Compositae: Madiinae). I. Cytogenetics of spontaneous hybrids. Evolution **35**:543–556.

Carr, G.D., and D.W. Kyhos. 1986. Adaptive radiation in the Hawaiian silversword alliance (Compositae: Madiinae). II. Cytogenetics of artificial and natural hybrids. Evolution **40**:959–976.

Carson, H.L. 1985. Unification of speciation theory in plants and animals. Syst. Bot. **10**:380–390.

Carulli, J.P., and D.E. Fairbrothers. 1988. Allozyme variation in three eastern United States species of *Aeschynomene* (Fabaceae), including the rare *A. virginica*. Syst. Bot. **13**:559–566.

Cavalier-Smith, T. 1982. Skeletal DNA and the evolution of genome sizes. Annu. Rev. Biophys. Bioeng. **11**:273–302.

Cavalier-Smith, T., Ed. 1985a. "The Evolution of Genome Size." New York: Wiley.

Cavalier-Smith, T. 1985b. Introduction: the evolutionary significance of genome size. In: T. Cavalier-Smith, ed. "The Evolution of Genome Size." New York: Wiley, pp. 1–36.

Cavalier-Smith, T. 1985c. Eukaryotic gene numbers, non-coding DNA, and genome size. In: T. Cavalier-Smith, ed. "The Evolution of Genome Size." New York: Wiley, pp. 69–103.

Cavalier-Smith, T. 1985d. Cell volume and the evolution of eukaryotic genome size. In: T. Cavalier-Smith, ed. "The Evolution of Genome Size." New York: Wiley, pp. 105–184.

Champion, A.B., E.M. Prager, and A.C. Wilson. 1974. Micro complement fixation. In: C.A. Wright, ed. "Biochemical and Immunological Taxonomy of Animals." New York: Academic Press, pp. 397–416.

Chan, P.H., and S.G. Wildman. 1972. Chloroplast DNA codes for the primary structure of the large subunit of fraction I protein. Biochem. Biophys. Acta **277**:677–680.

Chase, M.W., and R.G. Olmstead. 1988. Isozyme number in the subtribe Oncidiinae (Orchidaceae): an evaluation of polyploidy. Am. J. Bot. **75**:1080–1085.

Chen, K., and S.G. Wildman. 1981. Differentiation of fraction I protein in relation to age and distribution of angiosperm groups. Plant Syst. Evol. **138**:89–113.

Chen, K., S.D. Kung, J.C. Gray, and S.G. Wildman. 1975. Polypeptide composition of fraction I protein from *Nicotiana glauca* and from cultivars of *Nicotiana tabaccum*, including a male sterile line. Biochem. Genet. **13**:771–778.

Cherry, J.P., F.R.H. Katterman, and J.E. Endrizzi. 1970. Comparative studies of seed proteins of species of *Gossypium* by gel electrophoresis. Evolution **24**:431–447.

Cherry, J.P., F.R.H. Katterman, and J.E. Endrizzi. 1972. Seed esterases, leucine aminopeptidases and catalases of species of the genus *Gossypium*. Theor. Appl. Genet. **42**:218–226.

Chiapella, L., and G. Cristofolini. 1980. Sero-systematics of *Cytisus* sect. *Trianthocytisus* (Fabaceae). Plant Syst. Evol. **136**:209–216.

Choudhary, M., and R.S. Singh. 1987. Historical effective size and level of genetic diversity in *Drosophila melanogaster* and *Drosophila pseudoobscura*. Biochem. Genet. **25**:41–51.

Clarkson, R.B., and D.E. Fairbrothers. 1970. A serological and electrophoretic investigation of eastern North American *Abies* (Pinaceae). Taxon **19**:720–727.

Clausen, J. 1951. "Stages in the Evolution of Plant Species." Ithaca: Cornell University Press.

Clausen, J., D.D. Keck, and W. Heisey, 1940. Experimental studies on the nature of species. I. Effect of varied environments on western North American plants. Washington, DC: Publ. of Carnegie Inst: p. 520.

Clausen, J., D.D. Keck, and W.M. Heisey. 1941. Experimental taxonomy. Carnegie Inst. Wash. Yearbook **40**:160–170.

Clausen, J., D.D. Keck, and W.M. Heisey. 1947. Heredity of geographically and ecologically isolated races. Am. Naturalist **81**:114–133.

Clegg, M.T., J.R.Y. Rawson, and K. Thomas 1984. Chloroplast DNA variation in pearl millet and related species. Genetics **106**:449–461.

Clegg, M.T., K. Ritland, and G. Zurawski. 1986. Processes of chloroplast DNA evolution. In: S. Karlin and E. Nevo, eds. "Evolutionary Processes and Theory." New York: Academic Press, pp. 275–294.

Clifford, H., and W. Stephenson. 1975. "An Introduction to Numerical Classification." New York: Academic Press.

Coates, D., and C.A. Cullis. 1987. Chloroplast DNA variability among *Linum* species. Am. Jo. Bot. **74**:260–268.

Colless, D.H. 1980. Congruence between morphometric and allozyme data for *Menidia* species: a reappraisal. Syst. Zool. **29**:288–299.

Comas, C.I., J.H. Hunziker, and J.V. Crisci. 1979. Species relationships in *Bulnesia* as shown by seed protein electrophoresis. Biochem. Syst. Ecol. **7**:303–308.

Conkle, M.T., G. Schiller, and G. Grunwald. 1988. Electrophoretic analysis of diversity and phylogeny of *Pinus brutia* and closely related taxa. Syst. Bot. **13**:411–424.

Coyne, J.A. 1976. Lack of genic similarity between two sibling species of *Drosophila* as revealed by varied techniques. Genetics **84**:593–607.

Coyne, J.A., A.A. Felton, and R.C. Lewontin. 1978. Extent of genetic variation at a highly polymorphic esterase locus in *Drosophila pseudoobscura*. Proc. Natl. Acad. Sci. USA **7**:1816–1818.

Coyne, J.A., W.F. Fanes, J.A.M. Ramshaw, and R.E. Koehn. 1979. Electrophoretic heterogeneity of α-glycerophosphate dehydrogenase among many species of *Drosophila*. Syst. Zool. **28**:164–175.

Cracraft, J. 1987. DNA hybridization and avian phylogenetics. Evol. Biol. **21**:47–96.

Crawford, D.J. 1970. Systematic studies on Mexican *Coreopsis*, (sect. *Anathysana*), with special reference to the relationship between *C. cyclocarpa* and *C. pinnatisecta*. Bull. Torrey Bot. Club **97**:161–167.

Crawford, D.J. 1974. Variation in the seed proteins of *Chenopodium incanum*. Bull. Torrey Bot. Club. **101**:72–77.

Crawford, D.J. 1975. Systematic relationships in the narrow-leaved species of *Chenopodium* of the western United States. Brittonia **27**:279–288.

Crawford, D.J. 1976. Variation in the seed protein profiles of *Chenopodium fremontii*. Biochem. Syst. Ecol. **4**:169–172.

Crawford, D.J. 1977. A study of morphological variation in *Chenopodium incanum* (Chenopodiaceae) and recognition of two new varieties. Brittonia **29**:291–296.

Crawford, D.J. 1979a. Flavonoid chemistry and angiosperm evolution. Bot. Rev. **44**:431–456.

Crawford, D.J. 1979b. Allozyme variation in *Chenopodium incanum*: intraspecific variation and comparison with *Chenopodium fremontii*. Bull. Torrey Bot. Club **106**:257–261.

Crawford, D.J. 1983. Phylogenetic and systematic inferences from electrophoretic studies. In: S.D. Tanksley and T.J. Orton, eds. "Isozymes in Plant Genetics and Breeding," Part A. Amsterdam: Elsevier, pp. 257–287.

Crawford, D.J. 1985. Electrophoretic data and plant speciation. Syst. Bot. **10**:405–416.

Crawford, D.J. 1989. Enzyme electrophoresis and plant systematics. In: D.E. Soltis and P.S. Soltis, eds. "Isozymes in Plant Biology." Dioscorides Press, in press.

Crawford, D.J., and R.J. Bayer. 1981. Allozyme divergence in *Coreopsis cyclocarpa* (Compositae). Syst. Bot. **6**:373–386.

Crawford, D.J., and E.A. Julian. 1976. Seed protein profiles in the narrow-leaved species of *Chenopodium* of the western United States: taxonomic value and comparison with distribution of flavonoid compounds. Am. J. Bot. **63**:302–308.

Crawford, D.J., and R. Ornduff. 1989. Enzyme electrophoresis and evolutionary relationships among three species of *Lasthenia* (Asteraceae: Heliantheae) Am. J. Bot. **76**:289–296.

Crawford, D.J., and J.F. Reynolds. 1974. A numerical study of the common narrow-leaved taxa of *Chenopodium* occurring in the western United States. Brittonia **26**:398–410.

Crawford, D.J., and E.B. Smith. 1982a. Allozyme variation in *Coreopsis nuecensoides* and *C. nuecensis* (Compositae), a progenitor–derivative species pair. Evolution **36**:379–386.

Crawford, D.J., and E.B. Smith. 1982b. Allozyme divergence between *Coreopsis basalis* and *C. wrightii* (Compositae). Syst. Bot. **7**:359–364.

Crawford, D.J., and E.B. Smith. 1983a. The distribution of anthochlor floral pigments in North American *Coreopsis* (Compositae): taxonomic and phyletic interpretations. Am. Jo. Bot. **70**:355–362.

Crawford, D.J., and E.B. Smith. 1983b. Leaf flavonoid chemistry of North American *Coreopsis*: intra- and intersectional variation. Bot. Gaz. **144**:577–583.

Crawford, D.J., and E.B. Smith. 1984. Allozyme divergence and intraspecific variation in *Coreopsis grandiflora* (Compositae). Syst. Bot. **9**:219–225.

Crawford, D.J., and R. Whitkus. 1988. Allozyme divergence and the mode of speciation of *Coreopsis gigantea* and *C. maritima* (Compositae). Syst. Bot. **13**:256–264.

Crawford, D.J., and H.D. Wilson. 1977. Allozyme variation in *Chenopodium fremontii*. Syst. Bot. **2**:180–190.

Craword, D.J., and H.D. Wilson. 1979. Allozyme variation in several closely related diploid species of *Chenopodium* of the western United States. Am. Jo. Bot. **66**:237–244.

Crawford, D.J., E.B. Smith, and R.E. Pilatowski. 1984. Isozymes of *Coreopsis* section *Calliopsis* (Compositae): genetic variation within and divergence among the species. Brittonia **36**:375–381.

Crawford, D.J., R. Ornduff, and M.C. Vasey. 1985. Allozyme variation within and between *Lasthenia minor* and its derivative species *L. maritima* (Asteraceae). Am. Jo. Bot. **72**:1177–1184.

Crawford, D.J., E.B. Smith, M.L. Roberts, M. Benkowski, and M. Hoffman. 1990. Phylogenetic implications of differences in number of plastid phosphoglucose isomerase isozmyes in North American *Coreopsis* (Asteraceae: Heliantheae: Coreopsidinae). Am. Jo. Bot. in press.

Crawford, D.J., T.F. Stuessy, and M. Silva O. 1987. Allozyme divergence and the evolution of *Dendroseris* (Compositae: Lactuceae) on the Juan Fernandez Islands. Syst. Bot. **12**:435–443.

Crisci, J.V., J.H. Hunziker, R.A. Placios, and C.A. Naroajo. 1979. A numerical–taxonomic study of the genus *Bulnesia* (Zygophyllaceae): cluster analysis, ordination, and simulation of evolutionary trees. Am. Jo. Bot. **66**:133–140.

Cristofolini, C. 1980. Interpretation and analysis of serological data. In: F.A.

Bisby, J.G. Vaughan, and C.A. Wright, eds. "Chemosystematics: Principles and Practice." New York: Academic Press, pp. 269–288.

Cristofolini, C. 1981. Serological systematics of the Leguminosae. In: R. McPohill and P.H. Raven, eds. "Advances in Legume Systematics." Kew: Kew Royal Botanic Gardens, pp. 409–425.

Cristofolini, G., and L.F. Chiapella. 1977. Serological systematics of the tribe Genisteae (Fabaceae). Taxon **26**:43–56.

Cristofolini, G., and L. Chiapella. 1984. Serotaxonomy and systematic relationships of the genus *Adenocarpus* (Genisteae-Fabaceae). Nord. Jo. Bot. **4**:457–461.

Cristofolini, G., and P. Peri. 1983. Immunochemistry and phylogeny of selected Leguminosae tribes. In: U. Jensen and D.E. Fairbrothers, eds. "Proteins and Nucleic Acids in Plant Systematics." Heidelberg: Springer-Verlag, pp. 324–340.

Cristofolini, G., and P. Peri. 1984. Immunosystematical comparison of *Phaseolus coccineus* seed proteins with those of other legumes. Plant Syst. Evol. **144**:155–163.

Cronquist, A. 1973. Chemical plant taxonomy: a generalist's view of a promising specialty. In: G. Bendz and J. Santesson, eds. "Chemistry on Botanical Classification." New York: Academic Press, pp. 29–39.

Cronquist, A. 1976. The taxonomic significance of the structure of plant proteins: a classical taxonomist's view. Brittonia **28**:1–27.

Cronquist, A. 1980. Chemistry in plant taxonomy: an assessment of where we stand. In: F.A. Bisby, J.G. Vaughan, and C.A. Wright, eds. "Chemosystematics: Principles and Practice." New York: Academic Press, pp. 1–27.

Cronquist, A. 1981. "An Integrated System of Classification of Flowering Plants." New York: Columbia University Press.

Cronquist, A. 1984. A commentary on the definition of the order Myrtales. Ann. Missouri Bot. Gard. **71**:780–782.

Cronquist, A. 1987. A botanical critique of cladism. Bot. Rev. **53**:1–52.

Curtis, S.E., and M.T. Clegg. 1984. Molecular evolution of chloroplast DNA sequences. Mol. Biol. Evol. **1**:291–301.

Dahlgren, R. 1983. The importance of modern serological research for angiosperm classification. In: U. Jensen and D.E. Fairbrothers, eds. "Proteins and Nucleic Acids in Plant Systematics." Heidelberg: Springer-Verlag, pp. 373–394.

Dahlgren, R., and R.F. Thorne. 1984. The order Myrtales: circumscription, variation, and relationships. Ann. Missouri Bot. Gard. **71**:633–699.

D'Arcy, W.G. 1982. Combinations in *Lycopersicon* (Solanaceae). Phytologia **51**:240.

Davies, R.W. 1982. DNA sequencing. In: D. Rickwood and B.O. Haines, eds. "Gel Electrophoresis of Nucleic Acids, a Practical Approach." Oxford: IRL Press, pp. 117–172.

Davis, B.J. 1964. Disc-electrophoresis II. Methods and application to human serum proteins. Ann. NY Acad. Sci. **121**:404–427.

Davis, W.S. 1970. The systematics of *Clarkia bottae, C. cylindrica*, and a related new species, *C. rostrata*. Brittonia **22**:270–284.

Dean, C., P. Van den Elzen, S. Tamaki, P. Dursmuir, and J. Bedbrook. 1985a. Linkage and homology analysis divides the eight genes for the small subunit of petunia ribulose 1,5-bisphosphate carboxylase into three gene families. Proc. Natl. Acad. Sci. USA **82**:4964–4968.

Dean, D., P. van den Elzen, S. Tamaki, P. Dunsmuir, and J. Bedbrook. 1985b. Differential expression of the eight genes of the *Petunia* ribulose bisphosphate carboxylase small subunit multi-gene family. EMBO J. **4**:3055–3061.

Dean, C., P. van den Elzen, S. Tamaki, M. Black, P. Dunsmuir, and J. Bedbrook. 1987. Molecular characterization of the rboS multi-gene family of *Petunia* (Mitchell). Mol. Gen. Genet. **206**:465–474.

DeBry, R.W., and N.A. Slade. 1985. Cladistic analysis of restriction endonuclease cleavage maps within a maximum-likelihood framework. Syst. Zool. **34**:31–34.

Decker, D.S., and H.D. Wilson. 1987. Allozyme variation in the *Cucurbita pepo* complex: *C. pepo* var. *ovifera* vs. *C. texana*. Syst. Bot. **12**:263–273.

Dhaliwal, H.S. 1977. Genetic control of seed proteins in wheat. Theor. Appl. Genet. **50**:235–239.

Doebley, J.F., and H.H. Iltis. 1980. Taxonomy of *Zea* (Gramineae). I. A subgeneric classification with key to taxa. Am. Jo. Bot. **67**:982–993.

Doebley, J., and J.D. Wendel. 1989. Application of RFLPs to plant systematics. In: T. Helentjarus and B. Barr, eds. "Development and Application of Molecular Markers to Problems in Plant Genetics." New York: Cold Spring Harbor Laboratories, pp. 57–67.

Doebley, J.F., M.M. Goodman, and C.W. Stuber 1984. Isoenzymatic variation in *Zea* (Gramineae). Syst. Bot. **9**:203–218.

Doebley, J.F., M.M. Goodman, and C.W. Stuber. 1985. Isozyme variation in the races of maize from Mexico. Am. Jo. Bot. **72**:629–639.

Doebley, J.F., M.M. Goodman, and C.W. Stuber. 1986. Exceptional genetic divergence of northern flint corn. Am. Jo. Bot. **73**:64–69.

Doebley, J.F., W. Renfroe, and A. Blanton. 1987. Restriction site variation in the *Zea* chloroplast genome. Genetics **117**:139–147.

Doebley, J., J.D. Wendel, J.S.C. Smith, C.W. Stuber, and M.M. Goodman.

1988. The origin of cornbelt maize: the isozyme evidence. Econ. Bot. **42**:120–131.

Doolittle, R.F. 1981. Similar amino acid sequences: chance or common ancestry? Science **214**:149–159.

Dover, G.A. 1987. DNA turnover and the molecular clock. J. Mol. Evol. **26**:47–58.

Doyle, J.J. 1981. Biosystematic studies on the *Claytonia virginica* aneuploid complex [PhD thesis]. Bloomington: Indiana University.

Doyle, J.J. 1983. Flavonoid races of *Claytonia virginica* (Portulacaceae). Am. Jo. Bot. **70**:1085–1091.

Doyle, J.J., and R.N. Beachy. 1985. Ribosomal gene variation in soybean (*Glycine*) and its relatives. Theor. Appl. Genet. **70**:369–376.

Doyle, J.J., and E.L. Dickson. 1987. Preservation of plant samples for DNA restriction endonuclease analysis. Taxon **36**:715–722.

Doyle, J.J., and J.L. Doyle. 1987. A rapid DNA isolation procedure for small quantities of fresh leaf material. Phytochem. Bull. **19**:11–15.

Doyle, J.J., and J.L. Doyle. 1988. Natural interspecific hybridization in eastern North American *Claytonia*. Am. Jo. Bot. **75**:1238–1246.

Doyle, J.J., R.J. Beachy, and W.H. Lewis. 1984. Evolution of rDNA in *Claytonia* polyploid complexes. In: W.F. Grant, ed. "Plant Biosystematics." Orlando, FL: Academic Press, pp. 321–341.

Doyle, J.J., P.S. Soltis, and D.E. Soltis. 1985. Ribosomal RNA gene sequence variation: *Tolmiea, Tellima*, and their intergeneric hybrid. Am. Jo. Bot. **72**:1388–1391.

duCros, D.L., L.R. Joppa, and C.W. Wriglcy. 1983. Two-dimensional analysis of gliadin proteins associated with quality in durum wheat: chromosomal location of genes for their synthesis. Theor. Appl. Genet. **66**:297–302.

Dunbar, B.S. 1987. "Two Dimensional Electrophoresis and Immunology Techniques." New York: Plenum.

Dvorak, J., and R. Appels. 1982. Chromosome and nucleotide sequence differentiation in genomes of polyploid *Triticum* species. Theor. Appl. Genet. **63**:349–360.

Eckenwalder, J.E. 1976. Re-evaluation of Cupressaceae and Taxodiaceae: a proposed merger. Madrono **23**:237–256.

Edmonds, J.M., and S.M. Glidewell. 1977. Acrylamide gel electrophoresis of seed proteins from some *Solanum* (section *Solanum*) species. Plant Syst. Evol. **127**:277–291.

Ehrendorfer, F. 1959. Differentiation–hybridization cycles and polyploidy in *Achillea*. Cold Spring Harbor Symp. Quant. Biol. **24**:141–152.

Elisens, W.J. 1989. Genetic variation and evolution of the Galapagos shrub snapdragon. Natl. Geogr. Res. **5**:98–110.

Elisens, W.J., and D.J. Crawford. 1988. Genetic variation and differentiation in the genus *Mabrya* (Scrophulariaceae–Antirrhineae): systematic and evolutionary inferences. Am. Jo. Bot. **75**:85–96.

Elseth, G.D., and K.D. Baumgardner. 1984. "Genetics." Menlo Park: Addison–Wesley.

Engelke, D.R., P.A. Hoener, and F.S. Collins. 1988. Direct sequencing of enzymatically amplified human genomic DNA. Proc. Natl. Acad. Sci. USA **85**:544–548.

Ennos, R.A. 1986. Allozyme variation, linkage and duplication in the perennial grass *Cynosurus cristatus*. J. Hered. **77**:61–62.

Epes, D.A., and D.E. Soltis. 1984. An electrophoretic investigation of *Galax urceolata* (Diapensiaceae) [Abstract]. Am. J. Bot. **71**:165.

Erickson, L.R., N.A. Straus, and W.D. Beversdorf. 1983. Restriction patterns reveal origins of chloroplast genomes in *Brassica* amphiploids. Theor. Appl. Genet. **65**:201–206.

Esen, A., and K.W. Hilu. 1989. Immunological affinities among subfamilies of the Poaceae. Am. J. Bot. **76**:196–203.

Fairbrothers, D.E. 1967. Chemosystematics with emphasis on systematic serology. In: V.H. Heywood, ed. "Modern Methods in Plant Taxonomy." New York: Academic Press, pp. 141–174.

Fairbrothers, D.E. 1977. Perspectives in plant serotaxonomy. Ann. Missouri Bot. Gard. **64**:147–160.

Fairbrothers, D.E. 1983. Evidence from nucleic acid and protein chemistry, in particular serology, in angiosperm classification. Nord. J. Bot. **3**:35–41.

Fairbrothers, D.E., and F.P. Petersen. 1983. Serological investigation of the Annoniflorae (Magnoliiflorae, Magnoliidae). In: U. Jensen and D.E. Fairbrothers, eds. "Proteins and Nucleic Acids in Plant Systematics." Heidelberg: Springer-Verlag, pp. 301–310.

Fairbrothers, D.E., T.J. Mabry, R.L. Scogin, and B.L. Turner. 1975. The bases of angiosperm phylogeny: chemotaxonomy. Ann. Missouri Bot. Gard. **62**:765–800.

Farris, J.S. 1981. Distance data in phylogenetic analysis. In: V.A. Funk, and D.R. Brooks, eds. "Advances in Cladistics," Proc. First Meeting Willi Henig Soc. Meeting. New York: New York Bot. Gard, pp. 3–23.

Farris, J.S. 1985. Distance data revisted. Cladistics **1**:67–85.

Farris, J.S. 1986. Distances and statistics. Cladistics **2**:144–157.

Felsenstein, J. 1973. Maximum likelihood and minimum-steps methods for estimating evolutionary trees from data on discrete characters. Syst. Zool. **22**:240–249.

Felsenstein, J. 1978. Cases in which parsimony or compatibility methods will be positively misleading. Syst. Zool. **27**:401–410.

Felsenstein, J. 1982. Numerical methods for inferring evolutionary trees. Quart. Rev. Biol. **57**:379–404.

Felsenstein, J. 1983. Parsimony in systematics: biological and statistical issues. Annu. Rev. Ecol. Syst. **14**:313–333.

Felsenstein, J. 1984. Distance methods for inferring phylogenies: a justification Evolution **38**:16–24.

Felsenstein, J. 1986. Distance methods: reply to Farris. Cladistics **2**:130–143.

Felsenstein, J. 1988. Phylogenies from molecular sequences: inference and reliability. Annu. Rev. Genet. **22**:521–565.

Fink, W.L. 1986. Microcomputers and phylogenetic analysis. Science **234**:1135–1139.

Fitch, W.M., and C.H. Langley. 1976. Protein evolution and the molecular clock. Fed. Proc. **35**:2091–2097.

Fitch, W.M., and E. Margoliash. 1967. Construction of phylogenetic trees. Science **155**:279–284.

Flavell, R.B. 1980. The molecular characterization and organization of plant chromosomal DNA sequences. Annu. Rev. Plant Physiol. **31**:569–596.

Flavell, R.B. 1982. Sequence amplification, deletion and rearrangement: major sources of variation during species divergence. In: G.A. Dover and R.B. Flavell, eds. "Genome evolution." New York: Academic Press, pp. 301–323.

Flavell, R.B., and D.B. Smith. 1975. Genome organization in higher plants. Stadler Symp. **7**:47–69. Columbia, MI: University of Missouri.

Flavell, R.B., J. Rimpau, and D.B. Smith, 1977. Repeated sequence DNA relationships in four cereal genomes. Chromosoma **63**:205–222.

Flavell, R.B., M. O'Dell, and D.B. Smith. 1979. Repeated sequence DNA comparisons between *Triticum* and *Aegilops* species. Heredity **42**:209–222.

Flavell, R.B., J. Rimpau, D.B. Smith, M. O'Dell, and J.R. Bedbrook. 1980. The evolution of plant genome structure. In: C.J. Leaver, ed. "Genome Organization and Expression in Plants." New York: Plenum Press, pp. 35–47.

Flavell, R.B., M. O'Dell, P. Sharp, E. Nevo, and A. Beiles. 1986. Variation in the intergenic spacer of ribosomal DNA of wild wheat, *Triticum dicoccoides* in Israel. Mol. Biol. Evol. **3**:547–558.

Fluhr, R., P. Moses, G. Morelli, G. Coruzzi, and N. H. Chua. 1986. Expression dynamics of the pea rbcS multigene family and organ distribution of transcripts. EMBO J. **5**:2063–2071.

Frankel, O.H., W.L. Gerlach, and W.J. Peacock. 1987. The ribosomal RNA genes in synthetic tetraploids of wheat. Theor. Appl. Genet. **75**:138–143.

Frelin, C.H., and F. Vuilleumier. 1979. Biochemical methods and reasoning in systematics. Z. Zool. Syst. Evolutforsch. **17**:1–10.

Friday, A.E. 1980. The status of immunological distance data in the construction of phylogenetic classifications: a critique. In: F.A. Bisby, J.G. Vaughan, and C.A. Wright, eds. "Chemosystematics: Principles and Practice." New York: Academic Press, pp. 289–304.

Fristrom, J.W., and M.T. Clegg. 1988. "Principles of Genetics." 2nd ed. New York: Chiron Press.

Furnier, G.R., and W.T. Adams. 1986. Geographic patterns of allozyme variation in Jeffrey pine. Am. J. Bot. **73**:1009–1015.

Gallez, G.P., and L.D. Gottlieb. 1982. Genetic evidence for the hybrid origin of the diploid plant *Stephanomeria diegensis*. Evolution **36**:1158–1167.

Ganders, F.R., and K.M. Nagata. 1984. The role of hybridization in the evolution of *Bidens* in the Hawaiian Islands. In: W.F. Grant, ed. "Plant Biosystematics." Orlando, FL: Academic Press, pp. 179–194.

Gastony, G.J. 1986. Electrophoretic evidence for the origin of fern species by unreduced spores. Am. J. Bot. **73**:1563–1569.

Gastony, G.J. 1988. The *Pellaea glabella* complex: electrophoretic evidence for the derivations of the agamospermous taxa and a revised taxonomy. Am. Fern J. **78**:44–67.

Gastony, G.J., and D.C. Darrow. 1983. Chloroplastic and cytosolic isozymes of the homosporous fern *Athyrium felix-femina* L. Am. J. Bot. **70**:1409–1415.

Gastony, G.J., and L.D. Gottlieb. 1982. Evidence for genetic heterozygosity in a homosporous fern. Am. J. Bot. **69**:634–637.

Gastony, G.J., and L.D. Gottlieb. 1985. Genetic variation in the homosporous fern *Pellaea andromedifolia*. Am. J. Bot. **72**:257–267.

Gatenby, A.A., and E.C. Cocking. 1978a. The polypeptide composition of the subunits of fraction I protein in the genus *Lycopersicon*. Plant Sci. Lett. **13**:171–176.

Gatenby, A.A., and E.C. Cooking. 1978b. The evolution of fraction I protein and the distribution of the small subunit polypeptide coding sequences in the genus *Brassica*. Plant Sci. Lett. **12**:299–303.

Gatenby, A.A., and E.C. Cocking. 1978c. Fraction 1 protein and the origin of European potato. Plant. Sci. Lett. **12**:177–181.

Gell, P.G.H., J.G. Hawkes, and S.T.C. Wright. 1960. The application of immunological methods to the taxonomy of species within the genus *Solanum*. Proc. Roy. Soc. Lond. B. **151**:364–383.

Gerlach, W.L., and W.J. Peacock. 1980. Chromosomal locations of highly repeated DNA sequences in wheat. Heredity **44**:269–276.

Giannasi, D.E., and D.J. Crawford. 1986. Biochemical systematics II. A reprise. Evol. Biol. **20**:25–248.

Gillham, N.W., J.E. Boynton, and E.H. Harris. 1985. Evolution of plastid DNA. In: T. Cavallier-Smith, ed. "The Evolution of Genome Size." New York: Wiley, pp. 299–351.

Goldberg, R.B., W.P. Bemis, and A. Siegel. 1972. Nucleic acid hybridization studies within the genus *Cucurbita*. Genetics **72**:253–266.

Goodwin, T.W., and E.I. Mercer. 1983. "Introduction to Plant Biochemistry," 2nd ed. New York: Pergamon Press.

Gordon, A.H. 1975. "Electrophoresis of Proteins in Polyacrylamide and Starch Gels." New York: Elsevier.

Gottlieb, L.D. 1971. Evolutionary relationships in the outcrossing diploid annual species of *Stephanomeria* (Compositae). Evolution **25**:312–329.

Gottlieb, L.D. 1972. Levels of confidence in the analysis of hybridization in plants. Ann. Missouri Bot. Gard. **59**:435–446.

Gottlieb, L.D. 1973a. Genetic differentiation, sympatric speciation and the origin of a diploid species of *Stephanomeria*. Am. J. Bot. **60**:545–553.

Gottlieb, L.D. 1973b. Enzyme differentiation and phylogeny in *Clarkia franciscana*, *C. ribicunda* and *C. amoena*. Evolution **27**:205–214.

Gottlieb, L.D. 1973c. Genetic control of glutamate oxaloacetate transaminase isozymes in the diploid plant *Stephanomeria exigua* and its allotetraploid derivative. Biochem. Genet. **9**:97–107.

Gottlieb, L.D. 1974a. Genetic confirmation of the origin of *Clarkia lingulata*. Evolution **28**:244–250.

Gottlieb, L.D. 1974b. Gene duplication and fixed heterozygosity in the diploid plant *Clarkia franciscana*. Proc. Natl. Acad. Sci. USA **71**:1816–1818.

Gottlieb, L.D. 1977a. Electrophoretic evidence and plant systematics. Ann. Missouri Bot. Gard. **64**:161–180.

Gottlieb, L.D. 1977b. Phenotypic variation in *Stephanomeria exigua ssp. coronaria* (Compositae) and its recent derivative "Malheurensis." Am. J. Bot. **64**: 873–880.

Gottlieb, L.D. 1977c. Evidence for duplication and divergence of the structural gene for phosphoglucoisomerase in diploid species of *Clarkia*. Genetics **86**: 289–307.

Gottlieb, L.D. 1978. *Stephanomeria malheurensis* (Compositae), a new species from Oregon. Madrono **75**:44–46.

Gottlieb, L.D. 1979. The origin of phenotype in a recently evolved species. In:

O.T. Solbrig, S. Jain, G.B. Johnson, and P. Raven, eds. "Topics in Plant Population Biology." New York: Columbia University Press, pp. 264–286.

Gottlieb, L.D. 1981a. Electrophoretic evidence and plant populations. Prog. Phytochem. **7**:1–46.

Gottlieb, L.D. 1981b. Gene numbers in species of Astereae that have different chromosome numbers. Proc. Natl. Acad. Sci. USA **78**:3726–3729.

Gottlieb, L.D. 1982. Conservation and duplication of isozymes in plants. Science **216**:373–380.

Gottlieb, L.D. 1983. Isozyme number and plant phylogeny. In: U. Jensen and D.E. Fairbrothers, eds. "Proteins and Nucleic Acids in Plant Systematics." Heidelberg: Springer-Verlag, pp. 210–221.

Gottlieb, L.D. 1984a. Genetics and morphological evolution in plants. Am. Naturalist **123**:681–709.

Gottlieb, L.D. 1984b. Electrophoretic analysis of the phylogeny of the self-pollinating populations of *Clarkia xantiana*. Plant Syst. Evol. **147**:91–102.

Gottlieb, L.D. 1986. Genetic differentiation, speciation and phylogeny in *Clarkia* (Onagraceae). In: K. Iwatsuki, P.H. Raven, and W.J. Beck, eds. "Modern Aspects of Species." Tokyo: University of Tokyo Press, pp. 145–160.

Gottlieb, L.D. 1987. Phosphoglucomutase and isocitrate dehydrogenase gene duplications in *Layia* (Compositae). Am. J. Bot. **74**:9–15.

Gottlieb, L.D. 1988. Towards molecular genetics in *Clarkia*: gene duplications and molecular characterization of PGI genes. Ann. Missouri Bot. Gard. **75**:1169–1179.

Gottlieb, L.D., and V.S. Ford. 1987. Genetic and developmental studies of ray florets in *Layia discoidea*. In: H. Thomas and D. Grierson, eds. "Developmental Mutants in Higher Plants." New York: Cambridge University Press, pp. 1–17.

Gottlieb, L.D., and G. Pilz. 1976. Genetic similarity between *Gaura longiflora* and its obligately outcrossing derivative *G. demereei*. Syst. Bot. **1**:181–187.

Gottlieb, L.D., and N.F. Weeden. 1979. Gene duplication and phylogeny in *Clarkia*. Evolution **33**:1024–1039.

Gottlieb, L.D., S.I. Warwick, and V.S. Ford. 1985. Morphological and electrophoretic divergence between *Layia discoidea* and *L. glandulosa*. Syst. Bot. **10**:484–495.

Grant, V. 1981. "Plant Speciation," 2nd ed. New York: Columbia University Press.

Gray, J.C. 1980. Fraction I protein and plant phylogeny. In: F.A. Bisby, J.G. Vaughan, and C.A. Wright, eds. "Chemosystematics: Principles and Practice." New York: Academic Press, pp. 167–193.

Gray, J.C., S.D. Kung, and S.G. Wildman. 1974. Origin of *Nicotiana tabacum* L. detected by polypeptide composition of fraction I protein. Nature **252**:226–227.

Grund, C., J. Gilroy, T. Gleaves, U. Jensen, and D. Boulter. 1981. Systematic relationships of the Ranunculaceae based on amino acid sequence data. Phytochemistry **20**:1559–1565.

Hamby, R.K., and E.A. Zimmer. 1988. Ribosomal RNA sequences for inferring phylogeny within the grass family. Plant Syst. Evol. **160**:29–37.

Hames, B.D., and D. Rickwood, eds. 1981. "Gel Electrophoresis of Proteins, a Practical Approach." Oxford: IRL Press.

Hammel, B., and J.R. Reeder. 1979. The genus *Crypsis* (Gramineae) in the United States. Syst. Bot. **4**:267–280.

Hammond, H.D. 1955. Systematic serological studies in Ranunculaceae. Bull. Serological Museum, New Brunswick **14**:1–3.

Hamrick, J.L. 1987. Gene flow and distribution of genetic variation in plant populations. In: K.M. Urbanska, ed. "Differentiation Patterns in Higher Plants." Orlando, FL: Academic Press, pp. 53–67.

Hamrick, J.L., and M.D. Loveless. 1986. Isozyme variation in tropical trees: procedures and preliminary results. Biotropica **18**:201–207.

Hamrick, J.L., J.B. Mitton, and Y.B. Linhart. 1981. Levels of genetic variation in trees: influences of life history characteristics. In: M.T. Conkle, ed. "Proceedings of the Symposium on Isozymes of North American Forest Trees and Forest Insects." U.S. For. Serv. Gen. Tech. Rep. PSW-48, pp. 35–44.

Harborne, J.B., and B.L. Turner. 1984. "Plant Chemosystematics." Orlando, FL: Academic Press.

Hart, G.E. 1970. Evidence for triplicate genes for alcohol dehydrogenase in hexaploid wheat. Proc. Natl. Acad. Sci. USA **66**:1136–1141.

Hart, G.E. 1979. Evidence for a triplicate set of glucosephosphate isomerase structural genes in hexaploid wheat. Biochem. Genet. **17**:585–598.

Hartl, D.L., and A.G. Clark. 1989. "Principles of Population Genetics," 2nd ed. Sunderland, MA: Sinauer.

Haslett, B.G., T. Gleaves, and D. Boulter. 1977. N-terminal amino acid sequences of plastocyanins from various members of the Compositae. Phytochemistry **16**:363–365.

Hauber, D.P. 1986. Autotetraploidy in *Haplopappus spinulosus* hybrids: evidence from natural and synthetic tetraploids. Am. J. Bot. **73**:1595–1606.

Haufler, C.H. 1985. Enzyme variability and modes of evolution in *Bommeria* (Pteridaceae). Syst. Bot. **10**:92–104.

Haufler, C.H. 1987. Electrophoresis is modifying our concepts of evolution in homosporous pteridophytes. Am. Jo. Bot. **74**:953–966.

Haufler, C.H., and D.E. Soltis. 1986. Genetic evidence suggests that homosporous ferns with high chromosome numbers are diploid. Proc. Natl. Acad. Sci. USA **83**:4389–4393.

Hedrick, P.W. 1971. A new approach to measuring genetic similarity. Evolution **25**:276–280.

Hedrick, P.W. 1983. "Genetics of Populations." Boston: Science Books International.

Heiser, C.B. 1949. Study in the evolution of the sunflower species *Helianthus annuus* and *H. bolanderi*. Univ. Calif. Publ. Bot. **23**:157–196.

Heiser, C.B. 1956. Biosystematics of *Helianthus debilis*. Madrono **13**:145–176.

Helentjaris, T., M. Solcum, S. Wright, A. Schaefer, and J. Nienhuis. 1986. Construction of genetic linkage maps in maize and tomato using restriction fragment length polymorphisms. Theor. Appl. Genet. **72**:761–769.

Helenurm, K., and F.R. Ganders. 1985. Adaptive radiation and genetic differentiation in Hawaiian *Bidens*. Evolution **39**:753–765.

Heywood, J.S., and D.A. Levin. 1984. Allozyme variation in *Gaillardia pulchella* and *G. amblyodan* (Compositae): relation to morphological and chromosomal variation and to geographical isolation. Syst. Bot. **9**:448–457.

Heywood, V.H. 1973. The role of chemistry in plant systematics. Pure Appl. Chem. **34**:355–375.

Hill, R.J. 1977. Variability of soluble seed proteins in populations of *Mentzelia* L. (Loasaceae) from Wyoming and adjacent states. Bull. Torrey Bot. Club **104**:93–101.

Hillebrand, G.R., and D.E. Fairbrothers. 1969. A serological investigation of intrageneric relationships in *Viburnum* (Caprifoliaceae). Bull. Torrey Bot. Club **96**:556–567.

Hillis, D.M. 1987. Molecular versus morphological approaches to systematics. Annu. Rev. Ecol. Syst. **18**:23–42.

Hillis, D.M., and S.K. Davis. 1988. Ribosomal DNA: intraspecific polymorphisms, concerted evolution, and phylogeny reconstruction. Syst. Zool. **37**:63–66.

Hilu, K.W. 1984. The role of single gene mutations in the evolution of flowering plants. Evol. Biol. **16**:97–128.

Hirai, A., and K. Tsunewaki. 1981. Genetic diversity of the cytoplasm in *Triticum* and *Aegilops*. VII. Fraction I protein of 39 cytoplasms. Genetics **99**:487–493.

Hoffman, D.L., D.E. Soltis, F.J. Muehlbauer, and G. Ladizinsky. 1986. Isozyme polymorphism in *Lens* (Leguminosae). Syst. Bot. **11**:392–402.

Holsinger, K.E., and L.D. Gottlieb. 1988. Isozyme variability in the tetraploid

Clarkia gracilis (Onagraceae) and its diploid relatives. Syst. Bot. **13**:1–6.

Hunter, R.L., and C.L. Markert. 1957. Histochemical demonstration of enzymes separated by zone electrophoresis in starch gels. Science **125**:1294–1295.

Hutchinson, J. 1983. *In situ* mapping of plant chromosomes. In: P.E. Brandham and M.D. Bennett, eds. "Kew Chromosome Conference II." London: Allen and Unwin, pp. 27–34.

Hutchinson, J., V. Chapman, and T.E. Miller. 1980. Chromosome pairing at meiosis in hybrids between *Aegilops* and *Secale* species: a study by *in situ* hybridization using cloned DNA. Heredity **45**:245–254.

Hutchinson, J., R.B. Flavell, and J. Jones. 1981. Physical mapping of plant chromosomes by *in situ* hybridization. In: J. Setlow and A. Hollaender, eds. "Genetic Engineering," Vol. 3. New York: Plenum, pp. 207–222.

Hyun, J.O., O.P. Rajora, and L. Zsuffa. 1987. Genetic variation in trembling aspen in Ontario based on isozyme studies. Can. J. For. Res. **17**:1134–1138.

Jaaska, V. 1978. NADP-dependent aromatic alcohol dehydrogenase in polyploid wheats and their diploid relatives, on the origin and phylogeny of polyploid wheats. Theor. Appl. Genet. **53**:209–217.

Jacobs, B.F., C.R. Werth, and S.I. Guttman, 1984. Genetic relationships in *Abies* (fir) of eastern United States: an electrophoretic study. Can. J. Bot. **62**:609–616.

Jansen, R.K., and J.D. Palmer. 1987a. Chloroplast DNA from lettuce and *Barnadesia* (Asteraceae): structure, gene localization, and characterization of a large inversion. Curr. Genet. **11**:553–564.

Jansen, R.K., and J.D. Palmer. 1987b. A chloroplast DNA inversion marks an ancient evolutionary split in the sunflower family (Asteraceae). Proc. Natl. Acad. Sci. USA **84**:5818–5822.

Jansen, R.K., and J.D. Palmer. 1988. Phylogenetic implications of chloroplast DNA restriction site variation in the Mutisieae (Asteraceae). Am. J. Bot. **75**:753–766.

Jansen, R.K., E.B. Smith, and D.J. Crawford. 1987. A cladistic study of North American *Coreopsis* (Asteraceae: Heliantheae). Plant. Syst. Evol. **153**:73–84.

Jeffries, R.L., and L.D. Gottlieb. 1982. Genetic differentiation of the microspecies *Salicornia europaea* L. (sensa stricto) and *S. ramosissima*, J. Woods. New Phytol. **92**:123–129.

Jensen, U. 1968. Serologische Beitrage zur Systematik der Ranunculaceae. Bot. Jahrb. Syst. **88**:204–268.

Jensen, U. 1973. Interpretation of comparative serological results. In: G. Bendz and J. Santesson, eds. "Chemistry in Botanical Classification." New York: Academic Press, pp. 217–227.

Jensen, U. 1984. Legumin-like and vicilin-like proteins in *Nigella damaseua* (Ranunculaceae) and six other dicotyledonous species. J. Plant Physiol. **115**:161–170.

Jensen, U., and C. Buttner. 1981. The distribution of storage proteins in Magnoliophytina (angiosperms) and their serological similarities. Taxon **30**:404–419.

Jensen, U., and B. Greven. 1984. Serological aspects and phylogenetic relationships of the Magnoliidae. Taxon **33**:563–577.

Jensen, U., and B. Grumpe. 1983. Seed storage proteins. In: U. Jensen and D.E. Fairbrothers, eds. "Proteins and Nucleic Acids in Plant Systematics." Heidelberg: Springer-Verlag, pp. 238–254.

Jensen, U., and R. Penner. 1980. Investigation of serological determinants from single storage plant proteins. Biochem. Syst. Ecol. **8**:161–170.

Johnson, B.L. 1967. Tetraploid wheats: seed protein electrophoretic patterns of Emmer and Timopheevi groups. Science **158**:131–132.

Johnson, B.L. 1972a. Seed protein profiles and the origin of the hexaploid wheats. Am. J. Bot. **59**:952–960.

Johnson, B.L. 1972b. Protein electrophoretic profiles and the origin of the B genome of wheat. Proc. Natl. Acad. Sci. USA **69**:1398–1402.

Johnson, B.L. 1975. Identification of the apparent B-genome donor of wheat. Can. J. Genet. Cytol. **17**:21–39.

Johnson, B.L. 1976. Polyploid wheats and fraction I protein. Science **192**:1252.

Johnson, B.L., and M.M. Thien. 1970. Assessment of evolutionary affinities in *Gossypium* by protein electrophoresis. Am. J. Bot. **57**:1081–1092.

Johnson, G.B. 1976. Hidden alleles at the α-glycerophosphate dehydrogenase locus in colias butterflies. Genetics **83**:149–167.

Johnson, G.B. 1977a. Characterization of electrophoretically cryptic variation in the alpine butterfly *Colias meadii*. Biochem. Genet. **15**:665–693.

Johnson, G.B. 1977b. Assessing electrophoretic similarity: the problem of hidden heterogeneity. Annu. Rev. Ecol. Syst. **8**:309–328.

Jones, J.D.G., and R.B. Flavell. 1982. The structure, amount and chromosomal localisation of defined repeated DNA sequences in species of the genus *Secale*. Chromosoma **86**:613–641.

Jorgensen, R.A, and P.D. Cluster. 1988. Modes and tempos in the evolution of nuclear ribosomal DNA: new characters for evolutionary studies and new markers for genetic and population studies. Ann. Missouri Bot. Gard. **75**: 1238–1247.

Jorgensen, R.B. 1986. Relationships in the barley genus (*Hordeum*); an electrophoretic examination of proteins. Hereditas **104**:273–291.

Jukes, T.H. 1987. Transitions, transversions, and the molecular evolutionary clock. J. Mol. Evol. **26**:87–98.

Kawashima, N., and S.G. Wildman. 1972. Studies on fraction I protein. IV. Mode of inheritance of primary structure in relation to whether chloroplast or nuclear DNA contains the code for a chloroplast protein. Biochim. Biophys. Acta. **262**:42–49.

Kesseli, R.V., and S.K. Jain. 1984. New variation and biosystematic patterns detected by allozyme and morphological comparisons in *Limnanthes* sect. *Reflexae* (Limnanthaceae). Plant Syst. Evol. **147**:133–165.

Kim, S.E., L. Saur, and J. Mosse. 1979. Some features of the inheritance of avenins, the alcohol soluble proteins of oat. Theor. Appl. Genet. **54**:49–54.

Kimura, M. 1968. Evolutionary rate at the molecular level. Nature **217**:624.

Kimura, M. 1979. The neutral theory of molecular evolution. Sci. Am. **241**:98–126.

Kimura, M. 1983. The neutral theory of molecular evolution. In: M. Nei and R.K. Koehn, eds. "Evolution of Genes and Proteins." Sunderland, MA: Sinauer, pp. 208–233.

Kimura, M. 1984. "The Neutral Theory of Molecular Evolution." Cambridge: Cambridge University Press.

Kimura, M. 1987. Molecular evolutionary clock and the neutral theory. J. Mol. Evol. **26**:24–33.

Kitamura, K., C.S. Davies, and N.C. Nielson. 1984. Inheritance of alleles for Cgy_1 and Cy_4 storage proteins in soybean. Theor. Appl. Genet. **68**:253–257.

Kohne, D.E. 1968. Taxonomic applications of DNA hybridization techniques. In: J.G. Hawkes, ed. "Chemotaxonomy and Serotaxonomy." London: Academic Press, pp. 117–130.

Kraljevic-Balalic, M., D. Stajner, and O. Gasic. 1982. Inheritance of grain proteins in wheat. Theor. Appl. Genet. **63**:121–124.

Krebbers, E., J. Seurinck, L. Herdies, A.R. Cashmore, and M.P. Timko. 1988. Four genes in two diverged subfamilies encode the ribulose-1,5-bisphosphate carboxylase small polypeptides of *Arabidopsis thaliana*. Plant Mol. Biol. **11**:745–759.

Kung, S. 1976. Tobacco fraction I protein: a unique genetic marker. Science **191**:429–434.

Ladizinsky, G., and T. Hymowitz. 1979. Seed protein electrophoresis in taxonomic and evolutionary studies. Theor. Appl. Genet. **54**:145–151.

Ladizinsky, G., and G. Waines. 1982. Seed protein polymorphism in *Vicia sativa* agg. (Fabaceae) Plant Syst. Evol. **141**:1–5.

Lafiandra, D., D.D. Kasarda, and R. Morris. 1984. Chromosomal assignment of genes coding for the wheat gliadin protein components of the cultivars "Cheyenne" and "Chinese Spring" by two-dimensional (two-pH) electrophoresis. Theor. Appl. Genet. **68**:531–539.

Lander, E.S., and D. Botstein. 1989. Mapping Mendelian factors underlying quantitative traits using RFLP linkage maps. Genetics **121**:185–199.

Larkins, B.A. 1981. Seed storage proteins: characterization and biosynthesis. In: A. Marcus, ed. "The Biochemistry of Plants," Vol. 6. New York: Academic Press, pp. 449–489.

Lavin, M., A. Bruneau, J.J. Doyle, and J.D. Palmer. 1988. Loss of the chloroplast DNA inverted repeat as a tribal character in the Leguminosae. [Abstract] Am. J. Bot. **75**:189.

Lawrence, G.J., and K.W. Shepherd. 1981. Inheritance of glutenin protein subunits of wheat. Theor. Appl. Genet. **60**:333–337.

Lebacq, P., and F. Vedel. 1981. Sal I. restriction enzyme analysis of chloroplast and mitochondrial DNAs in the genus *Brassica*. Plant Sci. Lett. **23**:1–9.

Ledig, F.T., and M.T. Conkle. 1983. Gene diversity and genetic structure in a narrow endemic, Torrey pine (*Pinus torreyana* Parry ex Carr.). Evolution **37**:79–85.

Lee, Y.S. 1977. An immunoelectrophoretic comparison of three species of *Coffea*. Syst. Bot. **2**:169–179.

Leonard, A., C. Damerval, and D. De Vienne. 1988. Organ-specific variability and inheritance of maize proteins revealed by two-dimensional electrophoresis. Genet. Res. **52**:97–103.

Lester, R.N. 1979. The use of protein characters in the taxonomy of *Solanum* and other Solanaceae. In: J.G. Hawkes, R.N. Lester, and A.D. Skelding, eds. "The Biology and Taxonomy of the Solanaceae." New York: Academic Press, pp. 285–303.

Lester, R.N. 1984. Pre-absorption—a valuable tool in serotaxonomic research. Taxon **33**:578–580.

Lester, R.N., P.A. Roberts, and C. Lester. 1983. Analysis of immunotaxonomic data obtained from spur identification and absorption techniques. In: U. Jensen and D.E. Fairbrothers, eds. "Proteins and Nucleic Acids in Plant Systematics." Heidelberg: Springer-Verlag, pp. 275–300.

Leutwiler, L.S., B.R. Hough-Evans, and E.M. Meyerowitz. 1984. The DNA of *Arabidopsis thaliana*. Mol. Gen. Genet. **194**:15–23.

Levin, D.A. 1977. The organization of genetic variability in *Phlox drummondii*. Evolution **31**:477–494.

Levin, D.A. 1978. Genetic variation in annual *Phlox*: self-compatible versus self-incompatible species. Evolution **32**:245–263.

Levin, D.A. 1983. Polyploidy and novelty in flowering plants. Am. Naturalist **122**:1–25.

Levin, D.A., and B.A. Schaal. 1970. Reticulate evolution in *Phlox* as seen through protein electrophoresis. Am. J. Bot. **57**:977–987.

Levin, D.A, and B.A. Schaal. 1972. Seed protein polymorphism in *Phlox pilosa* (Polemoniaceae). Brittonia **24**:46–56.

Lewin, B. 1987. "Genes III." New York: Wiley.

Lewis, H. 1961. Experimental sympatric populations of *Clarkia*. Am. Naturalist **95**:155–168.

Lewis, H. 1962. Catastrophic selection as a factor in speciation. Evolution **16**:257–271.

Lewis, H. 1973. The origin of diploid neospecies in *Clarkia*. Am. Naturalist **107**:161–170.

Lewis, H. 1980. Mode of evolution in *Clarkia*. Second Int. Congress of Syst. and Evol. Biol., p. 73 of published abstracts. Vancouver, Canada.

Lewis, H., and C. Epling. 1959. *Delphinium gypsophyllum*, a diploid species of hybrid origin. Evolution **13**:511–525.

Lewis, H., and M.E. Lewis. 1955. The genus *Clarkia*. Univ. Calif. Publ. Bot. **20**:241–392.

Lewis, H., and P. Raven. 1958a. Rapid evolution in *Clarkia*. Evolution **12**:319–336.

Lewis, H., and P. Raven. 1958b. *Clarkia franciscana*, a new species from central California. Brittonia **10**:7–13.

Lewis, H., and M.R. Roberts. 1956. The origin of *Clarkia lingulata*. Evolution **10**:126–138.

Lewis, W.H., ed. 1980. "Polyploidy—Biological Relevance." New York: Plenum.

Lin, C.M., Z.Q. Liu, and S.D. Kung. 1986. *Nicotiana* chloroplast genome: X. Correlation between the DNA sequences and the isoelectric focusing patterns of the LS of Rubisco. Plant Mol. Biol. **6**:81–87.

Loukas, M., Y. Vergini, and C.B. Krimbas. 1983. Isozyme variation and heterozygosity in *Pinus halepensis* L. Biochem. Genet. **21**:497–509.

Loveless, M.D., and J.L. Hamrick. 1984. Ecological determinants of genetic structure in plant populations. Annu. Rev. Ecol. Syst. **15**:65–95.

Loveless, M.D., and J.L. Hamrick. 1988. Genetic organization and evolutionary history in two North American species of *Cirsium*. Evolution **42**:254–265.

Lowrey, T.K. 1986. A biosystematic revision of Hawaiian *Tetramolopium* (Compositae: Astereae). Allertonia **4**:203–265.

Lowrey, T.K., and D.J. Crawford. 1985. Allozyme divergence and evolution in *Tetramolopium* (Compositae: Astereae) on the Hawaiian Islands. Syst. Bot. **10**:64–72.

Lowrey, T.K., and D.J. Crawford. 1987. Electrophoretic confirmation of the intergeneric hybrid x *Ruttyruspolia* (Acanthaceae). Plant Syst. Evol. **158**:29–35.

Lumaret, R. 1986. Doubled duplication of the structural gene for cytosolic phosphoglucose isomerase in the *Dactylis glomerata* L. polyploid complex. Mol. Biol. Evol. **36**:499–521.

Mabry, T.J. 1977. The order Centrospermae. Ann. Missouri Bot. Gard. **64**:210–220.

MacIntyre, R.J., ed. 1985. "Molecular Evolutionary Genetics." New York: Plenum.

Maggini, F. 1975. Homologies of ribosomal RNA nucleotide sequences in monocots. J. Mol. Evol. **4**:317–322.

Maggini, F., R.I. DeDominicis, and G. Salvi. 1976. Similarities among ribosomal RNAs of angiospermae and gymnospermae. J. Mol. Evol. **8**:329–335.

Mahmoud, S.H., and J.A. Gatehouse. 1984. Inheritance and mapping of vicilin storage protein genes in *Pisum sativum* L. Heredity **53**:185–191.

Maniatis, T., E.F. Fritsch, and J. Sambrook. 1982. "Molecular Cloning: A Laboratory Manual." Cold Spring Harbor: Cold Spring Harbor Laboratory.

Manos, P.S., and D.E. Fairbrothers. 1987. Allozyme variation in populations of six northeastern American red oaks (Fagaceae: *Quercus* subg. *Erythrobalanus*). Syst. Bot. **12**:365–373.

Margoliash, E., S. Ferguson-Miller, C.H. Kange, and D.L. Brautigan. 1976. Do evolutionary changes in cytochrome c structure reflect functional adaptations? Fed. Proc. **35**:2124–2130.

Markert, C.L. 1977. Isozymes: the development of a concept and its application. In: M.C. Rattazz, J.G. Scandalios and G.S. Whitt, eds. "Isozymes: Current Topics in Biological and Medical Research," Vol. 1. New York: Alan R. Liss, pp. 1–17.

Markert, C.L., and F. Moller. 1959. Multiple forms of enzymes: tissue, ontogenetic and species-specific patterns. Proc. Natl. Acad. Sci. USA **45**:753–763.

Marshall, D.R., and A.H.D. Brown. 1975. Optimum sampling strategies in genetic conservation. In: O.H. Frankel and J.G. Hawkes, eds. "Genetic Resources for Today and Tomorrow." Cambridge: Cambridge University Press, pp. 53–80.

Martin, P.G., and J.M. Dowd. 1984a. The study of plant phylogeny using amino acid sequences of ribulose-1,5-bisphosphate carboxylase. III. Addition of Malvaceae and Ranunculaceae to the phylogenetic tree. Aust. J. Bot. **32**:283–290.

Martin, P.G., and J.M. Dowd. 1984b. The study of plant phylogeny using amino acid sequences of ribulose-1,5-bisphosphate carboxylase. IV. Proteaceae and Fagaceae and the rate of evolution of the small subunit. Aust. J. Bot. **32**:291–299.

Martin, P.G., and J.M. Dowd. 1984c. The study of plant phylogeny using amino acid sequences of ribulose-1,5-bisphosphate carboxylase. V. Magnoliaceae,

Polygonaceae and the concept of primitiveness. Aust. J. Bot. **32**:301–309.

Martin, P.G., and J.M. Dowd. 1986a. A phylogenetic tree for some monocotyledons and gymnosperms derived from protein sequences. Taxon **35**:469–475.

Martin, P.G., and J.M. Dowd. 1986b. Phylogenetic studies using protein sequences within the order Myrtales. Ann. Missouri Bot. Gard. **73**:442–448.

Martin, P.G., and A.C. Jennings. 1983. The study of plant phylogeny using amino acid sequences of ribulose-1,5-bisphosphate carboxylase. I. Biochemical methods and patterns of variability. Aust. J. Bot. **31**:395–409.

Martin, P.G., D. Boulter, and D. Penny. 1985. Angiosperm phylogeny studied using sequences of five macromolecules. Taxon **34**:393–400.

Martin, P.G., J.M. Dowd, and S.J.L. Stone. 1983. The study of plant phylogeny using amino acid sequences of ribulose-1,5-bisphosphate carboxylase. II. The analysis of small subunit data to form phylogenetic trees. Aust. J. Bot. **31**:411–419.

Martin, P.G., J.M. Dowd, C. Morris, and D.E. Symon. 1986. The study of plant phylogeny using amino acid sequences of ribulose-1,5-bisphosphate carboxylase. VI. Some *Solanum* and allied species from different continents. Aust. J. Bot. **34**:187–195.

Maxam, A.M., and W. Gilbert. 1980. Sequencing end-labeled DNA with base-specific chemical cleavages. Meth. Enzymol. **65**:499–560.

McDowell, R.E., and S. Prakash. 1976. Allelic heterogeneity within allozymes separated by electrophoresis in *Drosophila pseudoobscura*. Proc. Natl. Acad. Sci. USA **73**:4150–4153.

McLeod, M.J., S.I. Guttman, W.H. Eshbaugh, and R.E. Rayle. 1983. An electrophoretic study of evolution in *Capsicum* (Solanaceae). Evolution **37**:562–574.

McNeill, C.I., and S.K. Jain. 1983. Genetic differentiation studies and phylogenetic inference in the plant genus *Limnanthes* (section *Inflexae*). Theor. Appl. Genet. **66**:257–269.

Meacham, D.K., D.D. Kasarda, and C.O. Qualset. 1978. Genetic aspects of wheat gliadin proteins. Biochem. Genet. **16**:813–853.

Meinkoth, J., and G.M. Wahl. 1987. Nick translation. Meth. Enzymol. **152**:91–110.

Melchers, G., M.D. Sacristan, and A.A. Holder. 1978. Somatic hybrid plants of potato and tomato regenerated from fused protoplasts. Carlsberg Res. Comm. **43**:203–218.

Michaels, H.J., J.D. Palmer, and R.K. Jansen. 1987. Phylogenetic implications of chloroplast DNA restriction site variation in the Asteraceae [Abstract]. Am. J. Bot. **74**:744–745.

Mickevich, M.F. 1978. Taxonomic congruence. Syst. Zool. **27**:143–158.

Miege, M.N. 1982. Protein types and distribution. Encyclopedia Plant Physiol. new series. **14A**:291–345.

Miksche, J.P. 1968. Quantitative study of intraspecific variation of DNA per cell in *Picea glauca* and *Pinus banksiana*. Can. J. Genet. Cytol. **10**:590–600.

Miksche, J.P. 1971. Intraspecific variation of DNA per cell between *Picea sitchensis* (Bong.) Carr. provenances. Chromosoma **32**:343–352.

Millar, C.I., S.H. Strauss, M.T. Conkel, and R.D. Westfall. 1988. Allozyme differentiation and biosystematics of the Californian closed-cone pines (*Pinus* subsect. *Oocarpae*). Syst. Bot. **13**:351–370.

Mills, W.R., and K.W. Joy. 1980. A rapid method for isolation of purified, physiologically active chloroplasts, used to study the intracellular distribution of amino acids in pea leaves. Planta **148**:75–83.

Mitra, R., and C.R. Bhatia. 1986. Repeated DNA sequences and polyploidy in cereal crops. In: S.K. Dutta, ed. "DNA Systematics, Vol. II: Plants." Boca Raton, FL: CRC Press, Inc., pp. 21–43.

Mitton, J.B. 1983. Conifers. In: S.D. Tanksley and T.J. Orton, eds. "Isozymes in Plant Genetics and Breeding, Part. B. Amsterdam: Elsevier, pp. 443–472.

Morden, C.W., J. Doebley, and K.F. Schertz. 1987. "A Manual of Techniques for Starch Gel Electrophoresis of Sorghum Isozymes." Texas Agric. Exp. Station. MP-1635.

Mueller, L.D. 1979. A comparison of two methods for making statistical inferences on Nei's measure of genetic distance. Biometrics **35**:757–763.

Murdy, W.H., and M. Carter. 1985. Electrophoretic study of the allopolyploidal origin of *Talinum teretifolium* and the specific status of *T. appalachianum* (Portulaceae). Am. J. Bot. **72**:1590–1597.

Murphy, T.M., and W.F. Thompson. 1988. "Molecular Plant Development." Englewood Cliffs, J: Prentice Hall.

Murray, M.G., R.E. Cuellar, and W.F. Thompson. 1978. DNA sequence organization in the pea genome. Biochemistry **17**:5781–5790.

Narayan, R.K.J. 1982. Discontinuous DNA variation in the evolution of plant species: The genus *Lathyrus*. Evolution **36**:877–891.

Narayan, R.K.J. 1983. Chromosome changes in the evolution of *Lathyrus* species. In: P.E. Brandham, and M.D. Bennett, eds. "Kew Chromosome Conference II." London: Allen and Unwin, pp. 27–34.

Narayan, R.K.J. 1987. Nuclear DNA changes, genome differentiation and evolution in *Nicotiana*. Plant Syst. Evol. **157**:161–180.

Narayan, R.K.J., and H. Rees. 1976. Nuclear DNA variation in *Lathyrus*. Chromosoma 54:141–154.

Narayan, R.K.J., and H. Rees. 1977. Nuclear DNA divergence among *Lathyrus* species. Chromosoma **63**:101–107.

Nath, J., J.W. McNay, C.M. Paroda, and S.C. Gulati. 1983. Implication of *Triticum searsii* at the B-genome donor to wheat using DNA hybridizations. Biochem. Genet. **21**:745–760.

Nath, J., J.J. Hanzel, J.P. Thompson, and J.W. McNay. 1984. Additional evidence implicating *Triticum searsii* as the B-genome donor to wheat. Biochem. Genet. **22**:37–50.

Neale, D.B., M.A. Saghai-Maroof, R.W. Allard, Q. Zhang, and R.A. Jorgensen. 1988. chloroplast DNA diversity in populations of wild and cultivated barley. Genetics **120**:1105–1110.

Nei, M. 1972. Genetic distance between populations. Am. Naturalist **106**:283–293.

Nei, M. 1973. Analysis of gene diversity in subdivided populations. Proc. Natl. Acad. Sci. USA **70**:3321–3323.

Nei, M. 1978. Estimation of average heterozygosity and genetic distance from a small number of individuals. Genetics **89**:583–590.

Nei, M. 1987. "Molecular Evolutionary Genetics." New York: Columbia University Press.

Nelson, C.H., and G.S. VanHorn. 1975. A new simplified method for constructing Wagner networks and the cladistics of *Pentachaeta* (Compositae: Astereae). Brittonia **27**:362–372.

Nelson, G.J., and N.I. Platnick. 1981. "Systematics and Biogeography: Cladistics and Vicariance." New York: Columbia University Press.

Nesom, G.L. 1983. *Galax* (Diapensiaceae): geographic variation in chromosome number. Syst. Bot. **8**:1–14.

Nevo, E., D. Zohery, A.H.D. Brown, and M. Haber. 1979. Genetic diversity and environmental associations of wild barley, *Hordeum spontaneum*, in Israel. Evolution **33**:815–833.

Nevo, E., E. Golenberg, A. Beiles, A.H.D. Brown, and D. Zohary. 1982. Genetic diversity and environmental associations of wild wheat *Triticum dicoccoides*. Theor. Appl. Genet. **62**:241–254.

Nevo, E., A. Beiles, N. Storch, H. Doll, and B. Andersen. 1983. Microgeographic edaphic differentiation in hordein polymorphisms of wild barley. Theor. Appl. Genet. **64**:123–132.

Nickrent, D.L. 1986. Genetic polymorphism in the morphologically reduced dwarf mistletoes (*Arceuthobium*, Viscaceae): an electrophoretic study. Am. J. Bot. **73**:1492–1502.

Nickrent, D.L., S.I. Guttman, and W.H. Eshbaugh. 1984. Biosystematic and evolutionary relationships among selected taxa of New World *Arceuthobium*. In: F.G. Hawksworth and R.F. Sharp, eds. "Biology of Dwarf Mistletoes: Proceedings of the Symposium." U.S.D.A. For. Ser. Gen. Tech. Rep. RM-111, pp. 20–35.

Odrzykoski, I.J., and L.D. Gottlieb. 1984. Duplications of genes coding 6-phosphogluconate dehydrogenase in *Clarkia* (Onagraceae) and their phylogenetic implications. Syst. Bot. **9**:479–489.

Ogden, R.C., and D.A. Adams. 1987. Electrophoresis in agarose and acrylamide gels. Meth. Enzymol. **152**:61–87.

Ogihara, Y., and K. Tsunewaki. 1982. Molecular basis of the genetic diversity of the cytoplasm in *Triticum* and *Aegilops*. I. Diversity of the chloroplast genome and its lineage revealed by the restriction pattern of ct-DNAs. Jpn. J. Genet. **57**:361–396.

Ohri, D., and T.N. Khoshoo. 1986. Plant DNA: contents and systematics. In: S.K. Dutta, ed. "DNA Systematics, Vol. II: Plants." Boca Raton, FL: CRC Press, Inc., pp. 1–19.

Ohta, T. 1987. Very slightly deleterious mutations and the molecular clock. J. Mol. Evol. **26**:1–6.

Ohyama, K., H. Fukuzawa, T. Kohchi, H. Shirai, T. Sano, S. Sano, K. Umeso-no, Y. Shiki, M. Takeuchi, Z. Chang, S. Aota, H. Inokuchi, and H. Ozeki. 1986. Chloroplast gene organization deduced from complete sequence of liverwort *Marchantia polymorphia* chloroplast DNA. Nature **322**:572–574.

Ornduff, R. 1966. A biosystematic survey of the goldfield genus *Lasthenia* (Compositae: Helenieae). Univ. Calif. Publ. Bot. **40**:1–92.

Ornduff, R. 1969. The origin and relationships of *Lasthenia burkei* (Compositae). Am. J. Bot. **56**:1042–1047.

Ownbey, M. 1950. Natural hybridization and amphiploidy in the genus *Tragopogon*. Am. J. Bot. **37**:487–499.

Palmer, J.D. 1985a. Chloroplast DNA and molecular phylogeny. BioEssays **2**:263–267.

Palmer, J.D. 1985b. Comparative organization of chloroplast genomes. Annu. Rev. Genet. **19**:325–354.

Palmer, J.D. 1985c. Evolution of chloroplast and mitochondrial DNA in plants and algae. In: R.J. MacIntyre, ed. "Monographs in Evolutionary Biology, Molecular Evolutionary Genetics." New York: Plenum, pp. 131–240.

Palmer, J.D. 1986a. Isolation and structural analysis of chloroplast DNA. Meth. Enzymol. **118**:167–186.

Palmer, J.D. 1986b. Chloroplast DNA and phylogenetic relationships. In: S.K. Dutta, ed. "DNA Systematics, Vol. II: Plants," Boca Raton, FL: CRC Press, Inc., pp. xx–xx.

Palmer, J.D. 1987. Chloroplast DNA evolution and biosystematic uses of chloroplast DNA variation. Am. Naturalist **130**:S6–S29.

Palmer, J.D., and D. Zamir. 1982. Chloroplast DNA evolution and phylogenetic relationships in *Lycopersicon*. Proc. Natl. Acad. Sci. USA **79**:5006–5010.

Palmer, J.D., C.R. Sheilds, D.B. Cohen, and T.J. Orten. 1983. Chloroplast DNA evolution and the origin of *Brassica* species. Theor. Appl. Genet. **65**:181–189.

Palmer, J.D., B. Osorio, J.C. Watson, H. Edwards, T. Dodd, and W.F. Thompson. 1984. Evolutionary aspects of chloroplast genome expression and organization. In: J.P. Thornber, L.A. Staehelin and R. Hallick, eds. "Biosynthesis of the Photosynthetic Apparatus: Molecular Biology, Development and Regulation." UCLA Symposia on Mol. and Cell. Biol., New Series, Vol. 14. New York: Alan R. Liss, pp. 273–283.

Palmer, J.D., R.A. Jorgensen, and W.F. Thompson. 1985. Chloroplast DNA variation and evolution in *Pisum*: patterns of change and phylogenetic analysis. Genetics **109**:195–213.

Palmer, J.D., B. Osorio, J. Aldrich, and W.F. Thompson. 1987. Chloroplast DNA evolution among legumes: loss of a large inverted repeat occurred prior to other sequence rearrangements. Curr. Genet. **11**:275–286.

Palmer, J.D., R.K. Jansen, H.J. Michaels, M.W. Chase, and J.R. Manhart. 1988. Chloroplast DNA variation and plant phylogeny. Ann. Missouri Bot. Gard. **75**:1180–1206.

Paris, C.H., and M.D. Windham. 1988. A biosystematic investigation of the *Adiantum pedatum* complex in eastern North America. Syst. Bot. **13**:240–255.

Paterson, A.H., E.S. Lander, J.D. Hewitt, S. Peterson, S.E. Lincoln, and S.D. Tanksley. 1988. Resolution of quantitative traits into Mendelian factors by using a complete linkage map of restriction fragment length polymorphisms. Nature **335**:721–726.

Payne, P.I. 1986. Endosperm proteins. In: A.D. Blonstein and P.J. King, eds. "A Genetic Approach to Plant Biochemistry." Wien: Springer-Verlag.

Payne, P.I. 1987. Genetics of wheat storage proteins and the effect of allelic variation in bread-making quality. Ann. Rev. Plant Physiol. **38**:141–153.

Payne, P.I., K.G. Corfield, and J.A. Blackman. 1979. Identification of a high-molecular-weight subunit of glutenin whose presence correlates with bread-making quality in wheats of related parentage. Theor. Appl. Genet. **55**:153–159.

Payne, P.I., P.A. Harris, C.N. Law, L.M. Holt, and J.A. Blackman. 1980. The high-molecular-weight subunits of glutenin: structure, genetics and relationship to bread-making quality. Ann. Technol. Agric. **29**:309–320.

Payne, P.I., K.C. Corfield, L.M. Holt, and J.A. Blackman. 1981. Correlations between the inheritance of certain high-molecular weight subunits of glutenin and bread-making quality in progenies of six crosses of bread wheat. J. Sci. Food Agric. **32**:51–60.

Payne, W.W. 1976. Biochemistry and species problems in *Ambrosia* (Asteraceae–Ambrosieae). Plant Syst. Evol. **125**:169–178.

Peacock, D. 1981. Data handling for phylogenetic trees. In: H. Guffreund, ed. "Biochemical Evolution." Cambridge: Cambridge University Press, pp. 88–115.

Peacock, W.J., W.L. Gerlach, and E.S. Dennis. 1981. Molecular aspects of wheat evolution: repeated DNA sequences. In: L.T. Evans and W.J. Peacock, eds. "Wheat Science—Today and Tomorrow." New York: Cambridge University Press, pp. 41–60.

Petersen, F.P., and D.E. Fairbrothers. 1979. Serological investigation of selected amentiferous taxa. Syst. Bot. **4**:230–241.

Petersen, F.P., and D.E. Fairbrothers. 1983. A serotaxonomic appraisal of *Amphipterygium* and *Leitneria*—two amentiferous taxa of Rutiflorae (Rosidae). Syst. Bot. **8**:134–148.

Petersen, F.P., and D.E. Fairbrothers. 1985. A serotaxonomic appraisal of the "Amentiferae." Bull. Torrey Bot. Club **112**:43–52.

Pichersky, E., and L.D. Gottlieb. 1983. Evidence for duplication of the structural genes coding plastid and cytosolic isozymes of triose phosphate isomerase in diploid species of *Clarkia*. Genetics **105**:421–436.

Pichersky, E., L.D. Gottlieb, and R.C. Higgins. 1984. Hybridization between subunits of triose phosphate isomerase isozymes from different subcellular compartments of higher plants. Mol. Gen. Genet. **193**:158–161.

Pickering, J.L., and D.E. Fairbrothers. 1970. A serological comparison of Umbelliferae subfamilies. Am. J. Bot. **57**:988–992.

Piechura, J.E., and D.E. Fairbrothers. 1983. The use of protein-serological characters in the systematics of the family Oleaceae. Am. J. Bot. **70**:780–789.

Pinkas, R., D. Zamir, and G. Ladizinsky. 1985. Allozyme divergence and evolution in the genus *Lens*. Plant Syst. Evol. **151**:131–140.

Platnick, N.I. 1987. An empirical comparison of microcomputer parsimony programs. Cladistics **3**:121–144.

Poggio, L., and J.H. Hunziker. 1986. Nuclear DNA content variation in *Bulnesia*. J. Hered. **77**:43–48.

Prager, E.M., D.P. Fowler, and A.C. Wilson. 1976. Rates of evolution in conifers (Pinaceae). Evolution **30**:637–649.

Preparata, G., and C. Saccone. 1987. A simple quantitative model of the molecular clock. J. Mol. Evol. **26**:7–15.

Price, H.J. 1976. Evolution of DNA content in higher plants. Bot. Rev. **42**:27–52.

Price, H.J. 1988a. Plant genome size and the DNA C-value paradox. Plant Genet. Newsletter **4**:18–24.

Price, H.J. 1988b. DNA content variation among higher plants. Ann. Missouri Bot. Gard. **75**:1248–1257.

Price, H.J., and K. Bachmann. 1975. DNA content and evolution in the Microseridinae. Am. J. Bot. **62**:262–267.

Price, H.J., K. Bachmann, K.L. Chambers, and J. Riggs. 1980. Detection of intraspecific variation in nuclear DNA content in *Microseris douglasii*. Bot. Gaz. **141**:195–198.

Price, H.J., K.L. Chambers, and K. Bachmann. 1981a. Genome size variation in diploid *Microseris bigelovii* (Asteraceae). Bot. Gaz. **142**:156–159.

Price, H.J., K.L. Chambers, and K. Bachmann. 1981b. Geographic and ecological distribution of genomic DNA content variation in *Microseris douglasii* (Asteraceae). Bot. Gaz. **142**:415–426.

Price, H.J., K.L. Chambers, K. Bachmann, and J. Riggs. 1983. Inheritance of nuclear 2C DNA content variation in intraspecific and interspecific hybrids of *Microseris* (Asteraceae) Am. J. Bot. **70**:1133–1138.

Price, H.J., D.J. Crawford, and R.J. Bayer. 1984. Nuclear DNA contents of *Coreopsis nuecensoides* and *C. nuecensis* (Asteraceae), a progenitor–derivative species pair. Bot. Gaz. **145**:240–245.

Price, H.J., K.L. Chambers, K. Bachmann, and J. Riggs. 1986. Patterns of mean nuclear DNA content in *Microseris douglasii* (Asteraceae) populations. Bot. Gaz. **147**:496–507.

Price, R.A., and J.M. Lowenstein. 1989. An immunological comparison of the Sciadopityaceae, Taxodiaceae and Cupressaceae. Syst. Bot. **14**:141–149.

Price, R.A., J. Olsen-Stojkovich, and J.M. Lowenstein. 1987. Relationships among the genera of Pinaceae: an immunological comparison. Syst. Bot. **12**:91–97.

Quiros, C.F. 1982. Tetrasomic segregation for multiple alleles in alfalfa. Genetics **101**:117–127.

Quiros, C.F. 1983. Alfalfa, Luzerne (*Medicago sativa* L.). In: S.D. Tanksley and T.J. Orton, eds. "Isozymes in Plant Genetics and Breeding," Part B. Amsterdam: Elsevier, pp. 253–294.

Quiros, C.F., and N. McHale. 1985. Genetic analysis of isozyme variants in diploid and tetraploid potatoes. Genetics **111**:131–145.

Raina, S.N., and R.K.J. Narayan. 1984. Changes in DNA composition in the evolution of *Vicia* species. Theor. Appl. Genet. **68**:187–192.

Ramshaw, J.A.M. 1982. Structures of plant proteins. Encyclopedia Plant Physiol. new series. **14A**:229–290.

Ramshaw, J.A.M., D.L. Richardson, B.T. Meatyard, R.H. Brown, M. Richardson, E.W. Thompson, and D. Boulter, 1972. The time of origin of the flowering plants by using amino acid sequence data of cytochrome c. New Phytol. **71**:773–779.

Ramshaw, J.A.M., J.A. Coyne, and R.C. Lewontin. 1979. The sensitivity of gel

electrophoresis as a detector of genetic variation. Genetics **93**:1019–1037.

Ranker, T.A. 1987. Experimental systematics and population biology of the fern genera *Hemionitis* and *Gymnopteris* with reference to *Bommeria* [PhD thesis]. Laurence, KS: University of Kansas.

Ranker, T.A., and A.F. Schnabel. 1986. Allozymic and morphological evidence for a progenitor–derivative species pair in *Camassia* (Liliaceae). Syst. Bot. **11**:433–445.

Raven, P.H. 1976. Systematics and plant population biology. Syst. Bot. **1**:284–316.

Raven, P.H. 1979. A survey of reproductive biology of Onagraceae. NZ J. Bot. **17**:575–593.

Raven, P.H. 1980. Hybridization and the nature of species in higher plants. Can. Bot. Assoc. Bull. Suppl. **13**:3–10.

Raven, P.H., O.T. Solbrig, D.W. Kyhos, and R. Snow. 1960. Chromosome numbers in Compositae. I. Astereae. Am. J. Bot. **47**:124–132.

Rayburn, A.L., and B.S. Gill. 1985. Use of biotin-labeled probes to map specific DNA sequences on wheat chromosomes. J. Hered. **76**:78–81.

Rayburn, A.L., H.J. Price, J.D. Smith, and J.R. Gold. 1985. C-band heterochromatin and DNA content in *Zea mays*. Am. J. Bot. **72**:1610–1617.

Reddy, M.M., and E.D. Garber. 1971. Genetic studies of variant enzymes. III. comparative electrophoretic studies of esterases and peroxidases for species, hybrids and amphiploids in the genus *Nicotiana*. Bot. Gaz. **132**:158–166.

Reynolds, J.F., and D.J. Crawford. 1980. A quantitative study of variation in the *Chenopodium atrovirens–desiccatum–pratericola* complex. Am. J. Bot. **67**:1380–1390.

Richardson, B.J., P.R. Baverstock, and M. Adams. 1986. "Allozyme Electrophoresis—A Handbook for Animal Systematics and Population Studies." Orlando, FL: Academic Press.

Rick, C.M. 1979. Biosystematic studies in *Lycopersicon* and closely related species in *Solanum*. In: J.W. Hawkes, R.N. Lester, and A.D. Skalding, eds. "Biology and Taxonomy of the Solanaceae." New York: Academic Press, pp. 667–678.

Rick, C.M., and J.F. Fobes. 1975. Allozyme variation in the cultivated tomato and closely related species. Bull. Torrey Bot. Club **102**:376–384.

Rick, C.M., and S.D. Tanksley. 1981. Genetic variation in *Solanum pennellii*: comparisons with two other sympatric tomato species. Plant Syst. Evol. **139**:11–45.

Rick, C.M., E. Kesicki, J.F. Fobes, and M. Holle. 1976. Genetic and biosystematic studies of two new sibling species of *Lycopersicon* from interandean Peru. Theor. Appl. Genet. **47**:55–68.

Rick, C.M., J.F. Fobes, and M. Holle. 1977. Genetic variation in *Lycopersicon pimpinellifolium*: evidence of evolutionary change in mating systems. Plant Syst. Evol. **127**:139–170.

Rick, C.M., J.F. Fobes, and S.D. Tanksley. 1979. Evolution of mating systems in *Lycopersicon hirsutum* as deduced from genetic variation in electrophoretic and morphological characters. Plant Syst. Evol. **132**:279–298.

Rieseberg, L.H., and D.E. Soltis. 1987a. Allozymic differentiation between *Tolmiea menziesii* and *Tellima grandiflora*. Syst. Bot. **12**:154–161.

Rieseberg, L.H., and D.E. Soltis. 1987b. Phosphoglucomutase in *Helianthus debilis*: a polymorphism for isozyme number. Biochem Syst. Ecol. **5**:545–548.

Rieseberg, L.H., P.M. Paterson, D.E. Soltis, and C.R. Annable. 1987. Genetic divergence and isozyme number variation among four varieties of *Allium douglasii* (Alliaceae). Am. J. Bot. **74**:1614–1624.

Rieseberg, L.H., D.E. Soltis, and J.D. Palmer. 1988. A molecular reexaminimation of introgression between *Helianthus annuus* and *H. bolanderi* (Compositae). Evolution **43**:227–238.

Rimpau, J., D. Smith, and R. Flavell. 1978. Sequence organization analysis of the wheat and rye genomes by interspecies DNA/DNA hybridization. J. Mol. Biol. **123**:327–359.

Rimpau, J., D. Smith, and R. Flavell. 1979. Sequence organization in barley and oats chromosomes by interspecies DNA/DNA hybridization. Heredity **42**:131–149.

Ritland, K., and M.T. Clegg. 1987. Evolutionary analysis of plant DNA sequences. Am. Naturalist **130**:S74–S100.

Robbins, M.P., and J.G. Vaughan. 1983. Rubisco in the Brassicaceae. In: U. Jensen and D.E. Fairbrothers, eds. "Proteins and Nucleic Acids in Plant Systematics." Heidelberg: Springer-Verlag, pp. 191–240.

Roberts, M.L. 1983. Allozyme variation in *Bidens discoidea* (Compositae). Brittonia **35**:239–247.

Rodman, J.E., M.K. Oliver, R.R. Nakamura, J.U. McClammer, Jr., and A.H. Bledsoe. 1984. A taxonomic analysis and revised classification of Centrospermae. Syst. Bot. **9**:297–323.

Rogers, J.S. 1972. Measures of genetic similarity and genetic distance. Univ. Texas Publ. **7213**:145–153.

Rogers, S.O., and A.J. Bendich. 1987. Ribosomal RNA genes in plants: variability in copy number and in the intergenic spacer. Plant Mol. Biol. **9**:509–520.

Romero-Herrera, A.E., N. Lieska, M. Goodman, and E.L. Simons. 1979. The use of amino acid sequence analysis in assessing evolution. Biochimie **61**:767–779.

Roose, M.L., and L.D. Gottlieb. 1976. Genetic and biochemical consequences of

polyploidy in *Tragopogon*. Evolution **30**:818–830.

Roose, M.L., and L.D. Gottlieb. 1978. Stability of structural gene number in diploid species with different amounts of nuclear DNA and different chromosome numbers. Heredity **40**:159–163.

Roose, M.L., and L.D. Gottlieb. 1980. Alcohol dehydrogenase in the diploid plant *Stephanomeria exigua* (Compositae): gene duplication, mode of inheritance, and linkage. Genetics **95**:171–196.

Saghai-Maroof, M.A., K.M. Soliman, R.A. Jorgensen, and R.W. Allard. 1984. Ribosomal DNA spacer-length polymorphisms in barley: Mendelian inheritance, chromosomal location, and population dynamics. Proc. Natl. Acad. Sci. USA **81**:8014–8018.

Sahai, S., and R.S. Rana. 1977. Seed protein homology and elucidation of species relationships in *Phaseolus* and *Vigna* species. New Phytol. **79**:527–534.

Saidman, B.O., and J.C. Vilardi. 1987. Analysis of the genetic similarities among seven species of *Prosopis* (Leguminosae: Mimosoideae). Theor. Appl. Genet. **75**:109–116.

Sakai, R.K., S. Scharf, F. Faloona, K.B. Mullis, G.T. Horn, H.A. Evlich, and N. Arnheim. 1985. Enzymatic amplification of B-globin genomic sequences and restriction site analysis for diagnosis of sickle cell anemia. Science **230**:1350–1354.

Sakano, K., S.D. Kung, and S.G. Wildman, 1974. Identification of several chloroplast genes which code for the large subunit of *Nicotiana* fraction I proteins. Mol. Gen. Genet. **130**:91–97.

Salcedo, G., C. Aragoncillo, M.A. Rodriquez-Loperena, F. Carbonero, and F. Garcia–Olmedo. 1978. Differential allelic expression at a locus encoding an endosperm protein in tetraploid wheat (*Triticum turgidum*). Genetics **89**:147–156.

Salcedo, G., P. Fra-Mon, J.L. Molina-Cano, C. Aragoncillo, and F. Garcia-Olmedo. 1984. Genetics of CM-proteins (A-hordeins) in barley. Theor. Appl. Genet. **68**:53–59.

Sanger, F. 1981. Determination of nucleotide sequences in DNA. Science **214**:1205–1210.

Sanger, F., S. Nicklen, and A.R. Coulson. 1977. DNA sequencing with chain-terminating inhibitors. Proc. Natl. Acad. Sci. USA **74**:5463–5467.

Sarich, V.M., C.W. Schmid, and J. Marks. 1989. DNA hybridization as a guide to phylogenies: a critical analysis. Cladistics **5**:3–32.

Schaal, B.A., and G.H. Learn. 1988. Ribosomal DNA variation within and among plant populations. Ann. Missouri Bot. Gard. **75**:1202–1216.

Schaal, B.A., W.J. Leverich, and J. Nieto-Sotelo. 1987. Ribosomal DNA variation in the native plant *Phlox divaricata*. Mol. Biol. Evol. **4**:611–621.

Schoen, D.J. 1982. Genetic variation and the breeding system of *Gilia achillaeifolia*. Evolution **36**:361–370.

Scogin, R. 1981. Amino acid sequence studies and plant phylogeny. In: D.A. Young and D.S. Seiger, eds. "Phytochemistry and Angiosperm Phylogeny." New York: Praeger, pp. 19–42.

Schroeder, H.E., and A.H.D. Brown. 1984. Inheritance of legumin and albumin contents in a cross between round and wrinkled peas. Theor. Appl. Genet. **68**:101–107.

Scowcroft, W.R. 1979. Nucleotide polymorphism in chloroplast DNA of *Nicotiana dabneyi*. Theor. Appl. Genet. **55**:133–137.

Scribailo, R.W., K. Carey, and U. Posluszny, 1984. Isozyme variation and the reproductive biology of *Hydrocharis morsus-ranae* L. (Hydrocharitaceae). Bot. J. Linn. Soc. **89**:305–312.

Sealy, P.G., and E.M. Southern. 1982. Gel electrophoresis of DNA. In: D. Rickwood and B.O. Hames, eds. "Gel Electrophoresis of Nucleic Acids, a Practical Approach." Oxford: IRL Press, pp. 39–76.

Sheen, S. 1972. Isozymic evidence bearing on the origin of *Nicotiana tabacum*. Evolution **26**:142–154.

Shewry, P.R., H.M. Pratt, R.A. Finch, and B.J. Miflin. 1978. Genetic analysis of hordein polypeptides from single seeds of barley. Heredity **40**:463–468.

Shewry, P.R., A.J. Faulks, R.A. Pickering, I.T. Jones, R.A. Finch, and B.J. Miflin. 1980. The genetic analysis of barley storage proteins. Heredity **44**:383–389.

Shields, C.R., T.J. Orton, and C.W. Stuber. 1983. An outline of general resource needs and procedures for the electrophoretic separation of active enzymes from plant tissue. In: S.D. Tanksley and T.J. Orton, eds. "Isozymes in Plant Genetics and Breeding," Plant A. Amsterdam: Elsevier, pp. 443–468.

Shinozaki, K., M. Ohme, M. Tanaka, T. Wakasugi, N. Hayashida, T. Matsubayashi, N. Zaita, J. Chunwongse, J. Obokata, K. Yamaguchi-Shinozaki, C. Ohto, K. Tovazawa, B.Y. Meng, M. Sugita, H. Deno, T. Kamogashira, K. Yamata, J. Kusuda, F. Takaiwa, A. Kato, N. Tohdoh, H. Shimada, and M. Sugiuva. 1986. The complete nucleotide sequence of tobacco chloroplast genome: its gene organization and expression. EMBO J. **5**:2043–2049.

Sibley, C.G., J.E. Ahlquist, and E.H. Sheldon. 1987. DNA hybridization and avian phylogenetics, reply to Cracraft. Evol. Biol. **21**:97–125.

Singh, R.S. 1979. Genic heterogeneity within electrophoretic "alleles" and the pattern of variation among loci in *Drosophila pseudoobscura*. Genetics **93**:997–1018.

Singh, R.S., R.C. Lewontin, and A.A. Felton. 1976. Genetic heterogeneity within electrophoretic "alleles" of xanthine dehydrogenase in *Drosophila pseudoobscura*. Genetics **84**:609–629.

Smith, D.B., and R.B. Flavell. 1974. The relatedness and evolution of repeated nucleotide sequences in the DNA of some Gramineae species. Biochem. Genet. **12**:243–256.

Smith, D.B., and R.B. Flavell. 1977. Nucleotide sequence organization in the rye genome. Biochim. Biophys. Acta. **474**:82–97.

Smith, D.B., J. Rimpau, and R.B. Flavell. 1976. Interspersion of different repeated sequences in the wheat genome revealed by interspecies DNA/DNA hybridization. Nucleic Acids Res. **3**:2811–2825.

Smith, D.M., and D.A. Levin. 1963. A chromatographic study of reticulate evolution in the Appalachian *Asplenium* complex. Am. J. Bot. **50**:952–958.

Smith, E.B. 1973. A biosystematic study of *Coreopsis saxicola* (Compositae). Brittonia **25**:200–208.

Smith, E.B. 1974. *Coreopsis nuecensis* (Compositae) and a related new species from Southern Texas, Brittonia **26**:161–171.

Smith, E.B. 1975. The chromosome numbers of North American *Coreopsis* with phyletic interpretations. Bot. Gaz. **136**:78–86.

Smith, E.B. 1976. A biosystematic survey of *Coreopsis* in eastern United States and Canada. Sida **6**:123–215.

Smith, H.H., D. Hamill, E. Weaver, and K. Thompson. 1970. Multiple molecular forms of peroxidases, and esterases among *Nicotiana* species and amphiploids. Heredity **61**:203–212.

Smith, P.M. 1976. "The Chemotaxonomy of Plants." New York: Elsevier.

Smithies, O. 1955. Zone electrophoresis in starch gels: group variations in the serum proteins of normal human adults. Biochem. J. **61**:629–641.

Sneath, P.H.A., and R.S. Sokal. 1973. "Numerical Taxonomy—the Principle and Practice of Numerical Classifications." San Francisco: Freeman.

Solbrig, O.T., L.C. Anderson, D.W. Kyhos, P.H. Raven, and L. Rudenberg. 1964. Chromosome numbers in the Compositae V. Astereae II. Am. J. Bot. **51**:513–519.

Solbrig, O.T., L.C. Anderson, D.W. Kyhos, and P.H. Raven. 1969. Chromosome numbers in Compositae VII. Astereae III. Am. J. Bot. **56**:348–353.

Soltis, D.E. 1981a. Allozymic variability in *Sullivantia* (Saxifragaceae). Syst. Bot. **7**:26–34.

Soltis, D.E. 1981b. Variation in hybrid fertility among the disjunct populations and species of *Sullivantia* (Saxifragaceae). Can. J. Bot. **59**:1174–1180.

Soltis, D.E. 1984. Autopolyploidy in *Tolmiea menziesii* (Saxifragaceae). Am. J. Bot. **71**:1171–1174.

Soltis, D.E. 1985. Allozymic differentiation among *Heuchera americana, H. parviflora, H. pubescens*, and *H. villosa* (Saxifragaceae). Syst. Bot. **10**:193–198.

Soltis, D.E. 1986. Genetic evidence for diploidy in *Equisetum*. Am. J. Bot. **73**:908–913.

Soltis, D.E., and B.A. Bohm. 1986. Flavonoid chemistry of diploid and tetraploid cytotypes of *Tolmiea menziessii* (Saxifragaceae). Syst. Bot. **11**:20–25.

Soltis, D.E., and L.H. Rieseberg. 1986. Autopolyploidy in *Tolmiea menziesii* (Saxifragaceae): genetic insights from enzyme electrophoresis. Am. J. Bot. **73**:310–318.

Soltis, D.E., and P.S. Soltis. 1986a. Intergeneric hybridization between *Conimitella williamsii* and *Mitella stauropetala* (Saxifragaceae). Syst. Bot. **11**:293–297.

Soltis, D.E., and P.S. Soltis. 1986b. Electrophoretic evidence for inbreeding in the fern *Botrychium virginianum* (Ophioglossaceae). Am. J. Bot. **73**:588–592.

Soltis, D.E., and P.S. Soltis. 1988a. Electrophoretic evidence for tetrasomic segregation in *Tolmiea menziesii* (Saxifragaceae). Heredity **60**:375–382.

Soltis, D.E., and P.S. Soltis. 1988b. Are lycopods with high chromosome numbers ancient polyploids? Am. J. Bot. **75**:238–247.

Soltis, D.E., B.A. Bohm, and G.L. Nesom. 1983a. Flavonoid chemistry of cytotypes in *Galax* (Diapensiaceae). Syst. Bot. **8**:15–23.

Soltis, D.E., C.H. Haufler, D.C. Darrow, and G.J. Gastony. 1983b. Starch gel electrophoresis of ferns: a compilation of grinding buffers, gel and electrode buffers and staining schedules. Am. Fern. J. **73**:9–27.

Soltis, D.E., A.J. Gilmartin, L. Rieseberg, and S. Gardner. 1987. Genetic variation in the epiphytes *Tillandsia ionantha* and *T. recuruata* (Bromeliaceae). Am. J. Bot. **74**:531–537.

Soltis, D.E., P.S. Soltis, and B.D. Ness. 1989. High levels of chloroplast DNA variation and multiple origins of autopolyploidy in *Heuchera micrantha* (Saxifragaceae). Evolution **43**:650–656.

Soltis, P.S., and D.E. Soltis. 1986. Anthocyanin content in diploid and tetraploid cytotypes of *Tolmiea menziesii* (Saxifragaceae). Syst. Bot. **11**:32–34.

Soltis, P.S., and D.E. Soltis. 1988a. Electrophoretic evidence for genetic diploidy in *Psilotum nudum*. Am. J. Bot. **75**:1667–1671.

Soltis, P.S., and D.E. Soltis. 1988b. Genetic variation and population structure in *Blechnum spicant* (Blechnaceae) in western North America. Am. J. Bot. **75**:37–44.

Soltis, P.S., D.E. Soltis, and L.D. Gottlieb. 1987. Phosphoglucomutase gene duplications in *Clarkia* (Onagraceae) and their phylogenetic implications. Evolution **41**:667–671.

Song, K.M., T.C. Osborn, and P.H. Williams. 1988. *Brassica* taxonomy based on nuclear restriction fragment length polymorphisms (RFLPs). Theor. Appl. Genet. **75**:784–794.

Southern, E.M. 1975. Detection of specific sequences among DNA fragments separated by gel electrophoresis. J. Mol. Biol. **98**:503–517.

Stebbins, G.L. 1980. Polyploidy in plants: unresolved problems and prospects. In: W.H. Lewis, ed. "Polyploidy, Biological Relevance." New York: Plenum, pp. 495–520.

Steer, M.W., and D. Kernoghan. 1977. Nuclear and cytoplasmic genome relationships in the genus *Avena*. Analysis by isoelectric focusing of ribulose bisphosphate carboxylase subunits. Biochem. Genet. **15**:273–286.

Steer, M.W., and H. Thomas. 1976. Evolution of *Avena sativa*: origin of the cytoplasmic genome. Can. J. Genet. Cytol. **18**:769–771.

Stein, D.B., and W.F. Thompson. 1975. DNA hybridization and evolutionary relationships in three *Osmunda* species. Science **189**:888–890.

Stein, D.B., W.F. Thompson, and H.S. Belford. 1979. Studies on DNA sequences in the Osmundaceae. J. Mol. Evol. **13**:215–232.

Strauss, S.H., J.D. Palmer, G.T. Howe, and A.H. Doerksen. 1988. Chloroplast genomes of two conifers lack a large inverted repeat and are extensively rearranged. Proc. Natl. Acad. Sci. USA **85**:3898–3902.

Stuber, C.W., and M.M. Goodman. 1983. Inheritance, intracellular localization, and genetic variation of phosphoglucomatase isozymes in maize (*Zea mays* L.). Biochem. Genet. **21**:667–689.

Stucky, J., and R.C. Jackson. 1975. DNA content of seven species of Astereae and its significance to theories of chromosome evolution in the tribe. Am. J. Bot. **62**:508–518.

Stuessy, T.F., R.W. Sanders, and Mario Silva O. 1984. Phytogeography and evolution of the Juan Fernandez Islands: a progress report. In: P.A. Raven, F. Radousky, and S. Sohmer, eds. "Pacific Basin Biogeography." Stanford: Stanford University Press, pp. 55–69.

Sugita, M., T. Manzara, E. Pichersky, A. Cashmore, and W. Gruissem. 1987. Genomic organization, sequence analysis and expression of all five genes encoding the small subunit of ribulose-1,5-bisphosphate carboxylase/oxygenase from tomato. Mol. Gen. Genet **209**:247–256.

Swofford, D.L. 1981. On the utility of the distance Wagner procedure. In: V.A. Funk, and D.R. Brooks, eds. "Advances in Cladistics," Proc. First Willi Henig Soc. Meeting. New York: New York Botanic Garden, pp. 25–43.

Swofford, D.W., and S.H. Berlocher. 1987. Inferring evolutionary trees from gene frequency data under the principle of maximum parsimony. Syst. Zool. **36**:293–325.

Sytsma, K.J., and L.D. Gottlieb. 1986a. Chloroplast DNA evidence for the origin of the genus *Heterogaura* from a species of *Clarkia* (Onagraceae). Proc. Natl. Acad. Sci. USA **83**:5554–5557.

Sytsma, K.J., and L.D. Gottlieb. 1986b. Chloroplast DNA evolution and phylogenetic relationships in *Clarkia* sect. *Peripetasma* (Onagraceae). Evolution **40**:1248–1261.

Sytsma, K.J., and B.A. Schaal. 1985a. Genetic variation, differentiation, and evolution in a species complex of tropical shrubs based on isozymic data. Evolution **39**:582–593.

Sytsma, K.J., and B.A. Schaal. 1985b. Phylogenetics of the *Lisianthius skinneri* (Gentianaceae) species complex in Panama utilizing DNA restriction fragment analysis. Evolution **39**:594–608.

Sytsma, K.J., and J.F. Smith. 1988. DNA and morphology: comparisons in the Onagraceae. Ann. Missouri Bot. Gard. **75**:1217–1237.

Tanksley, S.D., and G.D. Kuehn. 1985. Genetics, subcellular localization and molecular characterization of 6-phosphogluconate dehydrogenase isozymes in tomato. Biochem. Genet. **23**:441–454.

Tanksley, S.D., and C.M. Rick. 1980. Genetics of esterases in species of *Lycopersicon*. Theor. Appl. Genet. **56**:209–219.

Tanksley, S.D., R. Bernatzky, N.L. Lapitan, and J.P. Prince. 1988a. Conservation of gene repertoire but not gene order in pepper and tomato. Proc. Natl. Acad. Sci. USA **85**:6419–6423.

Tanksley, S.D., J. Miller, A. Paterson, and R. Bernatzky. 1988b. Molecular organization of plant chromosomes. In: J.P. Gustafson and R. Appels, eds. "Chromosomes, Structure and Function," 18th Stadler Gent. Symp. New York: Plenum, pp. 157–173.

Tanksley, S.D., N.D. Young, A.H. Paterson, and M.W. Bonierbale. 1989. RFLP mapping in plant breeding: new tools for an old science. Biotechnology **7**:257–264.

Thompson, T.P., and J. Nath. 1986. Elucidation of the B-genome donor to *Triticum turgidum* by unique and repeated sequence DNA hybridizations. Biochem. Genet. **24**:39–50.

Thompson, W.F., and M.G. Murray. 1981. The nuclear genome: structure and function. In: E. Marcus, ed. "The Biochemistry of Plants," Vol. 6. New York: Academic Press, pp. 1–81.

Thompson, W.F., M.G. Murray, and R.E. Cuellar. 1980. Contrasting patterns of DNA sequence organization in plants. In: C.J. Leaver, ed. "Genome Organization and Expression in Plants," New York: Plenum, pp. 1–15.

Thorpe, J.P. 1982. The molecular clock hypothesis: biochemical evolution, genetic differentiation and systematics. Annu. Rev. Ecol. Syst. **13**:139–168.

Torres, M.A., and G.E. Hart. 1976. Developmental specificity and evolution of the acid phosphatase isozymes of *Triticum aestivum* and its progenitor species. Biochem. Genet. **14**:595–609.

Tsu, T.T., and L.A. Herzenberg. 1980. Solid phase radioimmune assays. In: B.B. Mishel and S.M. Shiigi, eds. "Selected Methods in Cellular Immunology." San Francisco: Freeman, pp. 373–397.

Tsunewaki, K., and Y. Ogihara. 1983. The molecular basis of genetic diversity among cytoplasms of *Triticum* and *Aegilops* species. II. On the origin of polyploid wheat cytoplasms as suggested by chloroplast DNA restriction fragment patterns. Genetics **104**:155–171.

Turner, B.L. 1977. Chemosystematics and its effect upon the traditionalist. Ann. Missouri Bot. Gard. **64**:235–242.

Turner, B.L., and D. Horne. 1964. Taxonomy of *Machaeranthera* sect. *Psilactis* (Compositae–Astereae). Brittonia **16**:316–331.

Turner, B.L., W.L. Ellison, and R.M. King. 1961. Chromosome numbers in the Compositae IV. North American species with phyletic interpretations. Am. J. Bot. **48**:216–223.

Uchimiya, H., and S.G. Wildman. 1978. Evolution of fraction I protein in relation to origin of amphidiploid *Brassica* species and other members of the Cruciferae. J. Hered. **69**:299–303.

Uchimiya, H., K. Chen, and S.G. Wildman. 1977. Polypeptide composition of fraction I protein as an aid in the study of plant evolution. Stadler Symposium **9**:83–99.

Uchimiya, H., K. Chen, and S.G. Wildman. 1979. Evolution of fraction I protein in the genus *Lycopersicon*. Biochem. Genet. **17**:333–341.

Vallejos, E. 1983. Enzyme activity staining. In: S.D. Tanksley and T.J. Orton, eds. "Isozymes in Plant Genetics and Breeding," Part A. Amsterdam: Elsevier, pp. 469–516.

Vaughan, J.G., and K.E. Denford. 1968. An acrylamide gel electrophoretic study of the seed proteins of *Brassica* and *Sinapis* species with special reference to their taxonomic value. J. Exp. Bot. **19**:724–732.

Vaughan, J.G., and A. Waite. 1967. Comparative electrophoretic studies of the seed proteins of certain amphidiploid species of *Brassica*. J. Exp. Bot. **18**:269–276.

Vaughan, J.G., A. Waite, D. Boulter, and S. Waiters. 1966. Comparative studies of the seed proteins of *Brassica campestris, Brassica oleracea*, and *Brassica nigra*. J. Exp. Bot. **17**:332–343.

Vaughan, J.G., K.E. Denford, and E.I. Gordon. 1970. A study of the seed proteins of synthesized *Brassica napus* with respect to its parents. J. Exp. Bot. **21**:892–898.

Vedel, F., F. Quetier, and M. Bayen. 1976. Specific cleavage of chloroplast DNA from higher plants by EcoRI restriction nuclease. Nature **263**:440–442.

Vedel, F., F. Quetier, F. Dosba, and G. Doussinault. 1978. Study of wheat phylogeny by EcoRI analysis of chloroplastic and mitochondrial DNAs. Plant Sci. Lett. **13**:97–102.

Villamil, C.B., and D.E. Fairbrothers. 1974. Comparative protein population investigation of the *Alnus serrulata–rugosa* complex. Biochem. Syst. Ecol. **2**:15–20.

Vittozzi, L., and V. Silano. 1976. The phylogenesis of protein 1-amylase inhibitors from wheat seed and the speciation of polyploid wheats. Theor. Appl. Genet. **48**:279–284.

Vodkin, M., and F.R.H. Katterman. 1971. Divergence of ribosomal RNA sequences within angiospermae. Genetics **69**:435–451.

Vogelmann, J.E., and G.J. Gastony. 1987. Electrophoretic analysis of North American and eastern Asian populations of *Agastache* sect. *Agastache* (Labiatae). Am. J. Bot. **74**:385–393.

Wagner, D.B., G.R. Furnier, M.A. Saghai-Maroof, S.M. Williams, B.P. Dancik, and R.W. Allard. 1987. Chloroplast DNA polymorphisms in lodgepole and jack pines and their hybrids. Proc. Natl. Acad. Sci. USA **84**:2097–2100.

Wagner, W.H. Jr. 1954. Reticulate evolution in the Appalachian Aspleniums. Evolution **8**:103–118.

Wahl, G.M., J.L. Meinkoth, and A.R. Kimmel. 1987. Northern and Southern blots. Meth. Enzymol. **52**:572–581.

Wain, R.P. 1982. Genetic differentiation in the Florida subspecies of *Helianthus debilis* (Asteraceae). Am. J. Bot. 1573–1578.

Wain, R.P. 1983. Genetic differentiation during speciation in the *Helianthus debilis* complex. Evolution **37**:1119–1127.

Wain, R.P., W.T. Haller, and D.F. Martin. 1983. Genetic relationships among the forms of *Cabomba*. J. Aquat. Plant Manag. **21**:96–98.

Walbot, V. 1979. Genome organization in plants. In: I. Rubenstein, C.E. Green, and B.G. Gregenbach, eds. "Molecular Biology of Plants." New York: Academic Press, pp. 31–72.

Walbot, V., and R. Goldberg. 1979. Plant genome organization and its relationship to classical plant genetics. In: T.C. Hall and J.W. Davies, eds. "Nucleic Acids in Plants." West Palm Beach, FL: CRC Press, pp. 3–40.

Walters, T.W. 1987. Electrophoretic evidence for the evolutionary relationship of the tetraploid *Chenopodium berlandieri* to its putative diploid progenitors. Selbyana **10**:36–55.

Walters, T.W. 1988. Relationship between isozymic and morphologic variation in the diploids *Chenopodium fremontii, C. neomexicanum, C. palmeri,* and *C. watsonii.* Am. J. Bot. **75**:97–105.

Warwick, S.I., and S.G. Aiken. 1986. Electrophoretic evidence for the recognition of two species in annual wild rice (Zizania, Poaceae). Syst. Bot. **11**:464–473.

Warwick, S.I., and L.D. Gottlieb. 1985. Genetic divergence and geographic speciation in *Layia* (Compositae). Evolution **39**:1236–1241.

Weber, J.C., and R.F Stettler. 1981. Isoenzyme variation among ten populations of *Populus trichocarpa* Torr. et Gray in the Pacific Northwest. Silvae Genet. **30**:82–87.

Weeden, N.F. 1983. Plastid isozymes. In: S.D. Tanksley, and T.J. Orton, eds. "Isozymes in Plant Genetics and Breeding," Part A. Amsterdam: Elsevier, pp. 139–156.

Weeden, N.F., and L.D. Gottlieb. 1979. Distinguishing allozymes and isozymes of phosphoglucoisomerase by electrophoretic comparisons of pollen and somatic tissues. Biochem. Genet. **17**:287–296.

Weeden, N.F., and L.D. Gottlieb. 1980a. The genetics of chloroplast enzymes. J. Hered. **71**:392–396.

Weeden, N.F., and L.D. Gottlieb. 1980b. Isolation of cytoplasmic enzymes from pollen. Plant Physiol. **66**:400–403.

Wells, E.F. 1979. Interspecific hybridization in eastern North American *Heuchera* (Saxifragaceae). Syst. Bot. **4**:319–338.

Wendel, J.F. 1989. New World tetraploid cottons contain Old World cytoplasm. Proc. Natl. Acad. Sci. USA **86**:4132–4136.

Wendel, J.F., and C.R. Parks. 1985. Genetic diversity and population structure in *Camellia japonica* L. (Theaceae). Am. J. Bot. **72**:52–65.

Wendel, J.F., and C.W. Stuber. 1984. Plant isozymes: enzymes studied and buffer systems for their electrophoretic resolution in starch gels. Isozyme Bull. **17**:4–11.

Werth, C.R. 1985. Implementing an isozyme laboratory at a field station. Virginia J. Sci. **36**:55–76.

Werth, C.R. 1989. The use of isozyme data for inferring ancestry of polyploid species of pteridophytes. Biochem Syst. Ecol. in press.

Werth, C.R., S.I. Guttman, and W.H. Eshbaugh. 1985a. Electrophoretic evidence of reticulate evolution in the Appalachian *Asplenium* complex. Syst. Bot. **10**:184–192.

Werth, C.R., S.I. Guttman, and W.H. Eshbaugh. 1985b. Recurring origin of allopolyploid species in *Asplenium*. Science **228**:731–733.

Whalen, M.D. 1979. Allozyme variation and evolution in *Solanum* section *Androceras*. Syst. Bot. **4**:203–222.

Whalen, M.D., D.E. Costich, and C.B. Heiser. 1981. Taxonomy of *Solanum* section *Lasiocarpa*. Genetes Herb. **12**:41–129.

Whalen, M.D., and E.E. Caruso. 1983. Phylogeny in *Solanum* sect. *Lasiocarpa* (Solanaceae): congruence of morphological and molecular data. Syst. Bot. **8**:369–380.

Wheeler, N.C., and R.P. Guries. 1982. Population structure, genic diversity and morphological variation in *Pinus contorta* Dougl. Can. J. For. Res. **12**:595–606.

Wheeler, N.C., R.P. Guries, and D.M. O'Malley. 1983. Biosystematics of the genus *Pinus*, subsection *Contortae*. Biochem Syst. Ecol. **11**:333–340.

Whitfeld, R.R., and W. Bottomley. 1983. Organization and structure of chloroplast genes. Ann. Rev. Plant Physiol. **34**:279–310.

Whitkus, R. 1985. FORTRAN program for computing genetic statistics from allelic frequency data. J. Hered. **76**:142.

Whitkus, R. 1988. Modified version of GENESTAT, a program for computing genetic statistics from allelic frequency data. Plant Genet. Newsletter **4**:10.

Wildman, S.G. 1981. Molecular aspects of wheat evolution: rubisco composition. In: L.T. Evans and W.J. Peacock, eds. "Wheat Science. Today and Tomorrow." New York: Cambridge University Press, pp. 61–74.

Wildman, S.G. 1983. Polypeptide composition of rubisco as an aid in studies of plant phylogeny. In: U. Jensen and D.E. Fairbrothers, eds. "Proteins and Nucleic Acids in Plant Systematics." Heidelberg: Springer-Verlag, pp. 182–208.

Wiley, E.O. 1981. "Phylogenetics." New York: Wiley.

Williams, S.M., R.W. DeBry, and J.L. Feder. 1988. A commentary on the use of ribosomal DNA in systematic studies. Syst. Zool. **37**:60–62.

Wilson, A.C., S.S. Carlson, and T.J. White. 1977. Biochemical evolution. Annu. Rev. Biochem. **46**:573–639.

Wilson, H.D. 1976. Genetic control and distribution of leucine aminopeptidase in the cultivated chenopods and related weed taxa. Biochem. Genet. **14**:913–919.

Wilson, H.D. 1981. Genetic variation among South American populations of tetraploid *Chenopodium* sect. *Chenopodium* subsect. *Cellulata*. Syst. Bot. **6**:380–398.

Wilson, H.D. 1988a. Allozyme variation and morphological relationships of *Chenopodium hircinum* (s.l.). Syst. Bot. **13**:215–228.

Wilson, H.D. 1988b. Quinoa biosystematics I: domesticated populations. Econ. Bot. **42**:461–477.

Wilson, H.D. 1988c. Quinoa biosystematics II: free-living populations. Econ. Bot. **42**:478–494.

Wilson, H.D. 1989. Free-living and domesticated *Cucurbita* populations of Mexico: contrasting patterns of electrophoretic and morphometric variation. Syst. Bot. in press.

Wilson, H.D., and C.B. Heiser, Jr. 1979. The origin and evolutionary relationships of "Huauzontle" (*Chenopodium nuttalliae* Safford). domesticated chenopod of Mexico. Am. J. Bot. **66**:198–206.

Wilson, H.D., S.C. Barber, and Terrence Walters. 1983. Loss of duplicate gene expression in tetraploid *Chenopodium*. Biochem. Syst. Ecol. **11**:7–13.

Wimpee, C.F., W.J. Stiekema, and E.M. Tobin. 1983. Sequence heterogeneity in the RuBP carboxylase small subunit family of *Lemma gibba*. Plant Mol. Biol. **2**:391–401.

Witter, M. 1986. Adaptive radiation and genetic differentiation in the Hawaiian silversword alliance (Compositae: Madiinae) [PhD thesis]. Honolulu: University of Hawaii, Chapter 3.

Witter, M., and G.D. Carr. 1988. Adaptive radiation and genetic differentiation in the Hawaiian silversword alliance (Compositae: Madiinae). Evolution **42**: 1278–1287.

Wolf, P.G., C.H. Haufler, and E. Sheffield. 1987. Electrohoretic evidence for genetic diploidy in the brackenfern (*Pteridium aquilinum*). Science **236**:947–949.

Wolfe, K.H., W.Li, and P.M. Sharp. 1987. Rates of nucleotide substitution vary greatly among plant mitochondrial, chloroplast and nuclear DNAs. Proc. Natl. Acad. Sci. USA **84**:9054–9058.

Yadava, J.S., J.B. Chowdhury, S.N. Kakar, and H.S. Nainawatee. 1979. Comparative electrophoretic studies of proteins and enzymes of some *Brassica* species. Theor. Appl. Genet. **54**:89–91.

Yatskievych, G., D.B. Stein, and G.J. Gastony. 1988. Chloroplast DNA evolution and systematics of *Phanerophlebia* (Drypteridaceae) and related fern genera. Proc. Natl. Acad. Sci. USA **85**:2589–2593.

Yeh, F.C., M.A.K. Khalil, Y.A. El-Kussaby, and D.C. Trust. 1986. Allozyme variation in *Picea mariana* from Newfoundland: genetic diversity, population structure, and analysis of differentiation. Can. J. For. Res. **16**:713–720.

Young, D.A. 1976. Flavonoid chemistry and the phylogenetic relationships of the Julianiaceae. Syst. Bot. **1**:149–162.

Zimmer, E.A., E.R. Jape, and V. Walbot. 1988. Ribosomal gene structure, variation and inheritance in maize and its ancestors. Genetics **120**:1125–1136.

Zuckerkandl, E. 1987. On the molecular evolutionary clock. J. Mol. Evol. **26**:34–46.

Zuckerkandl, E., and L. Pauling. 1965. Molecules as documents of evolutionary history. J. Theor. Biol. **8**:357–366.

Zurawski, G., and M.T. Clegg. 1987. Evolution of higher-plant chloroplast DNA-encoded genes: implications for structure–function and phylogenetic studies. Annu. Rev. Plant Physiol. **38**:391–418.